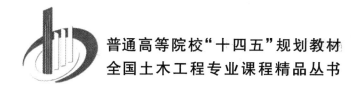

普通高等院校"十四五"规划教材
全国土木工程专业课程精品丛书

# 土木工程材料双语简明教程

主　审　赵庆新　薛才红
主　编　杜　森
副主编　林建军　张洋洋
　　　　赵德志　陈正心

中国建材工业出版社
北　京

图书在版编目(CIP)数据

土木工程材料双语简明教程/杜森主编. --北京：
中国建材工业出版社,2023.11
ISBN 978-7-5160-3701-0

Ⅰ.①土… Ⅱ.①杜… Ⅲ.①土木工程—建筑材料—
双语教学—高等学校—教材 Ⅳ.①TU5

中国国家版本馆CIP数据核字(2023)第008289号

## 内容提要

本书根据国家最新颁布的标准和规范，以教育部高等学校土木工程专业教学指导委员会制定的"土木工程材料"课程教学大纲为依据进行编写，同时兼顾建筑学专业及与土建类相关的建筑材料课程的需求，具有较宽的专业适用性。

本书介绍了土木工程材料的基本性质、无机气硬性胶凝材料、水泥、普通混凝土与建筑砂浆、沥青与沥青混合料、建筑钢材、其他工程材料等内容，可作为高等学校土木建筑类及相关专业的本科教材，也可供土木工程设计、施工、科研、管理和监理人员参考学习。

### 土木工程材料双语简明教程
TUMU GONGCHENG CAILIAO SHUANGYU JIANMING JIAOCHENG

主　审　赵庆新　薛才红
主　编　杜　森
副主编　林建军　张洋洋
　　　　赵德志　陈正心

出版发行：中国建材工业出版社
地　　址：北京市海淀区三里河路11号
邮　　编：100831
经　　销：全国各地新华书店
印　　刷：北京雁林吉兆印刷有限公司
开　　本：787mm×1092mm　1/16
印　　张：23
字　　数：650千字
版　　次：2023年11月第1版
印　　次：2023年11月第1次
定　　价：80.00元

本社网址：www.jccbs.com，微信公众号：zgjcgycbs
请选用正版图书，采购、销售盗版图书属违法行为
**版权专有，盗版必究**。本社法律顾问：北京天驰君泰律师事务所，张杰律师
举报信箱：zhangjie@tiantailaw.com　举报电话：(010)57811389
本书如有印装质量问题，由我社市场营销部负责调换，联系电话：(010)57811387

# 丛书序言

党的二十大报告明确指出，高质量发展是全面建设社会主义现代化国家的首要任务。我国经济社会已经进入高质量发展新阶段，从外延扩张型的平面发展走向更注重质量的立体深度发展。

如何实现土木工程专业的高质量发展？必须从人类文明形态进步的高度认识土木工程专业，树立新的专业发展观，重新认识我国土木工程专业的技术发展情况以及社会对其专业人才的培育要求，深刻推动国家的经济社会进步和长远的可持续发展。

土木工程作为一门古老而新兴的科学，从史前文明到现代文明，从石窟树洞到高楼大厦，各式建筑材料构建的建筑不仅记录着人类文明的发展史，也承载着大地上的人类建材发展史。

自改革开放以来，土木工程材料不断变革，新材料、新技术不断推出，对更新建材产品、升级施工机械、创新施工技术、发展建筑理论起到了明显的推动作用。尤其是钢材和混凝土出现以后，建筑领域的钢结构、混凝土结构、钢-混凝土组合结构和相应的新型施工技术与施工机械等领域围绕材料的应用迅速展开研究与探索，不仅带动了工业材料的发展和技术革新，在一定程度上还提高了对土木工程专业人才的培养要求。

2023年《政府工作报告》指出，深入实施"强基计划"和基础学科拔尖人才培养计划，接续推进世界一流大学和一流学科建设，不断夯实发展的人才基础。中国建材工业出版社利用其专业优势，和国内的高校、企业携手合作，在业内组织专家学者，撰写《普通高等院校"十四五"规划教材/全国土木工程专业课程精品丛书》，阐述土木工程专业人才应具备的职业基本素质以及应掌握的专业技能知识。

本套丛书旨在为我国培养一批高质量专业人才，对推动我国土木工程事业的高质量发展、构建新发展格局、增强我国经济实力具有重要意义，将在我国构建中国式现代化强国建设的进程中发挥重要作用。鉴于以上原因，特将此套丛书推荐给广大读者，相信广大读者一定会从这套丛书中获得收益。

中国工程院 院士
清 华 大 学 教授
2023 年 5 月

# 前　言

近年来，随着我国基础设施建设水平的大幅提升，许多工程单位开始走出国门，参与世界范围的建筑设计、施工与管理。"土木工程材料"是涉及土木建筑类相关专业的一门重要课程，也是基建从业人员必备的基础知识。在此背景下，编写土木工程材料双语教材，是专业人才国际化能力培养的迫切需求。

本书根据国家最新颁布的标准和规范，以教育部高等学校土木工程专业教学指导委员会制定的"土木工程材料"课程教学大纲为依据进行编写，同时兼顾建筑学专业及与土建类相关的建筑材料课程的需求，具有较宽的专业适用性。与同类教材相比，本教材对有关章节进行了调整，例如：考虑砂浆与混凝土的相似性，将砂浆与混凝土合并为一章；因天然石材、木材以及建筑功能材料在现代土木工程中的应用比例不突出，所以将它们统称为"其他工程材料"；将沥青与沥青混合料、建筑钢材设为独立的两章。各自章节的英文采用学界公认程度最高或者使用范围最广的专业术语进行编写，力求满足从业人员进行准确的行业交流及学术交流的需要。

本书由杜森担任主编，林建军、张洋洋、赵德志、陈正心担任副主编。参加编写的有燕山大学林建军（绪论、土木工程材料的基本性质），张洋洋（无机气硬性胶凝材料、水泥），杜森（普通混凝土与建筑砂浆中的混凝土部分），赵德志（沥青与沥青混合料），陈正心（普通混凝土与建筑砂浆中的建筑砂浆部分、建筑钢材、其他工程材料）。研究生肖天宇和周克凡参与了普通混凝土与建筑砂浆一章中部分插图的绘制工作。全书由杜森负责统稿，燕山大学赵庆新和薛才红主审。

本书在编写过程中参考了有关文献，在此向文献作者表示衷心的感谢！

由于编者水平有限，书中疏漏和不足之处在所难免，敬请广大读者批评指正。

编者

2023 年 3 月

# 目　　录

绪　论 ·················································································································· 1

## 第一章　土木工程材料的基本性质 ······································································ 4
第一节　土木工程材料的组成、结构与性质 ······················································· 4
第二节　材料的物理性质 ················································································· 6
第三节　材料的力学性质 ··············································································· 13
第四节　材料的耐久性 ··················································································· 16
复习思考题 ································································································· 17

## 第二章　无机气硬性胶凝材料 ············································································ 18
第一节　石灰 ······························································································· 18
第二节　石膏 ······························································································· 21
第三节　水玻璃 ···························································································· 24
复习思考题 ································································································· 25

## 第三章　水泥 ··································································································· 26
第一节　水泥概述 ························································································· 26
第二节　硅酸盐水泥和普通硅酸盐水泥 ··························································· 28
第三节　大掺量混合材料的硅酸盐水泥 ··························································· 38
第四节　其他品种水泥 ··················································································· 44
复习思考题 ································································································· 46

## 第四章　普通混凝土与建筑砂浆 ········································································ 48
第一节　混凝土概述 ······················································································ 48
第二节　普通混凝土的组成材料 ····································································· 50
第三节　新拌混凝土的工作性能 ····································································· 62
第四节　硬化混凝土的结构 ············································································ 67
第五节　硬化混凝土的力学性能 ····································································· 68
第六节　混凝土的质量控制与评定 ·································································· 78
第七节　混凝土的变形性能 ············································································ 80
第八节　混凝土的耐久性 ··············································································· 84
第九节　混凝土的配合比设计 ········································································· 88

第十节　建筑砂浆 ………………………………………………………… 96
　　复习思考题 ………………………………………………………………… 101

第五章　沥青与沥青混合料 ……………………………………………………… 103
　　第一节　沥青材料 ………………………………………………………… 103
　　第二节　沥青混合料 ……………………………………………………… 113
　　复习思考题 ………………………………………………………………… 123

第六章　建筑钢材 ………………………………………………………………… 124
　　第一节　钢材的化学成分及其对钢材性能的影响 ……………………… 124
　　第二节　钢材的技术性质 ………………………………………………… 126
　　第三节　钢材的冷加工、时效 …………………………………………… 129
　　第四节　土木工程常用钢种 ……………………………………………… 130
　　复习思考题 ………………………………………………………………… 133

第七章　其他工程材料 …………………………………………………………… 134
　　第一节　天然石材 ………………………………………………………… 134
　　第二节　木材 ……………………………………………………………… 138
　　第三节　防水材料 ………………………………………………………… 141
　　第四节　绝热材料 ………………………………………………………… 146
　　第五节　吸声材料和隔声材料 …………………………………………… 149
　　第六节　防火材料 ………………………………………………………… 151
　　复习思考题 ………………………………………………………………… 152

参考文献 …………………………………………………………………………… 153

英文版本 …………………………………………………………………………… 155

# 绪　　论

## 一、土木工程材料的定义及分类

土木工程包括建筑工程、道路工程、桥梁工程、岩土与地下工程、港口工程、水利工程及市政工程等。用于这些工程的各种材料统称为土木工程材料。从狭义上讲，土木工程材料一般指用于建筑物本身的各种建筑材料。从广义上讲，土木工程材料是指工程建造过程中所使用到的所有材料。土木工程材料是土木工程建设的物质基础。扫码了解常见的土木工程结构及其组成材料。

土木工程材料种类繁多、性能各异，其分类方法也有许多。例如，按材料的制造方法可分为天然材料和人工材料；按材料在建筑物中的功能可分为承重材料、保温和隔热材料、吸声和隔声材料、防水材料、耐热防火材料、装饰材料、防腐材料、采光材料等；按材料在建筑中的使用部位可分为基础材料、结构材料、墙体材料、屋面材料、地面材料、饰面材料等。此外，按材料的化学组成又可分为无机材料、有机材料和复合材料三大类，如图 0-1 所示。

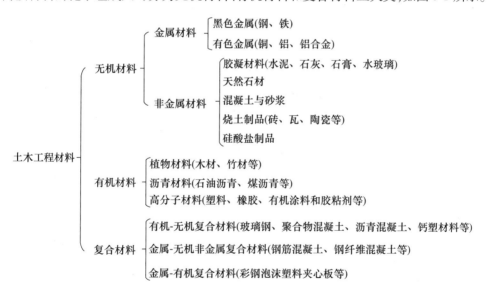

图 0-1　土木工程材料的分类

## 二、土木工程材料的发展历程及趋势

土木工程材料是伴随着人类社会的不断进步和社会生产力的不断发展而发展的。可以说，土木工程材料的发展史就是人类文明的编年史。在远古时代，人类居住于天然山洞或树巢中。距今 10000～6000 年前，人类学会了建造自己的居所。这一时期的房屋多为半地穴式,所使用的材料为天然的木、竹、苇、草、泥等。在距今约 6000 年的西安半坡遗址,考古学家发现当时的人类就已经开始采用木骨泥墙建房并有制陶窑场。随着人类取材能力的提高,人们开始利用天然石材建造房屋和纪念性结构物。最早利用大块石材的结构物当数公元前 2500 年前后建造的埃及金字塔。另外,人工烧土制品的出现使人类建造房屋的能力跃上了新的台阶。黏土砖是烧土制品的代表性材料,其强度高、耐水性好,同时外形规则、尺寸适中,易于砌筑。我国的秦汉时期,黏土砖作为最主要的房屋建筑材料被大量使用,因此有"秦砖汉瓦"之称。

水泥和钢材在土木工程中的应用掀开了建筑材料发展史的新篇章。1824 年,英国人 J. Aspding 将石灰石与黏土混合制成料浆,再经煅烧、磨细制成水泥并取得了发明专利。因其凝结后与英国波特兰岛的石灰石颜色相似,故称波特兰水泥(即我国的硅酸盐水泥)。钢材在土木工程中的应用也是 19 世纪的事。1823 年英国建成世界上第一条铁路;1889 年建造的法国巴黎埃菲尔铁塔高达 320m。钢材在使用过程中容易生锈;水泥混凝土则属于脆性材料,虽然抗压强度较高,但抗拉强度很低。在混凝土中放入钢筋既可使钢筋免于大气中有害介质的侵蚀,防止生锈;同时钢筋提高了构件的抗拉性能,于是出现了钢筋混凝土复合材料。1850 年法国人朗波制造了第一条钢筋混凝土小船;1872 年在纽约出现了第一座钢筋混凝土房屋。1887 年,M. Koenen 发表了钢筋混凝土梁的荷载计算方法。

进入 20 世纪,钢筋混凝土材料有了两次较大的飞跃。其一是 1908 年由 C. R. Steiner 提出了预应力钢筋混凝土的概念;1928 年法国的 E. Fregssinet 使用高拉力钢筋和高强度混凝土使预应力混凝土结构实用化。其二是 1934 年美国人发明了减水剂,在普通的混凝土中加入少量的减水剂可使材料的工作性、耐久性等大大提高。20 世纪以来,相继出现的高分子有机材料、新型金属材料和各种复合材料使建筑物的功能和外观发生了根本性变革,也使其可靠性和服役寿命得到了很大提高。

高性能化、多功能化、工业规模化和生态化成为当今土木工程材料的发展趋势。高性能建筑材料是指比现有材料的性能更为优异的建筑材料,具有高强、高耐久、高抗渗、力学性能稳定等特点及功能,可适用于各种超高、超长、超大型的建筑结构及各种严酷条件。为降低成本、控制质量,材料的生产要实现现代化、工业化,生产要标准化、大型化和商业化。此外,在资源与环境成为当今世界亟待解决的两大问题的背景下,减少环境污染、节约资源、废物资源化利用成为 21 世纪建材工业发展的一大战略需求。绿色建筑材料又称生态建筑材料或健康建筑材料,它是指采用清洁生产技术,不用或少用天然资源和能源,大量使用工农业或城市固态废弃物生产的无毒害、无污染、无放射性,达到使用周期后可回收利用,有利于环境保护和人体健康的建筑材料。例如,利用工业废渣(粉煤灰、矿渣等)作为掺合料制备混凝土等材料,利用废弃泡沫塑料生产保温墙体板材等。这些做法既利用了工业废料、减轻了环境污染,又可节约自然资源。

### 三、土木工程材料的标准化

土木工程材料的技术标准是生产单位和使用单位检验并确认产品质量是否合格的技术文件。为了保证材料的质量,必须对材料产品的技术要求制定统一的技术标准,其主要包括产品规格、分类、技术要求、检验方法、验收规则、标志、运输和储存等内容。我国的技术标准分为国家标准、行业标准、地方标准和企业标准四级。此外，标准还分为强制性标准和推荐性标准两类：强制性标准具有法律属性，在规定的适用范围内必须严格执行；推荐性标准具有技术上的权威性和指导性，是自愿执行的标准，它在合同或行政文件确认的范围内也具有法律属性。各级标准都有各自的部门代号，例如：GB为国家强制性标准；GB/T为国家推荐性标准；JGJ为建筑工程行业建设标准；JG为建筑工业行业标准；SL为水利行业标准；DB为地方标准；QB为企业标准等。标准的表示方法由标准名称、部门代号、编号和批准年份等组成。例如，国家推荐性标准《混凝土结构耐久性设计标准》（GB/T 50476—2019）。

世界各国均有自己的国家标准，例如"ANSI"代表美国国家标准（"ASTM"是美国材料试验与材料协会标准）、"JIS"代表日本国家标准、"BS"代表英国标准。另外，在世界范围内统一执行的标准称为国际标准，其代号为"ISO"。

标准是根据一个时期的技术水平制定的，随着建筑材料及科学技术的发展，技术标准也在不断变化，应根据技术发展的要求不断进行修订和完善。

## 四、本课程学习目的、方法与特点

土木工程材料是土木工程类专业的专业基础课。它是以数学、力学、化学等课程为基础，而又为学习建筑、结构、施工等后续专业课程提供材料学的基本知识，同时它还为今后从事的工程实践和科学研究打下必要的专业基础。书中对每一种土木工程材料的叙述，一般包括原材料、生产、组成、构造、性质、应用、检验、运输和贮存等方面的内容，以及现行的相关技术标准。本课程的学习重点是掌握材料的基本性质和合理选用材料。要达到这一点，就必须了解各种材料的特性，在学习时，不但要了解每一种材料具有哪些基本性质，还应对不同类属、品种的材料特性相互进行比较，只有掌握其特点，才能做到正确合理地选用材料。同时，还要知道材料之所以具有某种基本性质的基本原理以及影响其性质变化的外界条件。

试验课是本课程的重要教学环节。通过试验，一方面要学会各种常用土木工程材料的检验方法，能对材料的合格性进行准确地判断和验收；另一方面是提高实践技能，能对试验数据、试验结果进行正确地分析和判别，培养科学认真的态度和实事求是的工作作风。

本课程具有实践性强、综合性强等特点。所学内容接近工程实际，不像基础理论课程那样：对具体的现象或结构进行简化、抽象。另外，面对浩瀚的工程学，需要学生具有广泛的知识和综合能力，并要善于归纳总结。

# 第一章 土木工程材料的基本性质

土木工程材料的基本性质直接影响着工程结构的可靠性、耐久性和使用性。面对不同的环境和功能要求,材料须具有相适应的性质。例如,结构材料应具有良好的力学性能;墙体材料应具有绝热、隔声性能;屋面材料应具有抗渗、防水性能;地面和路面材料应具有防滑、耐磨损性能等。根据材料学的基本原理,材料的组成与结构决定了其性质。本章在分析材料组成、结构与性质关系的基础上,讨论了土木工程材料的物理性质、力学性质及耐久性能等。这些性质既是我们选择、应用和分析材料时的重要依据,也是工程技术人员在工程设计和施工过程中必须掌握的内容,因此,掌握土木工程材料性质和性能特点是土木工程专业学习、合理选择和使用材料的基础。

## 第一节 土木工程材料的组成、结构与性质

### 一、土木工程材料的组成

土木工程材料的组成通常可分为化学组成和矿物组成。

化学组成是指构成土木工程材料的化学元素和化合物的种类及其相对含量。矿物组成是指材料中的各种矿物的相对含量。矿物是组成地壳的基本物质单元,它是指地质作用下各种化学成分所形成的自然单质和化合物,具有相对稳定的化学成分和内部结构。矿物也可理解为构成材料的单质及化合物的特定结合形式。土木工程材料中引申了这一概念,通常将人造的无机非金属材料中具有特定晶体结构、物理力学性能且与天然矿物相似的组织也称为矿物。

土木工程材料的化学组成相同,其矿物组成不一定相同。例如,化学组成同为碳元素的金刚石和石墨的矿物性能差别即明显不同。不同矿物组成的土木工程材料,其化学组成有可能相同。例如,半水石膏有 $\alpha$、$\beta$、$\gamma$ 等多种矿物相,但其化学组成均可表示为 $CaSO_4 \cdot 0.5H_2O$。土木工程材料的矿物组成相同,其化学组成一定相同。另外,土木工程材料的化学组成不同,其性能会有明显差异。例如,金属材料的导电性能明显优于非金属材料,有机材料的保温性能明显优于无机材料。有时,化学组成上的微小变化也会引起材料性能的显著不同。例如,纯铁、钢和生铁三者的主要成分都是铁元素,但纯铁强度相对较低且较柔软,钢较坚韧,生铁则硬脆。形成这种差异的主要原因之一就是它们的含碳量不同。

由上述可知,土木工程材料的组成直接影响材料性能,在生产和使用时应根据结构性能的要求来确定材料组成及所占比例。

## 二、土木工程材料的结构

土木工程材料的结构和构造决定了材料性能。通常可将材料的结构分为宏观结构、细观结构和微观结构。

(一)宏观结构

宏观结构是指用肉眼或放大镜能够分辨的粗大组织。

**1. 按材料孔隙特征分类**

(1)致密结构:指具有不吸水或吸水性很小的材料的结构,如金属材料、玻璃、塑料、橡胶等。

(2)多孔结构:指具有粗大孔隙材料的结构,如加气混凝土、泡沫混凝土、泡沫塑料及人造轻质多孔材料等。

(3)微孔结构:指具有微细孔隙材料的结构,如石膏制品、低温烧结黏土制品等。

**2. 按材料组织构造特征分类**

(1)堆聚结构:指由骨料和具有胶凝性或黏结性物质胶结而成的结构,如水泥混凝土、砂浆、沥青混合料等。

(2)纤维结构:指由天然或人工合成纤维物质构成的结构,如木材、玻璃纤维等。

(3)层状结构:指具有叠合结构的层状结构,如胶合板、纸面石膏板等各种叠合成层状的板材。

(4)散粒结构:指呈松散颗粒状构造的材料,如砂、石及粉状或颗粒状的材料(膨胀珍珠岩、膨胀蛭石、粉煤灰等)。

(5)纹理结构:指天然材料在生长或形成过程中自然造就的天然纹理,如木材、大理石等。

(二)细观结构

细观结构(原称亚微观结构)是指用光学显微镜所能观察到的材料结构,其尺寸范围在 $10^{-3} \sim 10^{-6}$ mm。细观结构只能针对某种具体材料来进行分类研究。例如,对混凝土可分为基相、骨料相、界面;对木材可分为木纤维、导管髓线、树脂道。

(三)微观结构

微观结构是指材料在分子、离子、原子甚至亚原子层次上的组成形式。材料的许多性质,如弹塑性、硬度、强度等都与材料的微观结构有着密切关系。土木工程材料的使用状态均为固态,其微观结构可分为晶体结构和非晶体结构。

**1. 晶体结构**

晶体结构指由离子、原子或分子在空间上按照特定规则呈周期性排列而成的结构,如石英、胆矾、冰糖、水晶等。晶体组成的每个晶粒具有各向异性,但它们排列起来组成的晶体材料却是各向同性的。晶体中离子、原子或分子的密集程度和它们之间的相互作用力,以及晶粒的外形都将影响材料的性质。晶体中质点的密集程度越高,材料的塑性变形能力越大。晶粒越小,分布越均匀,材料的强度越高。在使用材料时,人们常用改变晶粒粗细和结构的方法来改善材料的性质。如对钢材进行的冷加工和热处理,分别使晶粒细化和晶粒扭曲及滑移,以改善钢材的强度性能。

**2. 非晶体结构**

非晶体结构指结构无序或者近程有序而长程无序的物质,即组成物质的分子(或原子、离子)不呈空间有规则周期性排列的固体,如塑料、石蜡、沥青、橡胶等。玻璃体是典型的非晶体,所以非晶态又称为玻璃态。在熔融物冷却凝固过程中,如果冷却速度较快,质点来不及按一定规则排列,便形成玻璃体。玻璃体材料各向同性,破坏时没有解理面,无固定熔点,只是出现软化现象,将开始软化的起始温度称为软化温度或软化点。熔融物在急冷过程中,质点间的能量以内能的形式存储起来,使得玻璃体具有化学不稳定性,有时表现出一定的化学活性。

晶体和非晶体之间在一定条件下可以相互转化。例如,把石英晶体熔化并迅速冷却可以得到石英玻璃。对于高炉矿渣,如果在空气中自然冷却,则形成稳定的晶体结构,其化学活性很低,如果用压力水冲熔融的高炉矿渣,其内部质点未来得及定向排列,而形成大量的玻璃体结构,烘干磨细后制得的磨细矿渣则是一种活性较高的混凝土掺合料。

# 第二节 材料的物理性质

## 一、材料的密度、视密度、表观密度和堆积密度

(一)材料的密度

密度是指材料在绝对密实状态下单位体积的质量。密度可用下式表示:

$$\rho = m/V \tag{1-1}$$

式中 $\rho$——材料的密度($g/cm^3$);

$m$——材料在绝干状态下的质量(g);

$V$——材料在绝对密实状态下的体积($cm^3$)。

材料在绝对密实状态下的体积是指不包括材料内部孔隙的固体物质本身的体积,亦称实体积。建筑材料中除钢材、玻璃、沥青等外,绝大多数材料均含有一定的孔隙。测定含孔材料的密度时,需将材料磨成细粉(粒径小于0.20mm),经干燥后用李氏瓶测得其实体积。

(二)材料的视密度

视密度是指材料在包含闭口孔隙条件下单位体积的质量。视密度可用下式表示:

$$\rho' = m/V' \tag{1-2}$$

式中 $\rho'$——材料的视密度($g/cm^3$);

$m$——材料在绝干状态下的质量(g);
$V'$——材料在包含闭口孔隙条件下的体积(cm³)。

通常将材料在包含闭口孔隙条件下的体积称为视体积,其等于材料的实体积和闭口孔隙的体积之和。对于如钢材、玻璃等密实材料,视密度与密度十分接近,因此,也称为近似密度。

(三)材料的表观密度

材料在自然状态下单位体积的质量,亦称为体积密度。其公式为：

$$\rho_0 = m/V_0 \tag{1-3}$$

式中 $\rho_0$——材料的表观密度(kg/m³);
$m$——材料的质量(kg);
$V_0$——材料在自然状态下的表观体积(m³)。

材料在自然状态下的表观体积是指材料的实体积与材料内所含全部孔隙体积之和。对于外形规则的材料,其表观密度测定很简便,只要测得材料的质量和体积(用尺量测)即可算得。不规则材料的体积要采用排水法求得,但材料表面应预先涂上蜡,以防水分渗入材料内部。

材料表观密度的大小与含水状态有关,因此,测定材料表观密度时,需注明其含水状态。一般来说,散粒状材料的含水状态可归纳为以下4种,即干燥状态、气干状态、饱和面干状态和湿润状态,如图1-1所示。通常材料的表观密度是指气干状态下的表观密度。材料在干燥状态下的表观密度称为绝干表观密度。

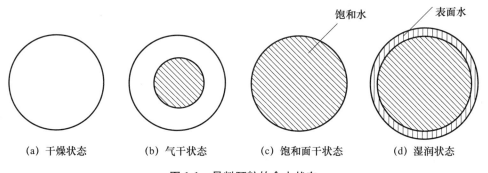

图1-1 骨料颗粒的含水状态

(四)材料的堆积密度

散粒材料在自然堆积状态下单位体积的质量,可用下式表示：

$$\rho_0' = m/V_0' \tag{1-4}$$

式中 $\rho_0'$——散粒材料的堆积密度(kg/m³);
$m$——散粒材料的质量(kg);
$V_0'$——散粒材料在自然堆积状态下的体积(m³)。

散粒材料的堆积体积既包含材料的实体积,又包含颗粒内部的孔隙以及颗粒之间空隙的体积。扫码可获取密度、视密度、表观密度和堆积密度试验的相关内容。

## 二、材料的孔隙率与密实度

大多数土木工程材料的内部都含有孔隙,材料中含有孔隙的多少常用孔隙率表征。孔隙率通常指材料内部孔隙体积($V_P$)占材料表观体积($V_0$)的百分率,可用下式表示:

$$P = \frac{V_P}{V_0} \times 100\% = \frac{V_0 - V}{V} \times 100\% = \left(1 - \frac{\rho_0}{\rho}\right) \times 100\% \tag{1-5}$$

与孔隙率相对应的是材料的密实度,即材料内部固体物质的实体积占材料表观体积的百分率,可用下式表示:

$$D = \frac{V}{V_0} \times 100\% = \frac{\rho_0}{\rho} \times 100\% = 1 - P \tag{1-6}$$

孔隙率的大小反映材料的致密程度,直接影响材料的力学性能、热学性能及耐久性能等。孔隙率相同的材料,它们的孔隙特征(即孔隙孔径与构造)也可以不同。不同尺寸、不同特征的孔隙对材料性能的影响程度不同。例如,封闭孔隙有利于提高材料的保温隔热性,在一定范围内对材料的抗冻性有利;开放或连通的孔隙则降低材料的保温性和抗渗性;孔径较大的孔隙对材料的强度极为不利;孔径在20nm以下的凝胶孔对强度几乎没有任何影响。

根据孔隙对材料性能的影响,可分为有害孔、无害孔和有益孔等;根据孔径尺寸大小,可分为微孔(纳米级孔)、细孔(微米级孔)和大孔(毫米级及以上的孔)等;根据孔隙构造,可分为开口孔$V_K$(连通孔)和闭口孔$V_B$(封闭孔),并且开口孔和闭口孔的体积之和等于材料的总孔隙。另外,通常定义开口孔隙率为$P_K = V_K/V_0$,闭口孔隙率$P_B = V_B/V_0$。

## 三、材料的空隙率与填充率

散粒材料颗粒间的空隙含量常用空隙率表示,其等于散粒材料颗粒间的空隙体积$V_V$占堆积体积$V_0'$的百分率。可按下式表示:

$$P' = \frac{V_V}{V_0'} \times 100\% = \frac{V_0' - V_0}{V_0'} \times 100\% = \left(1 - \frac{\rho_0'}{\rho_0}\right) \times 100\% \tag{1-7}$$

与空隙率相对应的是填充率,即颗粒的自然状态体积占堆积体积的百分率,可按下式计算:

$$D' = \frac{V_0}{V_0'} \times 100\% = \frac{\rho_0'}{\rho_0} \times 100\% = 1 - P' \tag{1-8}$$

空隙率反映散粒材料堆积体积内颗粒之间的相互填充状态,是衡量砂、石等粒状材料颗粒级配好坏、进行混凝土配合比设计的重要数据。在进行混凝土配合比设计时,通常根据骨料的堆积密度、空隙率等指标计算水泥浆用量及砂率等。

## 四、材料与水有关的性质

(一) 亲水性与憎水性

当材料与水接触时,有些材料能被水润湿,有些材料不能被水润湿,前者称为亲水性材料,后者称为憎水性材料。材料产生亲水性的原因是当其与水接触时,材料分子与水分子之间的亲和力大于水本身分子间的内聚力时,则表现为材料的亲水性;反之,当材料与水接触时,材料分子与水分子之间的亲和力小于水本身分子间的内聚力时,则表现为材料的憎水性。

材料被水润湿的情况可用润湿角 $\theta$ 表示。如图1-2所示,当材料与水接触时,在材料、水、空气三相的交点处沿水滴表面作切线,此切线与材料和水接触面的夹角 $\theta$ 即为润湿角。$\theta$ 的值越小,表明材料越易被水润湿。如图1-2(a)所示,当 $\theta \leqslant 90°$ 时,材料表面容易吸附水,材料能被水润湿而表现出亲水性。如图1-2(b)所示,当 $\theta > 90°$ 时,材料表面不易吸附水,表现为憎水性。当 $\theta = 0°$ 时,表明材料完全被水润湿。

亲水性材料易被水润湿,且水能通过毛细管作用而被吸入材料内部。憎水性材料则能阻止水分渗入毛细管中,从而降低材料的吸水性,故常被用作防水材料。土木工程材料大多为亲水性材料,如水泥、混凝土、砂、石、砖、木材等。少数材料为憎水性材料,如沥青、石蜡及某些塑料等。

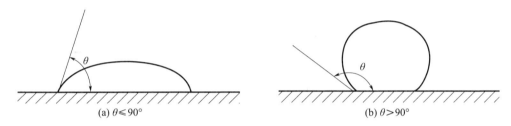

图1-2 材料浸润示意图

(二)材料的吸水性与吸湿性

**1. 吸水性**

材料在水中吸收水分的性质称为吸水性,常用吸水率表示。吸水率有以下两种表示方法:

(1)质量吸水率

质量吸水率指材料在吸水饱和时,其内部所吸水分的质量占材料绝干质量的百分率。质量吸水率用下式表示:

$$W_m = \frac{m_{sw}}{m} \times 100\% \tag{1-9}$$

式中 $W_m$——材料的质量吸水率(%);
$m_{sw}$——材料吸收水的质量(g);
$m$——材料的绝干质量(g)。

(2)体积吸水率

体积吸水率指材料在吸水饱和时,其内部所吸水分的体积占干燥材料表观体积的百分率。体积吸水率用下式表示:

$$W_V = \frac{V_{sw}}{V_0} \times 100\% \tag{1-10}$$

式中 $W_V$——材料的体积吸水率(%);
$V_{sw}$——材料吸收水的体积($cm^3$);
$V_0$——材料的表观体积($cm^3$)。

根据式(1-9)和式(1-10),质量吸水率与体积吸水率存在下列关系:

$$W_V = W_m \cdot \rho_{0d} \tag{1-11}$$

式中 $\rho_{0d}$——材料在干燥状态下的表观密度($g/cm^3$)。

材料中所吸水分是通过开口孔隙吸入的,因此,开口孔隙率越大,材料的吸水量越多。材料吸水达到饱和时的体积吸水率可视为材料的开口孔隙率。材料的吸水性与材料的孔隙率、孔隙特征有关。对于细微连通孔隙,孔隙率越大,则吸水率越大。闭口孔隙水分不能进去,而开口大孔虽然水分易进入,但不能存留,只能润湿孔壁,所以吸水率仍然较小。各种材料的吸水率差异很大,如花岗岩的吸水率只有 0.5%~0.7%,混凝土的吸水率为 2%~3%,烧结黏土砖的吸水率达 8%~20%,而木材的吸水率可超过 100%。

扫码可获取材料吸水率试验的相关内容。

**2. 吸湿性**

材料在潮湿空气中吸收水分的性质称为吸湿性,可用含水率表示。含水率是指材料内部所含水的质量占材料干燥状态下的质量的百分率。含水率用下式表示:

$$W_h = \frac{m_s}{m} \times 100\% \tag{1-12}$$

式中 $W_h$——材料的含水率(%);
$m_s$——材料在吸湿状态下的水的质量(g);
$m$——材料的绝干质量(g)。

材料的吸湿性随空气的湿度和环境温度的变化而改变,当空气湿度较大且温度较低时,材料的含水率就大,反之则小。材料中所含水分与空气的湿度相平衡时的含水率,称为平衡含水率。具有微小开口孔隙的材料吸湿性特别强,如木材及某些绝热材料,这是由于这类材料的内表面积大,吸附水分的能力强所致。

材料的吸水性和吸湿性均会对材料的性能产生不利影响。材料吸水后会导致其自重增大、绝热性降低,强度和耐久性将产生不同程度的下降。不过,利用材料的吸湿效应可起除湿作用,用于保持环境的干燥。

**(三)材料的耐水性**

材料长期在水的作用下不破坏,强度不显著降低的性质称为耐水性,材料的耐水性常用软化系数表示。软化系数用下式表示:

$$K_R = \frac{f_b}{f_g} \times 100\% \tag{1-13}$$

式中 $K_R$——材料的软化系数;
$f_b$——材料在饱水状态下的抗压强度(MPa);
$f_g$——材料在干燥状态下的抗压强度(MPa)。

$K_R$ 的大小表明材料在浸水饱和后强度降低的程度。一般来说,材料被水浸湿后,强度均会有所降低。这是因为水分被组成材料的微粒表面吸附,形成水膜,削弱了微粒间的结合力所致,$K_R$值越小,表示材料吸水饱和后强度下降越大。材料的软化系数 $K_R$ 在 0~1 之间。不同材料的 $K_R$ 值相差颇大,如黏土 $K_R=0$,而金属 $K_R=1$。土木工程材料中将 $K_R>0.85$ 的称为耐水的材料。在设计长期处于水中或潮湿环境中的重要结构时,必须选用 $K_R>0.85$ 的材料。对用于受潮较轻或次要结构物的材料,其 $K_R$ 值不宜小于 0.75。

(四)材料的抗渗性

材料抵抗压力水渗透的性质称为抗渗性或不透水性,常用渗透系数表示。渗透系数的物理意义是:一定厚度的材料,在单位压力水头作用下,单位时间内透过单位面积的水量。渗透系数用下式表示:

$$K_S = \frac{Qd}{AHt} \tag{1-14}$$

式中　$K_S$——材料的渗透系数(cm/h);

$Q$——渗透水量($cm^3$);

$d$——材料的厚度(cm);

$A$——渗水面积($cm^2$);

$t$——渗水时间(h);

$H$——静水压力水头(cm)。

$K_S$ 的值越大表示材料渗透的水量越多,即抗渗性越差。材料的抗渗性可用抗渗等级表示。抗渗等级是以规定的试件、在规定的条件和标准试验方法下所能承受的最大水压力来确定,以符号"P$n$"表示,其中 $n$ 为该材料所能承受的最大水压力的 10 倍值,如 P4 表示材料最大能承受 0.4MPa 的水压力而不渗水。

材料的抗渗性与其孔隙率、孔隙特征有关。细微连通的孔隙水易渗入,故这种孔隙越多,材料的抗渗性越差。闭口孔隙水不能渗入,因此,闭口孔隙率大的材料,其抗渗性仍然良好;开口大孔水最易渗入,其抗渗性最差。

(五)材料的抗冻性

材料在水饱和状态下经受多次冻融循环作用而不破坏,强度也不显著降低的性质称为抗冻性。材料的抗冻性可用抗冻等级表示。抗冻等级是以规定的方法对规定的试件进行冻融循环试验,测得其强度损失不超过规定值,并无明显损伤和剥落时所能经受的最大循环次数来确定,用符号"F$n$"表示,其中 $n$ 即为最大冻融循环次数,如 F25 等。

材料受冻融破坏主要是因其孔隙中的水结冰所致。水结冰时体积增大约 9%,若材料孔隙中充满水,则结冰膨胀对孔壁产生很大应力,当此应力超过材料的抗拉强度时,孔壁将产生局部开裂。随着冻融次数的增多,材料破坏加重。所以材料的抗冻性取决于其孔隙率、孔隙特征及充水程度。如果孔隙不充满水,即远未达饱和,具有足够的自由空间,则即使受冻也不致产生很大的冻胀应力。极细的孔隙,虽可充满水,但因孔壁对水的吸附力极大,吸附在孔壁上的水其冰点很低,它在很大负温条件下才会结冰。闭口孔隙水分不能渗入,而毛细管孔隙既易充满水分,又能结冰,故其对材料的冰冻破坏作用最大。因此,闭口孔隙比例越高,材料抗冻性越好。工程中常利用这一原理改善材料的抗冻性,例如,采用引气剂提高混凝土中闭口孔的比例,从而提高其抗冻性能。材料的变形能力大、强度高、软化系数大时,其抗冻性较高。

从外界条件看,材料受冻融破坏的程度与冻融温度、结冰速度、冻融频繁程度等因素有关。环境温度越低、降温越快、冻融越频繁,材料受冻破坏越严重。

### 五、材料的热工性质

土木工程材料中常用的热工性能有导热性、热容量、比热等。

(一)导热性

当材料两侧存在温度差时,热量将由温度高的一侧通过材料传递到温度低的一侧,材料的这种传导热量的能力称为导热性,可用导热系数来表示。导热系数的物理意义是:厚度为1m的材料,当温度改变1K(热力学温度单位开尔文)时,在1s时间内通过$1m^2$面积的热量。导热系数用下式表示:

$$\lambda = \frac{Q \cdot d}{A \cdot \Delta T \cdot t} \tag{1-15}$$

式中 $\lambda$——材料的导热系数[W/(m·K)];
$Q$——传导的热量(J);
$d$——材料的厚度(m);
$A$——材料传热的面积($m^2$);
$t$——传热时间(s);
$\Delta T$——材料两侧温度差(K)。

材料的导热系数越小,表示其绝热性能越好。各种材料的导热系数差别很大,如泡沫塑料$\lambda = 0.03W/(m·K)$,而大理石$\lambda = 3.48W/(m·K)$。工程中通常把$\lambda < 0.23W/(m·K)$的材料称为绝热材料。材料的导热系数不仅取决于材料的组成,还与材料内部孔隙含量、孔隙特征以及含水状态等有关。比如,空气的导热系数[0.025W/(m·K)]很小,而水的导热系数[0.6W/(m·K)]较大,如果材料内部含有大量封闭的微小孔隙,同时保持干燥状态,孔隙内部充满空气,可有效地降低材料的导热系数;但如果多孔材料吸收大量水分,将使导热系数增大,就会降低其保温效果。

(二)热容量与比热

热容量是指材料受热时吸收热量和冷却时放出热量的性质,其值可通过材料的比热算得。比热的物理意义是:质量为1kg的材料在温度改变1K时所吸收或放出的热量。热容量与比热的关系可用下式表示:

$$C = c \cdot m \tag{1-16}$$

式中 $C$——材料的热容量(kJ/K);
$c$——材料的比热[kJ/(kg·K)];
$m$——材料的质量(kg)。

材料的导热系数和热容量是设计建筑物围护结构(墙体、屋盖等)进行热工计算时的重要参数,设计时应选用导热系数较小而热容量较大的材料,以使建筑物保持室内温度的稳定性。

# 第三节　材料的力学性质

材料在外力作用下抵抗变形或破坏的性质称为力学性质,它是选用材料时首要考虑的基本性质。工程结构设计时通常以能够承受最大荷载,同时具有最小的变形为选择原则。

## 一、材料的强度、比强度与理论强度

当材料受外力作用时,材料内部会产生应力,随着外力增加,应力也相应增大,当应力达到极限值时,材料发生破坏,这个极限应力值即为材料的强度。换言之,材料强度是指材料在外力作用下不破坏时能承受的最大应力。由于外力作用的形式不同,破坏时的应力形式也不同,工程中最基本的外力作用形式如图 1-3 所示,相应的强度分别为抗拉强度、抗压强度、抗弯强度和抗剪强度。

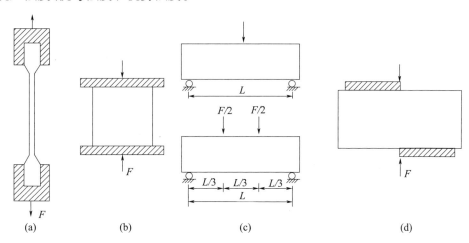

图 1-3　材料所受外力示意图

材料的抗拉强度、抗压强度和抗剪强度可用式(1-17)计算:

$$f = \frac{P}{A} \tag{1-17}$$

式中　$f$——材料的抗拉(或抗压或抗剪)强度(MPa);
　　　$P$——试件破坏时的最大荷载(N);
　　　$A$——试件受力面积(mm$^2$)。

材料的抗弯试验一般选用矩形截面试件,抗弯强度计算有两种情况。
一种是试件在两支点的中间受一集中荷载作用,计算式为:

$$f_{tm} = \frac{3PL}{2bh^2} \tag{1-18}$$

式中 $f_{tm}$——抗弯(折)强度(MPa);
 $P$——试件破坏时的最大荷载(N);
 $L$——两支点之间距离(mm);
 $b、h$——试件截面的宽度和高度(mm)。

另一种是在试件两支点的三分点处作用两个相等的集中荷载,计算式如下:

$$f_{tm} = \frac{PL}{bh^2} \quad (1\text{-}19)$$

影响材料强度的因素很多,除了材料的组成外,材料的孔隙率增加,强度将降低;材料含水率增加,温度升高,一般强度也会降低;另外,试件尺寸大的比小的强度低;棱柱体试件比立方体试件强度低;涂油试件比不涂油试件强度低;加载速度较慢或表面不平等因素均会使所测材料的强度值偏低。

由此可知,材料强度是在特定条件下测定的数值。为了使试验结果准确且具有可比性,各国都制定了统一的材料试验标准,在测定材料强度时,必须严格按照规定的试验方法进行。

承重的结构材料除了承受外荷载,尚需承受自身重力。不同材料的强度比较,常采用比强度。比强度是材料的强度与其表观密度之比,它是衡量材料轻质高强的指标。以常用钢材、木材、混凝土的抗压强度来比较,三者的比强度分别为 0.054、0.069、0.017,可见木材为轻质高强材料,而混凝土属于重质低强材料。努力促进普通混凝土这一当代最重要的结构材料向轻质、高强方向发展,是一项十分重要且紧迫的工作。

上述材料强度是通过试验测定的,故称实际强度。材料的实际强度远低于其理论强度。理论强度是指从材料结构的理论上分析得到的材料所能承受的最大应力。理论强度是克服固体内部质点间的结合力,形成两个新表面时所需的力。材料受力破坏主要是因为外力致使材料质点间产生拉裂或位移所致。计算固体材料理论强度的公式有很多,一般可用下式(奥洛旺公式)计算:

$$\sigma_L = \sqrt{\frac{E\gamma}{d}} \quad (1\text{-}20)$$

式中 $\sigma_L$——理论抗拉强度(Pa);
 $E$——弹性模量(N/m$^2$);
 $\gamma$——固体表面能(J/m$^2$);
 $d$——原子间距离(m),平均为 $2 \times 10^{-10}$ m。

实际工程中所用的材料,其内部结构中均存在一定的缺陷,主要是晶格缺陷(位错、杂原子)或微裂缝等。晶格缺陷的存在能致使材料在较小应力下就发生晶格位移。微裂缝的存在使材料受力时易在裂缝尖端处出现应力集中,致使裂缝不断扩大、延伸、相互连通,从而严重降低材料的强度。例如,钢的理论抗拉强度为 30000MPa,但实际上普通碳素钢的抗拉强度仅为 400MPa 左右(高强钢丝为 1800MPa)。

**二、材料的弹性与塑性**

材料在外力作用下产生变形,当外力卸除后能完全恢复到原始形状的性质称为弹性。

材料的这种可恢复的变形称为弹性变形。如果应力与应变呈直线关系,如式(1-21)所示,则该物体为虎克弹性体,其中,比例常数 $E$ 称为该材料的弹性模量。

$$\sigma = E \cdot \varepsilon \tag{1-21}$$

式中　$\varepsilon$——应变,即单位长度产生的变形量;

　　　$\sigma$——应力(MPa);

　　　$E$——弹性模量(MPa)。

弹性模量是衡量材料抵抗变形能力的一个指标。$E$ 值越大,材料越不易变形,刚度越好。弹性模量是结构设计时的重要参数。

材料在外力作用下产生变形,当外力去除后,有一部分变形不能恢复,这种性质称为塑性,这种变形称为材料的塑性变形,塑性变形是不可逆的变形。

实际上纯弹性变形的材料是没有的,通常一些材料在受力不大时,表现为弹性变形,而当外力达一定值时,则呈现塑性变形,如低碳钢等。另外,许多材料在受力时弹性变形和塑性变形会同时发生,当外力取消后,弹性变形会恢复,而塑性变形不会消失,如混凝土等。

### 三、材料的脆性与韧性

(一)脆性

材料受外力作用,当外力达到一定值时,材料发生突然破坏,且破坏时无明显的塑性变形,这种性质称为脆性,具有这种性质的材料称脆性材料。脆性材料的抗压强度远大于其抗拉强度,可高达数倍甚至数十倍,所以脆性材料不宜用于受拉部位和承受冲击荷载,只适用于作承压构件。土木工程材料中大部分无机非金属材料均为脆性材料,如天然岩石、陶瓷、玻璃、普通混凝土等。

(二)韧性

材料在冲击或振动荷载作用下,能吸收较大的能量,同时产生较大的变形而不破坏的性质称为韧性。与石材、混凝土等脆性材料相比,建筑钢材的韧性较高,因此工程中经常受冲击荷载作用的构件、有抗震要求的构件通常采用钢结构。

### 四、材料的硬度与耐磨性

(一)硬度

硬度是指材料表面抵抗硬物压入或刻划的能力。材料的硬度越大,其强度越高、耐磨性越好。

测定材料硬度的方法有多种,如刻划法、压入法和回弹法。刻划法常用于测定天然矿物的硬度,按硬度递增顺序分为10级,即滑石、石膏、方解石、萤石、磷灰石、正长石、石英、黄玉、刚玉、金刚石。钢材、木材及混凝土等的硬度常用压入法测定,比如布氏硬度就是以单位压痕面积上所受压力来表示的。回弹法常用于测定混凝土构件表面的硬度,并以此估算混凝土的抗压强度。

(二) 耐磨性

材料表面抵抗磨损的能力叫做耐磨性,可用磨损率或磨耗率表示。

$$N = \frac{m_1 - m_2}{A} \tag{1-22}$$

式中　$N$——材料的磨损率或磨耗率($g/cm^2$);

　　$m_1$、$m_2$——试件磨损前和磨损后的质量(g);

　　$A$——试件的受磨面积($cm^2$)。

材料的耐磨性与材料的组成成分、结构、强度、硬度等有关。在水利工程中,滚水坝的溢流面、闸墩和闸底板等部位经常受到夹砂的高速水流的冲刷作用或水底夹带石子的冲击作用而遭受破坏,用于这些部位的材料要求具有抵抗磨损的能力;建筑工程中楼梯的踏面、地面,道路工程中的路面等材料也要求具有较高的耐磨性。

## 第四节　材料的耐久性

材料的耐久性是指用于土木工程材料在环境的多种因素作用下能经久不改变其原有性质、不破坏,长久地保持其使用性能的性质。

### 一、材料经受的环境作用

在建筑物使用过程中,除材料内在原因使其组成、构造、性能等发生变化外,还要长期受到使用条件及各种自然因素的作用,这些作用可概括为以下几方面:

(1)物理作用。包括环境温度、湿度的交替变化,即冷热、干湿、冻融等循环作用。材料在经受这些作用后,将发生膨胀、收缩,或产生内应力,长期的反复作用将使材料渐遭破坏。

(2)化学作用。包括大气和环境水中的酸、碱、盐等溶液或其他有害物质对材料的侵蚀作用,以及日光、紫外线等对材料的作用。

(3)机械作用。包括荷载的持续作用,交变荷载对材料引起的疲劳、冲击、磨损、磨耗等。

(4)生物作用。包括菌类、昆虫等的侵害作用,导致材料发生腐朽、虫蛀等破坏。

### 二、材料的耐久性测定

对材料耐久性最可靠的判断是将其放在使用条件下进行长期的观察和测定,但这需要很长的时间。为此,多采用快速检验法,这种方法的实质是模拟实际使用条件,将材料在实验室进行相关的快速试验,根据测定结果对材料的耐久性给出对比性判定,比如干湿循环、冻融循环、碳化等。

## 复习思考题

1. 材料的组成、结构变化对其性能有何影响?
2. 材料的密度、表观密度与孔隙率之间有何关系?
3. 材料的孔隙对其性能影响如何?
4. 影响材料抗冻性的因素有哪些? 如何改善材料的抗冻性?
5. 何谓导热系数和比热容? 影响材料导热系数的因素有哪些?
6. 何谓材料的强度? 为什么需要按照标准方法测定材料的强度?
7. 脆性材料与韧性材料有何区别? 在使用时应注意哪些问题?
8. 何谓材料的耐久性? 包括哪些内容? 如何确定不同类型材料的耐久性?
9. 一块普通黏土砖的尺寸为 240mm×115mm×53mm,烘干后的质量为 2420g,吸水饱和后为 2640g,将其烘干磨细后取 50g,用李氏瓶测其体积为 19.2cm$^3$,求该砖的开口孔隙率及闭口孔隙率。

# 第二章 无机气硬性胶凝材料

通常,经过一系列物理、化学作用,能将散粒材料(如砂、石子)或块状材料(如砖、石块)黏结为整体的材料,统称为胶凝材料。胶凝材料按其化学成分可分为无机胶凝材料和有机胶凝材料两大类,前者如水泥、石灰、石膏、菱苦土、水玻璃等,后者如沥青、有机高分子聚合物等。其中无机胶凝材料在工程上应用更为广泛,用量也较大。

无机胶凝材料按其硬化条件的不同又可分为气硬性胶凝材料和水硬性胶凝材料两类。所谓气硬性胶凝材料是指只能在空气中硬化,也只能在空气中保持或继续发展其强度的胶凝材料,如石灰、石膏、菱苦土、水玻璃等。水硬性胶凝材料是指不仅能在空气中硬化,而且能更好地在水中硬化,并保持或继续发展其强度的胶凝材料,如水泥。因此,气硬性胶凝材料只适用于地上或干燥环境,不宜用于潮湿环境,更不可用于水中,而水硬性胶凝材料既可用于地上,也可用于地下或水中环境。

## 第一节 石 灰

石灰是在建筑工程上使用较早的胶凝材料之一,是具有不同化学成分和物理形态的生石灰、消石灰、水硬性石灰的统称。因其原材料分布广泛、生产工艺简单、成本低廉,故至今仍被广泛应用于土木工程中。

### 一、石灰的原材料与生产

制备石灰的主要原料是以碳酸钙为主要成分的石灰石、白云石、白垩、贝壳等天然岩石,通过煅烧,碳酸钙分解为氧化钙,即为生石灰。其化学反应式如下:

$$CaCO_3 \xrightarrow{900 \sim 1000℃} CaO + CO_2 \uparrow \quad -178 kJ/mol$$

由于石灰石的致密程度、块体大小、杂质含量不同,并考虑到热损失,因此为了使 $CaCO_3$ 充分煅烧,煅烧温度常控制在 1000~1100℃。生石灰呈块状,也称块灰。由于生产原料中常含有一些碳酸镁,因而经煅烧生成的生石灰中常伴有次要成分的氧化镁。根据我国建材行业标准《建筑生石灰》(JC/T 479—2013),氧化镁含量≤5%时,称为钙质石灰;氧化镁含量>5%时,称为镁质石灰。镁质石灰熟化较慢,但硬化后强度稍高。

将煅烧成的块状生石灰经过不同的加工,还可得到石灰的另外三种产品:

生石灰粉:由块状生石灰磨细而成。

消石灰粉:将生石灰用适量水经消化和干燥而成的粉末,主要成分为 $Ca(OH)_2$,亦称熟石灰。

石灰膏:将块状生石灰用过量水(约为生石灰体积的3~4倍)消化,或将消石灰粉与水拌和,所得到一定稠度的膏状物,主要成分为 $Ca(OH)_2$ 和水。

## 二、石灰的水化与硬化

### (一) 石灰的水化

生石灰的水化又称为熟化或消化,它是指生石灰与水发生水化反应生成熟石灰 $[Ca(OH)_2]$ 的过程。其反应式如下:

$$CaO + H_2O = Ca(OH)_2 + 64.9 kJ/mol$$

生石灰的消化反应为放热反应,消化时不但水化热大,而且放热速率也快。同时伴有显著的体积膨胀,体积膨胀 1.5~2 倍。

由于煅烧时火候不匀,石灰中常含有欠火石灰和过火石灰。欠火石灰是由于煅烧温度过低、时间不足或窑温不均匀导致,此时石灰石未完全分解,生成的石灰孔隙多,比表面积大,内部有未分解的 $CaCO_3$ 内核。过火石灰是由于煅烧时间过长或温度过高导致的,其内部结构致密,CaO 晶粒粗大,活性极低,与水反应的速率极慢,当石灰变硬后才开始熟化,产生体积膨胀,引起已变硬石灰体的隆起鼓包或开裂。为了防止过火石灰体积膨胀引起的隆起和开裂,石灰浆应在储灰坑中存放两星期以上,使其充分水化,该过程称为石灰的"陈伏"。"陈伏"期间,石灰浆表面应保持一层水分,隔绝外界空气,避免石灰浆碳化。

### (二) 石灰的硬化

水化石灰浆体在空气中逐渐硬化,是由两个同时进行的过程来完成的。

结晶过程:石灰浆体在干燥过程中,游离水分蒸发,使 $Ca(OH)_2$ 从饱和溶液中逐渐结晶析出。

碳化过程:$Ca(OH)_2$ 与空气中的 $CO_2$ 和水反应,形成不溶于水的碳酸钙晶体,析出的水分则逐渐被蒸发,其反应式为:

$$Ca(OH)_2 + CO_2 + nH_2O \longrightarrow CaCO_3 + (n+1)H_2O$$

由于碳化作用主要发生在与空气接触的表层,且生成的 $CaCO_3$ 膜层较致密,阻碍了空气中 $CO_2$ 的渗入,也阻碍了内部水分向外蒸发,因此硬化缓慢。

## 三、石灰的特性与技术要求

### (一) 石灰的特性

**1. 可塑性和保水性好**

生石灰熟化后形成的石灰浆,是高度分散的胶体,生成的氢氧化钙颗粒极细(直径约为 $1\mu m$),其比表面积大,可吸附大量水,表面附有较厚的水膜,降低了颗粒之间的摩擦力,具有良好的塑性和保水性,易铺摊成均匀的薄层。利用这一性质,在水泥砂浆中加入石灰浆,可使其可塑性和保水性显著提高。

**2. 凝结硬化慢,强度低**

石灰浆的硬化只能在空气中进行,由于空气中 $CO_2$ 含量少,使碳化作用进行缓慢,加之已硬化的表层对内部的硬化起阻碍作用,所以石灰浆的硬化过程较长。已硬化的石灰强度很低,以1∶3配成的石灰砂浆,28d强度通常只有0.2~0.5MPa。

**3. 耐水性差**

由于石灰浆硬化慢,强度低,当其受潮后,其中尚未碳化的$Ca(OH)_2$易产生溶解,硬化石灰体遇水会产生溃散,故石灰不宜用于潮湿环境。

**4. 硬化时体积收缩大**

石灰浆体黏结硬化过程中,蒸发出大量水分,由于毛细管失水收缩,引起体积收缩,使硬化石灰体产生裂纹,故石灰浆不宜单独使用,通常会在工程施工时掺入一定量的骨料(砂子)或纤维材料(麻刀、纸筋等)。

(二)石灰的技术要求

建筑生石灰的技术要求应符合《建筑生石灰》(JC/T 479—2013)的规定,扫码了解详细内容。

建筑消石灰的技术要求应符合《建筑消石灰》(JC/T 481—2013)的规定,扫码了解详细内容。

(三)石灰的应用与存放

石灰是建筑工程中面广量大的建筑材料之一,其常见的用途有:

**1. 制作石灰乳涂料**

熟石灰粉或石灰膏掺加大量水,可配成石灰乳涂料,是一种廉价的涂料,施工方便且颜色洁白,可用于内墙及顶棚的粉刷。

**2. 拌制建筑砂浆**

消石灰浆和消石灰粉可以单独或与水泥一起配制成砂浆,前者称石灰砂浆,后者称混合砂浆。石灰砂浆可用作砖墙和混凝土基层的抹灰,混合砂浆则用于砌筑,也常用于抹灰。

**3. 加固含水的软土地基**

生石灰块可直接用来加固含水的软土地基。它是在桩孔内灌入生石灰块,利用生石灰吸水熟化时体积膨胀的性能产生膨胀压力,从而使地基加固。

**4. 生产硅酸盐制品**

磨细生石灰(或消石灰粉)和硅质材料(粉煤灰、粒化高炉矿渣、炉渣等)加水拌和,经成型、蒸养或蒸压养护等工序而成的建筑材料,统称为硅酸盐制品。如灰砂砖、粉煤灰砖、粉煤灰砌块、硅酸盐砌块等。

**5. 配制石灰土和石灰三合土**

熟石灰粉与黏土配合成为灰土,再加入砂即成三合土。灰土或三合土在夯实或压实下,密实度大大提高,而且在潮湿环境中,黏土颗粒表面的少量活性氧化硅和氧化铝与$Ca(OH)_2$发生反应,生成水硬性的水化硅酸钙和水化铝酸钙,使黏土的抗渗能力、抗压强度、耐水性得到改善。三合土和灰土主要用于建筑物基础、路面和地面的垫层。

**6. 磨细生石灰粉**

大量采用磨细生石灰来代替石灰膏和消石灰粉配制灰土或砂浆,或直接用于制造硅酸盐制品。

应用石灰时应注意存放,块状生石灰放置太久,会吸收空气中的水分熟化成熟石灰粉,再与空气中的二氧化碳作用而成为碳酸钙,失去胶结能力。所以最好存放在封闭严密的仓库中,防潮防水。另外,存期不宜过长,如需长期存放,可熟化成石灰膏后用砂子铺盖防止碳化。块状生石灰在运输时,应尽量用带棚的车或用帆布盖好,防止水淋自行熟化,放热过高引起火灾。

# 第二节 石 膏

石膏是以硫酸钙为主要成分的气硬性胶凝材料。石膏是一种传统的胶凝材料,具有许多优良的建筑性能,且原料来源丰富、生产能耗低,因而在建筑材料领域中得到了广泛的应用。石膏胶凝材料品种很多,有建筑石膏、高强度石膏、无水石膏水泥、高温煅烧石膏等。

## 一、石膏的原料、生产与品种

生成石膏的主要原料为天然二水石膏、天然无水石膏。化学工业副产物的石膏废渣(如磷石膏、氟石膏、硼石膏)的成分也是二水石膏,也可作为生产石膏的原料。采用化工石膏时应注意,如废渣(液)中含有酸性成分时,必须预先用水洗涤或用石灰中和后才能使用。

石膏按其生产时煅烧的温度不同,分为低温煅烧石膏与高温煅烧石膏。

(一)低温煅烧石膏

低温煅烧石膏是在低温下(110~170℃)煅烧天然石膏所获得的产品,其主要成分为半水石膏($CaSO_4 \cdot 0.5H_2O$)。因为在此温度下,二水石膏脱水,转变为半水石膏,即

$$CaSO_4 \cdot 2H_2O =\!=\!= CaSO_4 \cdot 0.5H_2O + 1.5H_2O$$

属于低温煅烧石膏的产品有建筑石膏和高强石膏。

**1. 建筑石膏**

建筑石膏也称熟石膏或β型半水石膏($\beta\text{-}CaSO_4 \cdot 0.5H_2O$),是天然石膏在常压下在炉窑中进行加热煅烧生成。β型半水石膏晶体呈不规则的片状,是由细小的单个晶粒组成的次生颗粒,其结晶度较差,分散度较大。

**2. 高强度石膏**

高强度石膏是天然石膏在压蒸条件下(0.13MPa,125℃)蒸炼而成,又称α型半水石膏($\alpha\text{-}CaSO_4 \cdot 0.5H_2O$)。α型半水石膏是致密的、完整的、粗大的原生颗粒,其结晶比较完整,分散度较低。相较于β型半水石膏,α型半水石膏的水化速度慢、水化热低、需水量

小、硬化体的强度高。

### (二)高温煅烧石膏

高温煅烧石膏是天然石膏在800～1000℃下煅烧后经磨细而得到的产品。高温时,二水石膏不但完全脱水成无水硫酸钙($CaSO_4$),并且部分硫酸钙分解成氧化钙,少量的氧化钙是无水石膏与水进行反应的激发剂。

高温煅烧石膏与建筑石膏比较,凝结硬化慢,但耐水性和强度高,耐磨性好,用它调制抹灰、砌筑及制造人造大理石的砂浆,可用于铺设地面,也称地板石膏。

## 二、建筑石膏的凝结与硬化

建筑石膏与适量的水拌和后,最初是具有可塑性的石膏浆体,然后逐渐变稠失去可塑性,但尚无强度,这一过程称为"凝结",以后浆体逐渐变成具有一定强度的固体,这一过程称为"硬化"。

建筑石膏在凝结硬化过程中,与水进行水化反应,即

$$CaSO_4 \cdot 0.5H_2O + 1.5H_2O \Longrightarrow CaSO_4 \cdot 2H_2O$$

半水石膏加水后首先进行的是溶解,然后产生上述的水化反应,生成二水石膏。由于二水石膏在水中的溶解度比半水石膏在水中的溶解度小(仅为半水石膏的1/5),故二水石膏不断从过饱和溶液中沉淀而析出胶体颗粒。二水石膏析出,破坏了原有半水石膏的平衡浓度,这时半水石膏会进一步溶解。如此循环进行半水石膏的溶解和二水石膏的析出,直到半水石膏完全溶解。这一过程进行得较快,大约为7～12min。

随着水化的进行,二水石膏生成量不断增加,水分逐渐减少,浆体开始失去可塑性,这称为"初凝"。而后浆体逐渐变稠,颗粒之间的摩擦力、黏结力增加,并开始产生结构强度,表现为"终凝"。石膏终凝后,其晶体颗粒仍在逐渐长大、连生和互相交错,使其强度不断增长,直到剩余水分完全蒸发后,强度才停止发展,这就是石膏的硬化过程(图2-1)。

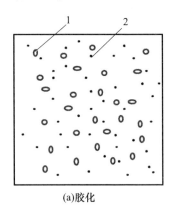
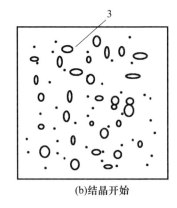
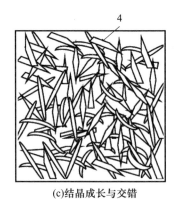

(a)胶化　　　　　　　　　(b)结晶开始　　　　　　　　(c)结晶成长与交错

图2-1　建筑石膏凝结硬化示意图

1—半水石膏;2—二水石膏胶体颗粒;3—二水石膏晶体;4—长大并交错的二水石膏晶体

## 三、建筑石膏的特性、技术性质及应用

（一）建筑石膏的特性

**1. 凝结硬化快**

建筑施工加水拌和后的浆体初凝时间不小于6min,终凝时间不早于30min,一星期左右完全硬化。初凝时间较短会使施工成型困难,为延缓其凝结时间,可以掺入缓凝剂,使半水石膏的溶解度降低或者降低其溶解速度,使水化速度减慢。常用的缓凝剂有动物胶、亚硫酸盐、酒精废液,也可以用硼砂、柠檬酸等。

建筑石膏硬化较快,如一等品石膏1d强度为5~8MPa,7d可达最大强度（为8~12MPa）。

**2. 硬化时体积微膨胀**

石灰和水泥等胶凝材料硬化时往往会产生收缩,而建筑石膏却微有膨胀（膨胀率为0.05%~0.15%）,这能使石膏制品表面光滑饱满、棱角清晰,干燥时不开裂。

**3. 硬化后空隙率较大,表观密度和强度较低**

建筑石膏在使用时,为获得良好的流动性,加入的水量往往比水化所需的水分多。理论需水量为18.6%,而实际加水量约为60%~80%。这些多余的自由水蒸发后会留下许多空隙,故建筑石膏硬化后的表观密度小,强度低。

**4. 隔热、吸声性良好**

石膏硬化体空隙率高,且均为微细的毛细孔,故导热系数小,具有良好的绝热能力;石膏的大量微孔,尤其是表面微孔使声音传导或反射的能力显著下降,从而具有较强的吸声能力。

**5. 防火性能良好**

遇火时,石膏硬化后的主要成分二水石膏中的结晶水蒸发并吸收热量,制品表面形成蒸汽幕,能有效阻止火的蔓延。

**6. 具有一定的调温调湿性**

建筑石膏的热容量大、吸湿性强,故能对室内温度和湿度起到一定的调节作用。

**7. 耐水性和抗冻性差**

建筑石膏的吸湿、吸水性大,故在潮湿环境中,建筑石膏晶体粒子间的黏合力会被削弱,在水中还会使二水石膏溶解而引起溃散,故建筑石膏的耐水性差。另外,建筑石膏中的水分受冻结冰后会产生崩裂,故抗冻性差。

**8. 加工性能好**

石膏制品可锯,可刨,可钉,可打眼。

（二）建筑石膏的技术要求

建筑石膏色白,密度为2.60~2.75g/cm³,堆积密度为0.8~1.1kg/cm³。根据国家强制性标准《建筑石膏》（GB 9776—2022）规定,建筑石膏按其凝结时间和强度分为4.0、3.0

和 2.0 三个等级(表 2-1)。扫码获取《建筑石膏》(GB 9776—2022)的详细内容。

表 2-1 建筑石膏的物理力学性能

| 等级 | 凝结时间(min) | | 强度(MPa) | | | |
| --- | --- | --- | --- | --- | --- | --- |
| | | | 2h 湿强度 | | 干强度 | |
| | 初凝 | 终凝 | 抗折 | 抗压 | 抗折 | 抗压 |
| 4.0 | ≥3 | ≤30 | ≥3.0 | ≥6.0 | ≥7.0 | ≥15.0 |
| 3.0 | | | ≥2.0 | ≥4.0 | ≥5.0 | ≥12.0 |
| 2.0 | | | ≥1.6 | ≥3.0 | ≥4.0 | ≥8.0 |

(三)建筑石膏的应用

石膏在建筑中的应用十分广泛,可用来制作石膏板、各种建筑艺术配件及建筑装饰、彩色石膏制品、石膏砖、空心石膏砌块、石膏混凝土、粉刷石膏、人造大理石等。另外,石膏作为重要的外加剂,广泛应用于水泥、水泥制品及硅酸盐制品中。

**1. 制备粉刷石膏**

粉刷石膏是由建筑石膏或由建筑石膏和 $CaSO_4$ 二者混合后再掺入外加剂、细骨料等,制成的气硬性胶凝材料。

**2. 建筑石膏制品**

建筑石膏制品的种类很多,如纸面石膏板、空心石膏条板、纤维石膏板、石膏砌块和装饰石膏板等,主要用于分室墙、内隔墙、吊顶和装饰等。

建筑石膏配以纤维增强材料、胶粘剂等,还可制成石膏角线、线板、角花、罗马柱、雕塑等艺术装饰石膏制品。

建筑石膏在储存中需要注意防水防潮。储存期一般不得超过三个月,过期或受潮都会使施工制品的强度显著降低。

# 第三节 水玻璃

水玻璃俗称泡花碱,是一种能溶于水的硅酸盐,由不同比例的碱金属氧化物和二氧化硅组成,化学通式为 $R_2O \cdot nSiO_2$。其中,$n$ 是二氧化硅与碱金属氧化物之间的摩尔比,也是水玻璃的模数,一般在 1.5~3.5。常见的水玻璃有硅酸钠($Na_2 \cdot nSiO_2$)和硅酸钾($K_2 \cdot nSiO_2$)等,建筑上常用的是硅酸钠水玻璃。

## 一、水玻璃的硬化

水玻璃在空气中吸收二氧化碳,析出无定型的二氧化硅凝胶,并逐渐干燥脱水成为氧化硅并硬化,反应式如下:

$$Na_2 \cdot nSiO_2 + CO_2 + mH_2O \longrightarrow Na_2CO_3 + nSiO_2 \cdot mH_2O$$

由于上述过程进行得非常缓慢,为了加速硬化,常加入适量的促硬剂氟硅酸钠($Na_2SiF_6$),以加速硅酸凝胶析出:

$$2(Na_2 \cdot nSiO_2) + Na_2SiF_6 + mH_2O \longrightarrow 6NaF + (2n+1)SiO_2 \cdot mH_2O$$

氟硅酸钠的适宜掺量为水玻璃的12%～15%,如果用量太少,硬化速度就会变慢、强度也会变低,且未反应的水玻璃易溶于水,导致水玻璃的耐水性差;用量过多,则凝结就会过快,易造成施工困难,且渗透性大,强度也会变低。

### 二、水玻璃的特性

(1)水玻璃具有良好的胶结能力,硬化时析出的硅酸凝胶有堵塞毛细孔、防止水渗透的作用。水玻璃的模数越大,胶体组分就越多,越难溶于水,水玻璃的黏结能力越强。同一模数的水玻璃,浓度越高,则密度越大,黏结力越强。工程中常用的水玻璃模数为2.6～2.8,密度为1.3～1.4g/cm³。

(2)水玻璃不燃烧,在高温下硅凝胶干燥得很快,强度甚至有所增加。

(3)水玻璃具有高度的耐酸性能,能抵抗大多数无机酸(氢氟酸除外)和有机酸对其的破坏作用。

### 三、水玻璃的应用

利用水玻璃的上述性能,其在建筑工程中主要有以下几方面的用途:

(1)水玻璃耐热混凝土。水玻璃耐热混凝土是由水玻璃、氟硅酸钠、磨细掺合料及粗细骨料按一定配合比例组成。这种混凝土耐热度为600～1200℃,强度等级为C10～C20,高温强度为9.0～20MPa,最高使用温度可达1000～1200℃。

(2)水玻璃耐酸混凝土。水玻璃耐酸混凝土是由水玻璃、氟硅酸钠、辉绿石粉、花岗石、石英砂或石英石等原料配制成的一种耐蚀耐磨材料,耐酸性强,能耐各种浓度的三酸、铬酸、醋酸(除氢氟酸、热磷酸、氟硅酸外)及有机溶剂等介质的腐蚀。

(3)碱激发材料。碱激发材料是将水玻璃与粒化高炉矿渣粉、赤泥、粉煤灰、偏高岭土等材料混合发生碱激发反应,生成具有胶凝能力的水化硅酸钙凝胶。

# 复习思考题

1. 何谓气硬性胶凝材料和水硬性胶凝材料?如何正确使用这两类材料?
2. 古代石灰比现代石灰更加耐水,请运用所学知识进行原因分析。
3. 生石灰熟化时必须进行"陈伏",什么是"陈伏"?有哪些注意事项?磨细的生石灰为什么不经"陈伏"就可以直接使用?
4. 石灰既然不耐水,为何石灰土或三合土可用于基础垫层等潮湿部位?
5. α型半水石膏与β型半水石膏有何区别?
6. 为什么说建筑石膏是一种很好的内装饰材料?为什么它不适用于室外?
7. 什么是水玻璃的模数?请简述水玻璃的凝结硬化过程。

# 第三章 水　　泥

水泥是一种粉末状矿物胶凝材料，它与水混合后形成浆体，经过一系列的物理化学变化，由可塑性浆体变成坚硬的石状体，能将散粒材料胶结成为一个整体。水泥属于水硬性胶凝材料，它不仅能在空气中凝结硬化，而且能更好地在水中凝结硬化，并保持强度增长。

水泥的品种繁多，目前生产和使用的水泥达200余种。按化学成分分，可分为硅酸盐水泥、铝酸盐水泥、硫铝酸盐水泥等。其中硅酸盐水泥是最基本的水泥，应用最广泛。

本章以硅酸盐系列水泥为主要内容，在此基础上介绍其他品种水泥。

## 第一节　水泥概述

### 一、水泥的发展历程

水泥的发展历史代表了胶凝材料的发展历史。早在公元前3000年，古埃及人就开始采用煅烧石膏作为建筑胶凝材料，而古希腊人则是将煅烧石灰石后制得的石灰作为建筑的胶凝材料。公元前146年，古罗马人对石灰的使用工艺进行了改进，在石灰中不仅掺入砂子，还掺入磨细的火山灰或磨细的碎砖，组成了性能更好并且具有部分水硬性的"石灰—火山灰—砂子"三组分砂浆，即"罗马砂浆"。到18世纪，英国为满足航海灯塔的建设需要，出现了含黏土的石灰石制成的水硬性石灰，后来又将含黏土的石灰石经高温煅烧后，磨细制成了"罗马水泥"。1824年，英国工程师约瑟夫·阿斯普丁（Joseph·Aspdin）发明的"波特兰水泥"（即Portland Cement，我国将之称为硅酸盐水泥）获得专利，这标志着现代水泥的诞生。

我国胶凝材料的发展历史十分悠久，早在5000年前，就有人开始使用二氧化硅含量较高的石灰石磨细制成的"白石灰"；公元前7世纪的周朝，开始出现石灰；公元5世纪的南北朝时期，出现了由石灰、黏土和细砂组成的"三合土"，"三合土"与"罗马砂浆"性质相近；自秦汉，我国就出现用掺糯米浆的石灰砌筑砖石以及将"石灰—桐油""石灰—血料"等无机与有机材料相结合的胶凝材料，这在当时的胶凝材料发展中处于世界领先地位。由于种种原因，近代中国胶凝材料的发展却远远落后于世界发展水平，19世纪初才开始现代水泥的生产。改革开放以来，我国的水泥工业获得了巨大的生机和活力，从此进入了重要的历史新阶段，水泥品种也从新中国成立初的几个发展到目前的近百个。2021年我国水泥年生产总量23.63亿t，位居世界第一。

但水泥行业也是重要的碳排放贡献者，生产1t水泥排放0.8～0.9t的$CO_2$。目前，水泥行业碳排放已占工业碳排放总量的26%，人类生活碳排放总量的7%。$CO_2$是全球全球变暖的因素之一，直接影响人类自身、动物以及整个社会的生存和发展。国家主席习近平在第七十五届联合国大会一般性辩论上，向世界作出实现"双碳"目标的中国承诺。目前我国水泥的生产平稳发展,国家鼓励水泥行业大力推广应用低碳绿色水泥基材料。

## 二、硅酸盐水泥的生产

生产硅酸盐水泥的主要原料是石灰质原料和黏土质原料,石灰质原料主要提供 CaO,常采用石灰石、白垩、石灰质凝灰岩等;黏土质原料主要提供 $SiO_2$、$Al_2O_3$ 及 $Fe_2O_3$,常采用黏土、黏土质页岩、黄土等。有时两种原料的化学成分不能满足要求,还需加入少量校正原料来调整,校正材料常采用黄铁矿生产硫酸时产生的废渣——铁矿粉等。为了改善煅烧条件,提高熟料质量,常加入少量矿化剂(如萤石、石膏等)。

生产水泥时首先要将原料按适当比例混合再磨细成生料,然后将制成的生料入窑(回转窑或立窑)进行高温煅烧;再将烧好的熟料配以适当的石膏和混合材料在粉磨机中磨成细粉,即可得到水泥,如图 3-1 所示。

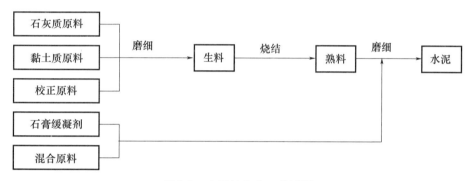

**图 3-1 水泥的生产工艺过程**

硅酸盐水泥的生产有三大环节,即生料制备、熟料烧成和水泥制成。其生产过程常形象地被概括为"两磨一烧"。在水泥生产工艺中,生料制备时加水制成料浆的方法称为湿法生产,干磨成料粉的方法称为干法生产。生料在煅烧过程中要经过干燥、预热、分解、烧成和冷却五个环节,通过一系列物理、化学变化,生成水泥矿物,形成水泥熟料。为使生料能充分反应,窑内烧成温度要达到 1450℃。目前,我国水泥熟料的煅烧主要有以悬浮预热和窑外分解技术为核心的新型干法回转窑生产工艺、传统的干法或湿法回转窑生产工艺和立窑生产工艺等几种。由于新型干法回转窑生产工艺具有规模大、质量好、消耗低、效率高等特点,已经成为发展方向和主流,而传统的回转窑和立窑生产工艺由于技术落后、能耗高、效率低而正逐渐被淘汰。

在硅酸盐水泥生产中,需加入适量石膏和混合材料,加入石膏的作用是延缓水泥的凝结时间,以满足使用的要求;加入混合材料则是为了改善其品种和性能,扩大其使用范围,降低水泥成本、增加水泥产量。

## 三、硅酸盐水泥的组成

硅酸盐水泥一般由硅酸盐水泥熟料、石膏和混合材料三部分组成。

**1. 硅酸盐水泥熟料**

以适当成分的生料煅烧至部分熔融,能得到以硅酸钙为主要成分的产物,该产物被称为硅酸盐水泥熟料。生料中的主要成分是 CaO、$SiO_2$、$Al_2O_3$ 及 $Fe_2O_3$，经高温煅烧后，反应生成硅酸盐水泥熟料中的四种主要矿物：硅酸三钙（$3CaO \cdot SiO_2$，简写式 $C_3S$），占 37%～60%；硅酸二钙（$2CaO \cdot SiO_2$，简写式 $C_2S$），占 15%～37%；铝酸三钙（$3CaO \cdot Al_2O_3$，简写式 $C_3A$），占 7%～15%；铁铝酸四钙（$4CaO \cdot Al_2O_3 \cdot Fe_2O_3$，简写式 $C_4AF$），占 10%～18%。

硅酸盐水泥熟料除上述主要组成外，尚含有少量以下成分：

（1）游离氧化钙。它是在煅烧过程中没有全部化合而残留下来呈游离状态的氧化钙，其含量过高将造成水泥安定性不良，危害很大。

（2）游离氧化镁。若其含量过高、晶粒过大时，也会导致水泥安定性不良。

（3）含碱矿物以及玻璃体等。含碱矿物及玻璃体的 $Na_2O$ 和 $K_2O$ 含量高的水泥，当遇有活性骨料时，易产生碱-骨料反应。

**2. 石膏**

石膏是硅酸盐水泥中必不可少的组成材料，主要作用是调节水泥的凝结时间，常采用天然的或合成的二水石膏（$CaSO_4 \cdot 2H_2O$），也可用含有 $CaSO_4 \cdot 2H_2O$ 的化工废渣。

**3. 混合材料**

混合材料是硅酸盐水泥生产中经常采用的组成材料，按其性能不同，可分为活性与非活性两大类。常用的混合材料有活性类的粒化高炉矿渣、火山灰质材料（沸石、火山灰）、粉煤灰及烧煤矸石等，非活性类的有石灰石、石英砂、钢渣、慢冷矿渣等。

### 四、通用硅酸盐水泥的定义及分类

根据国家标准《通用硅酸盐水泥》（GB 175—2007）规定：以硅酸盐水泥熟料和适量的石膏及规定的混合材料制成的水硬性胶凝材料称为通用硅酸盐水泥（Common Portland Cement）。通用硅酸盐水泥按混合材料的品种和掺量分为硅酸盐水泥、普通硅酸盐水泥、矿渣硅酸盐水泥、火山灰质硅酸盐水泥、粉煤灰硅酸盐水泥和复合硅酸盐水泥。

## 第二节　硅酸盐水泥和普通硅酸盐水泥

### 一、硅酸盐水泥的组分

国家标准《通用硅酸盐水泥》（GB 175—2007）规定：硅酸盐水泥由硅酸盐水泥熟料、0～5%石灰石或粒化高炉矿渣和适量石膏组成。硅酸盐水泥又分为两种类型：不掺加混合材料的称Ⅰ型硅酸盐水泥，其代号为 P·Ⅰ；在硅酸盐水泥熟料粉磨时掺入不超过水泥质量5%的石灰石或粒化高炉矿渣的称Ⅱ型硅酸盐水泥，其代号为 P·Ⅱ。

### 二、硅酸盐水泥的水化和凝结硬化

水泥加水拌和后，最初会形成具有可塑性的水泥浆体，随着水泥水化反应的进行，水泥浆体逐渐变稠失去可塑性，称为凝结。随着水化反应的继续，浆体逐渐变为具有一定强度的坚硬的固体水泥石,称为硬化。

(一)硅酸盐水泥的水化

硅酸盐水泥熟料由四种主要矿物组成,这些矿物的水化、硬化性质决定了水泥的性质。硅酸盐水泥与水拌和后,其熟料中各矿物立即单独与水发生水化反应,生成水化产物,各矿物的水化反应如下：

**1. 硅酸三钙的水化**

硅酸三钙是水泥熟料的主要矿物,其水化作用、产物和凝结硬化对水泥的性能有着重要影响。在常温下硅酸三钙的水化反应为：

$$2(3CaO \cdot SiO_2) + 6H_2O =\!=\!= 3CaO \cdot 2SiO_2 \cdot 3H_2O + 3Ca(OH)_2$$

可简写为

$$2C_3S + 6H =\!=\!= 3C\text{-}S\text{-}H + 3CH$$

式中,C-S-H 为水化硅酸钙。

硅酸三钙的水化产物为水化硅酸钙和氢氧化钙。水化硅酸钙为凝胶体,显微结构是纤维状；氢氧化钙为晶体,易溶于水。硅酸三钙的水化速度很快,水化放热量大,生成的硅酸钙凝胶构成具有高强度的空间网格结构,是水泥强度的主要来源,其凝结时间正常,早期和后期强度都较高。

**2. 硅酸二钙的水化**

硅酸二钙的水化与硅酸三钙相似,但水化速度慢很多,其水化反应如下：

$$2(2CaO \cdot SiO_2) + 4H_2O =\!=\!= 3CaO \cdot 2SiO_2 \cdot 3H_2O + Ca(OH)_2$$

可简写为

$$2C_2S + 4H =\!=\!= 3C\text{-}S\text{-}H + CH$$

在硅酸二钙的水化产物中,水化硅酸钙在形貌方面都与硅酸三钙的水化产物无大的区别,因此水化硅酸钙也称为硅酸钙凝胶；而氢氧化钙的生成量较硅酸三钙的少,且结晶比较粗大。在硅酸盐水泥熟料矿物中,硅酸二钙水化速度最慢,但后期增长,水化放热量小；其早期强度低,后期强度增长,可接近甚至超过硅酸三钙的强度,是保证水泥后期强度增长的主要因素。

**3. 铝酸三钙的水化**

铝酸三钙水化产物通称为水化铝酸钙,其水化反应如下：

$$2(3CaO \cdot Al_2O_3) + 6H_2O =\!=\!= 3CaO \cdot Al_2O_3 \cdot 6H_2O$$

可简写为

$$2C_3A + 6H =\!=\!= 3C\text{-}A\text{-}H$$

在硅酸盐水泥熟料矿物中,铝酸三钙水化速度最快,水化放热量大且放热速度快。其早期强度增长快,但强度值并不高,后期几乎不再增长,对水泥的早期(3d 以内)强度有一定的影响。水化铝酸钙凝结速度快,会使水泥瞬间凝结。为了控制 $C_3A$ 的水化和凝结硬化速度,就必须在水泥中掺入适量石膏。这样,$C_3A$ 水化后的产物将与石膏反应,生成高硫型水化硫铝酸钙($3CaO \cdot Al_2O_3 \cdot 3CaSO_4 \cdot 31H_2O$,又称钙矾石)。首先,钙矾石的形成反应速度比纯 $C_3A$ 的反应慢；其次,水泥颗粒表面析出钙矾石晶体构成阻碍层,形成"半透膜",延缓了水泥颗粒的水化,因此避免了闪凝或假凝。石膏完全消耗后,一部分钙矾石将转变为单硫型水化硫铝酸钙。但石膏掺量不能过多,否则不仅缓凝作用不大,还会引起水泥的体积安定性不良。

**4. 铁铝酸四钙的水化**

铁铝酸四钙的水化反应及产物与 $C_3A$ 相似,生成水化铝酸钙与水化铁酸钙的固溶体,其反应式如下:

$$4CaO \cdot Al_2O_3 \cdot Fe_2O_3 + 7H_2O = 3CaO \cdot Al_2O_3 \cdot 6H_2O + CaO \cdot Fe_2O_3 \cdot H_2O$$

可简写为 $C_4AF + 27H = C_3AH_6 + CFH$

铁铝酸四钙的水化速度较快,仅次于 $C_3A$,水化热不高,凝结正常,其强度值较低,但抗折强度相对较高。提高 $C_4AF$ 的含量,可降低水泥的脆性,有利于道路等有振动交变荷载作用的场合使用。

上述各单熟料矿物水化与凝结硬化表现出的特性见表 3-1 和图 3-2。

表 3-1 硅酸盐水泥主要矿物组成及其特性

| 性能指标 | | 熟料矿物 | | | |
| --- | --- | --- | --- | --- | --- |
| | | $C_3S$ | $C_2S$ | $C_3A$ | $C_4AF$ |
| 密度(g/cm³) | | 3.25 | 3.28 | 3.04 | 3.77 |
| 水化反应速率 | | 快 | 慢 | 最快 | 快 |
| 水化放热量 | | 大 | 小 | 最大 | 中 |
| 强度 | 早期 | 高 | 低 | 低 | 低 |
| | 后期 | | 高 | | |
| 收缩 | | 中 | 中 | 大 | 小 |
| 抗硫酸盐腐蚀性 | | 中 | 最好 | 差 | 好 |

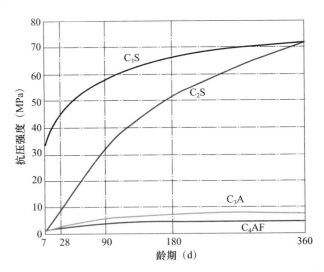

图 3-2 水泥熟料矿物不同龄期的抗压强度

硅酸盐水泥与水作用后生成的主要水化产物有:水化硅酸钙凝胶、水化铁酸钙凝胶、氢氧化钙晶体、水化铝酸钙晶体、水化硫铝酸钙晶体。在完全水化的水泥中,水化硅酸钙约占 70%,氢氧化钙约占 20%,钙矾石和单硫型水化硫铝酸钙约占 7%。

### (二)硅酸盐水泥的凝结硬化过程

水泥的凝结和硬化是人为划分的,实际上,凝结和硬化是一个连续的复杂的物理、化学变化过程。一般按水化反应速率和水泥浆体的结构特征划分,硅酸盐水泥的凝结硬化过程可分为诱导期、凝结期、硬化期三个阶段。

水泥强度发展的一般规律是:3~7d 内强度增长最快,28d 内强度增长较快,超过 28d 后强度将继续发展而增长较慢。

### (三)影响硅酸盐水泥凝结硬化的主要因素

**1. 水泥矿物组成的影响**

熟料的矿物组成直接影响着水泥的水化和凝结硬化。如图 3-3 所示,$C_3S$ 最初反应较慢,但之后反应很快;$C_3A$ 和 $C_3S$ 刚好相反,反应速度先快后慢;$C_4AF$ 开始反应速率比 $C_3S$ 快,以后变慢;$C_2S$ 的水化最慢,但后期稳步增长。

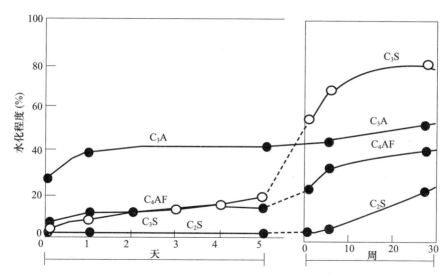

图 3-3 水泥熟料矿物不同龄期的水化程度

**2. 石膏掺量的影响**

石膏的掺入主要是为降低 $C_3A$ 的水化速度。随着石膏掺量的增加,水泥放热速度减慢,放热峰延后出现,但石膏掺量过多则对水泥的凝结硬化影响不大。

**3. 水泥细度的影响**

水泥颗粒的粗细直接影响水泥的水化、凝结硬化、强度、干缩及水化热等。但水泥颗粒过细,易与空气中的水分及二氧化碳反应,致使水泥不宜久存,过细的水泥在硬化时产生的收缩亦较大,而且磨制过细的水泥耗能大,成本高。一般认为水泥颗粒小于 40μm 时就具有较高的活性,大于 90μm 时活性较小。

**4. 养护湿度和温度的影响**

通常,提高温度可加速硅酸盐水泥的早期水化,使早期强度能较快发展,但对后期强度反而可能有所降低。相反,在较低温度下硬化时,虽然硬化速率慢,但水化产物较致密,可获得较高的最终强度。保持环境的温度和湿度,使水泥石强度不断增长的措施,称为养护。

**5. 拌和加水量的影响**

拌和水泥浆体时,为使浆体具有一定塑性和流动性,所加入的水量通常要大大超过水泥充分水化时所需的水量,多余的水在硬化的水泥石内形成毛细孔。因此拌和水越多,硬化水泥石中的毛细孔越多,当水灰比为 0.40 时,完全水化后水泥石的空隙率为 29.6%,而水灰比为 0.70 时,水泥石的空隙率高达 50.3%。水泥石的强度随其毛细孔隙率的增加呈线性下降。因此,在熟料矿物组成大致相近的情况下,拌和水泥浆的用水量是影响水泥石强度的主要因素。

**6. 养护龄期的影响**

水泥的水化和硬化是一个较长时期不断进行的过程,随着水泥颗粒内各熟料矿物水化程度的提高,凝胶体不断增加,毛细孔隙相应减少,从而随着龄期的增长使水泥石的强度逐渐提高。水泥在 3~14d 内强度增长较快,28d 后增长缓慢。

**7. 调凝外加剂的影响**

加入促凝剂($CaSO_4$、$CaCl_2$)能促进水泥水化、硬化,提高早期强度。相反,掺加缓凝剂(木钙、糖类等)可延缓水泥的水化、硬化,影响水泥早期强度的发展。

**8. 水泥受潮与久存**

水泥受潮后,因表面已水化而结块,从而丧失胶凝能力,严重降低其强度。而且,即使在良好的储存条件下,水泥也不可储存过久,因为水泥会吸收空气中的水分和二氧化碳,产生缓慢水化和碳化作用,经三个月后水泥强度降低约 10%~20%,6 个月后降低约 15%~30%,一年后降低约 25%~40%。

## 三、硅酸盐水泥的技术性质

国家标准《通用硅酸盐水泥》(GB 175—2007)对硅酸盐水泥的主要技术性质作出下列规定。扫码获取该标准的详细内容。

**(一)细度**

细度是指水泥颗粒的粗细程度。水泥颗粒细度影响水泥的水化活性和凝结硬化速度,水泥颗粒太粗,水化活性就低,不利于凝结硬化;虽然水泥越细,凝结硬化越快,早期强度会越高,但是水化放热速度也快,水泥收缩也越大,对水泥石性能会产生不利影响。

水泥的细度可用筛析法和比表面积法检验。筛析法是采用边长为 80μm 的方孔筛对水泥试样进行筛析试验,以筛余百分数表示水泥的细度。普通水泥的筛余要求不得超过 10.0%。扫码可了解水泥细度筛析试验内容。

比表面积法是用透气法测定水泥的比表面积(单位质量的粉末所具有的总表面积),以 $m^2/kg$ 为单位。比表面积法与筛析法相比,能较好地反映水泥粗细颗粒的分配情况,是较为合理的方法。标准规定,硅酸盐水泥的比表面积应不小于 $300m^2/kg$。

## (二)凝结时间

凝结时间是指水泥从加水开始,到水泥浆失去塑形所需的时间。凝结时间分初凝时间和终凝时间,初凝时间是指从水泥加水到水泥浆开始失去可塑性的时间,终凝时间是指从水泥加水到水泥浆开始凝固并具有强度的时间。标准规定,硅酸盐水泥的初凝时间不小于45min,终凝时间不大于390min,凡初凝时间不符合规定者为废品,终凝时间不符合规定者为不合格品。

水泥凝结时间的测定,是以标准稠度的水泥净浆在规定温度和湿度下凝结所需要的时间。凝结时间用测定仪来测定。标准稠度是指水泥净浆达到规定稠度时所需的拌和水量,以占水泥质量的百分率来表示。扫码可了解水泥标准稠度用水量和凝结时间试验的内容。硅酸盐水泥的标准稠度用水量,一般在24%~30%。水泥熟料矿物成分不同时,其标准稠度用水量亦有所差别,磨得越细的水泥,标准稠度用水量越大。

水泥的凝结时间对水泥混凝土和砂浆的施工有重要的意义。初凝时间不宜过短,以便施工时有足够的时间来完成混凝土和砂浆拌和物的运输、浇捣或砌筑等操作;终凝时间不宜过长,这是为了使混凝土和砂浆在浇捣或砌筑完毕后能尽快凝结硬化,以利于下一道工序的及早进行。

## (三)体积安定性

体积安定性是指水泥石硬化后抵抗体积变化而不发生延迟性膨胀破坏的性能。体积安定性不良的水泥会使混凝土构件因膨胀而产生裂缝,降低工程质量。

引起水泥体积安定性不良的原因是水泥熟料中存在过多的游离氧化钙和游离氧化镁,或者水泥熟料中石膏掺量过多。游离氧化钙和氧化镁均是过烧的、熟化很缓慢,在水泥硬化并产生一定强度后才开始水化。由于它们在熟化过程中体积膨胀,引起水泥石不均匀的体积变化,使水泥石产生裂缝;当水泥熟料中的石膏掺量过多,水泥硬化后石膏还会与固态的水化铝酸钙反应生成高硫型水化硫铝酸钙,体积膨胀约1.5倍以上,导致水泥体积安定性不良,致使水泥石开裂。

对于因过量的游离氧化钙引起的水泥体积安定性不良,国家标准规定采用沸煮法检验,因为沸煮法可以加速游离氧化钙的熟化。沸煮法又分为试饼法和雷氏法两种,当这两种方法发生争议时,以雷氏法为准。扫码可了解水泥体积安定性试验的具体内容。

由于游离氧化镁比游离氧化钙熟化更为缓慢,因此,沸煮法对游离氧化镁无效果,一般会采用压蒸法来测定游离氧化镁的体积安定性。因石膏掺量过多引起水泥的体积安定性不良,需要长期浸在常温水中才能发现。由此可知,游离氧化镁和石膏掺量过多引起的水泥体积安定性不良均不能快速检测,因此,国家标准规定,硅酸盐水泥中的氧化镁含量不得超过5.0%,三氧化硫含量不得超过3.5%,以保证水泥的体积安定性。

国家标准规定水泥体积安定性必须合格,安定性不良的水泥视为废品,不能用在工程中。

## (四)强度和强度等级

水泥的强度是水泥的主要技术指标。一般以28d的抗压强度来表征硅酸盐水泥的强度等级。

目前水泥强度的测定采用《水泥胶砂强度检验方法(ISO法)》(GB/T 17671—2021)规定的方法,扫码可了解水泥胶砂强度试验的内容。《通用硅酸盐水泥》(GB 175—2007)规定,根据3d、28d的抗折与抗压强度结果将硅酸盐水泥分为42.5,42.5R,52.5,52.5R,62.5和62.5R六个强度等级。其中代号R属于早强型水泥。不同类型各强度等级的硅酸盐水泥各龄期强度不得低于表3-2中规定的数值。

表3-2 硅酸盐水泥各龄期的强度要求(GB 175—2007)

| 强度等级 | 抗压强度(MPa) | | 抗折强度(MPa) | |
| --- | --- | --- | --- | --- |
| | 3d | 28d | 3d | 28d |
| 42.5 | ≥17.0 | ≥42.5 | ≥3.5 | ≥6.5 |
| 42.5R | ≥22.0 | | ≥4.0 | |
| 52.5 | ≥23.0 | ≥52.5 | ≥4.0 | ≥7.0 |
| 52.5R | ≥27.0 | | ≥5.0 | |
| 62.5 | ≥28.0 | ≥62.5 | ≥5.0 | ≥8.0 |
| 62.5R | ≥32.0 | | ≥5.5 | |

(五)碱含量

水泥中的碱含量按$Na_2O + 0.658K_2O$计算的质量百分率来表示。当混凝土骨料中含有活性二氧化硅时,混凝土骨料会与水泥中的碱相互作用形成碱硅酸盐凝胶,由于碱硅酸盐凝胶体积膨胀可引起混凝土开裂,造成结构的破坏,这种现象称为碱-骨料反应。它是影响混凝土耐久性的一个重要因素。为防止发生碱-骨料反应,《通用硅酸盐水泥》(GB 175—2007)对碱含量作出了相应规定,将碱含量定为任选要求。当用户有要求时,由供需双方商定;但有用户要求提供低碱水泥时,水泥中的碱含量不得大于0.60%。

(六)水化热

水泥在水化过程中放出的热量称为水化热。水化热的多少与放热速度快慢取决于水泥熟料的矿物成分与水泥的细度,还与混合材料、外加剂的品种及掺量有关。

水泥的水化热对于混凝土工程既有利也有弊,水泥的水化热有利于混凝土的冬期施工,但是对于大体积混凝土是不利的。由于硅酸盐水泥放热量大而且集中,对于大型基础、大坝和桥墩等大体积混凝土,其内部散热缓慢,使内部温度常常达到50~60℃。而大体积混凝土外部散热较快,温度较低,内外温差达到一定值时,就会形成温差裂缝,降低混凝土强度和耐久性。

## 四、水泥石的腐蚀与防腐

硅酸盐水泥硬化后,在一般使用条件下具有较高的耐久性。但是当水泥石长期处于某些腐蚀性介质的环境中时,会使其强度和耐久性降低,甚至导致混凝土结构破坏,这种现象称为水泥石的腐蚀。下面介绍几种典型介质腐蚀类型。

## (一)软水侵蚀(溶出性侵蚀)

$Ca(OH)_2$晶体是水泥的主要水化产物之一,水泥的其他水化产物也需在一定浓度的$Ca(OH)_2$溶液中才能稳定存在,而$Ca(OH)_2$又是易溶于水的,特别易溶于含碳酸氢钙很少的软水。若水泥石中的$Ca(OH)_2$晶体被溶解流失,其浓度低于水化产物稳定所需要的最低要求时,水泥的其他水化产物就会被溶解或分解,从而造成水泥石的破坏,这就是溶出性腐蚀。

雨水、雪水、蒸馏水、冷凝水、含碳酸盐较少的河水和湖水等都是软水,当水泥石长期处在软水中时,氢氧化钙因其溶解度最大首先被溶出,并很快达到其饱和溶液。在静止水或无压水中,氢氧化钙的溶出仅限于水泥石表面,因此,对水泥石性能影响不大。但在流动水或者有压水中,水泥石中的氢氧化钙会溶出并被水带走,水泥石中的氢氧化钙浓度会不断降低,这一方面使水泥石孔隙率增大,密实度和强度下降,水更易向内部渗透;另一方面,水泥石的碱度不断降低,引起水化产物分解,最终变成胶结能力很差的产物,使水泥石结构受到破坏。而每流失1%的CH,基体强度损失5%~7%。

## (二)盐类侵蚀

### 1. 硫酸盐侵蚀

在海水、湖水、盐沼水、地下水和某些工业污水及流经高炉矿渣或煤渣的水中,常含有钠、钾和铵的硫酸盐,它们与水泥石中的氢氧化钙发生复分解反应,产物硫酸钙沉积在已硬化的水泥石表面孔隙内,结晶膨胀导致水泥石开裂。如生成的硫酸钙较多,且新生成的硫酸钙活性高,则容易与水泥石中固态水化铝酸钙反应,反应式如下:

$$3CaO \cdot Al_2O_3 \cdot 6H_2O + 3(CaSO_4 \cdot 2H_2O) + 19H_2O = 3CaO \cdot Al_2O_3 \cdot 3CaSO_4 \cdot 31H_2O$$

上述反应生成了带有31个结晶水的高硫型水化硫铝酸钙,膨胀体积为原体积的1.5倍以上,对水泥石结构造成的破坏会更为严重。由于高硫型的水化硫铝酸钙是针状晶体,因此称之为"水泥杆菌"。

值得注意的是,在生产硅酸盐水泥时,为调节硅酸盐水泥的凝结时间,会适量加入石膏。石膏与水化铝酸钙反应也生成了高硫型的水化硫铝酸钙,但是它是在水泥浆尚具有一定可塑性时生成的,因此不具有破坏作用。

### 2. 镁盐侵蚀

在海水、地下水和盐沼水中,常含有大量的以硫酸镁和氯化镁为主的镁盐,它们与水泥石中的氢氧化钙发生复分解反应:

$$MgSO_4 + Ca(OH)_2 + 2H_2O = CaSO_4 \cdot 2H_2O + Mg(OH)_2$$
$$MgCl_2 + Ca(OH)_2 = CaCl_2 + Mg(OH)_2$$

产物中的氢氧化镁松散没有胶结能力,而氯化钙易溶于水,二水石膏则会受到硫酸盐的腐蚀。因此镁盐对水泥石具有双重破坏作用。

### 3. 酸类侵蚀

(1)碳酸的腐蚀

在工业废水和地下水中,常常溶解有较多的二氧化碳,水泥石中的氢氧化钙和二氧化

碳作用,生成不溶于水的碳酸钙:
$$Ca(OH)_2 + CO_2 + H_2O = CaCO_3 + 2H_2O$$
生成的碳酸钙再与含碳酸的水作用生成易溶于水的碳酸氢钙,此反应为可逆反应:
$$CaCO_3 + CO_2 + 2H_2O = Ca(HCO_3)_2$$

由于水中溶解的二氧化碳较多,使上述反应向右进行,水泥石中氢氧化钙转化成易溶于水的碳酸氢钙而流失。当水泥石中氢氧化钙浓度降低到一定值时,其他水化产物相继分解,使水泥石结构遭到破坏。

(2) 一般酸的腐蚀

工业废水、地下水和沼泽水中常含有无机酸和有机酸。各种酸对水泥石腐蚀程度不同,它们与水泥石中的氢氧化钙发生反应,其产物或是易溶于水,或是结晶膨胀,均会降低水泥石的强度。无机酸中对水泥石腐蚀最严重的是盐酸、氢氟酸、硫酸和硝酸等,而有机酸中的醋酸、蚁酸和乳酸等对水泥石的腐蚀作用最为严重。例如盐酸与硫酸分别与水泥石中的氢氧化钙反应,反应式如下:
$$2HCl + Ca(OH)_2 = CaCl_2 + 2H_2O$$
$$H_2SO_4 + Ca(OH)_2 = CaSO_4 \cdot 2H_2O$$

反应生成的氯化钙易溶于水,硫酸钙在水泥石的孔隙中结晶膨胀。如硫酸钙生成量较多,还会与水泥石中的固态水化铝酸钙反应,生成高硫型的水化硫铝酸钙,结晶膨胀1.5倍以上,对水泥石的危害更大。

**4. 强碱的腐蚀**

硅酸盐水泥水化后,由于水化产物中的氢氧化钙存在,使水泥石显碱性(pH值一般为12.5~13.5),所以一般碱浓度不高时对水泥石不产生腐蚀。如果水泥中铝酸盐含量较高时,遇到强碱介质(如氢氧化钠),两者反应生成易溶于水的铝酸钠,其反应为:
$$3CaO \cdot Al_2O_3 + 6NaOH = 3NaO \cdot Al_2O_3 + 3Ca(OH)_2$$

当水泥石被氢氧化钠浸透后,在空气中,氢氧化钠与二氧化碳反应生成碳酸钠,碳酸钠会在水泥石毛细孔中结晶膨胀,导致水泥石产生裂缝,强度降低。其反应式如下:
$$2NaOH + CO_2 = Na_2CO_3 + H_2O$$

(三) 水泥石腐蚀的防止

根据以上对侵蚀作用的分析可以看出,水泥石被腐蚀的基本原因为:①水泥石中存在易被腐蚀的成分,如氢氧化钙和水化铝酸钙;②水泥石本身不致密,有很多毛细孔通道,侵蚀性介质易于进入其内部。因此,针对以上情况可采取下列措施防止水泥石的腐蚀。

(1) 根据侵蚀环境特点,合理选择水泥品种。水泥石中易引起腐蚀的组分是氢氧化钙和水化铝酸钙,如选用水化产物中氢氧化钙含量少的水泥(如掺有混合材料的水泥),可有效地防止软水和镁盐等介质的侵蚀;如选用铝酸盐含量较少的水泥(如抗硫酸盐水泥),可显著地提高水泥石抵抗硫酸盐侵蚀的能力。

(2) 提高水泥石的密实度。水泥实际拌和用水量为30%~60%,远大于其化学反应的理论需水量(23%),导致多余水分蒸发,在水泥石中形成许多孔隙。因此在实际工程中一般采取降低水灰比,选择级配良好的骨料,掺入混合材料和外加剂,改善施工工艺等方法来提高混凝土的密实度。另外,还可以在混凝土表面进行碳化处理或者氟硅酸处理,提高混凝土表面的密实度以阻止侵蚀性介质侵入内部。

(3)设置隔离层或保护层。在混凝土表面加做耐腐蚀性高且不透水的保护层,如工程中常采用耐酸陶瓷、耐酸石料、玻璃、塑料等作为耐酸保护层。

## 五、硅酸盐水泥的特性与应用

水泥的特性与其应用是相适应的,硅酸盐水泥具有以下特性:

(1)凝结硬化快,早期强度与后期强度均高

这是因为硅酸盐水泥中熟料多,即水泥中 $C_3S$ 多。因此,适用于现浇混凝土工程、预制混凝土工程、冬期施工混凝土工程、预应力混凝土工程、高强混凝土工程等。

(2)抗冻性好

由于硅酸盐水泥凝结硬化快,早期强度高,而且其拌和物不易发生泌水,密实度高,适用于寒冷地区和严寒地区遭受反复冻融的混凝土工程。

(3)抗碳化性能好

由于硅酸盐水泥凝结硬化后,水化产物中氢氧化钙浓度高,水泥石的碱度高,再加上硅酸盐水泥混凝土的密实度高,开始碳化生成的碳酸钙填充混凝土表面的孔隙,使混凝土表面更密实,有效地阻止了进一步碳化。因此,硅酸盐水泥抗碳化性能高,可用于有碳化要求的混凝土工程中。

(4)耐腐蚀性差

由于硅酸盐熟料中硅酸三钙和铝酸三钙含量高,其水化产物中易腐蚀的氢氧化钙和水化铝酸三钙含量高,因此耐腐蚀性差,不宜在受流动水和压力水作用的工程长期使用,也不适用于受海水及其他侵蚀性介质作用的工程。

(5)水化热高

由于硅酸盐水泥熟料中的硅酸三钙和铝酸三钙含量较高,水化热高,因此不宜用于大体积混凝土工程中,但可应用于冬期施工的工程中。

(6)耐热性差

硅酸盐水泥混凝土在温度不高时(一般为100~250℃),尚存的游离水可使水化继续进行,混凝土的密实度进一步增加,强度有所提高;当温度高于250℃时,水泥中的水化产物氢氧化钙分解为氧化钙,如再遇到潮湿的环境,氧化钙熟化体积膨胀,使混凝土遭到破坏;另外,水泥受热约300℃时,体积收缩,强度开始下降;温度达700℃时,强度降低很多,甚至完全破坏。因此,硅酸盐水泥不宜应用于有耐热性要求的混凝土工程中。

(7)耐磨性好

由于硅酸盐水泥混凝土的强度高,因此其耐磨性好,可应用于路面和机场跑道等混凝土工程中。

## 六、普通硅酸盐水泥

按《通用硅酸盐水泥》(GB 175—2007)的规定:普通硅酸盐水泥由硅酸盐水泥熟料、

5%~20%混合材料及适量石膏组成,代号P·O。

由普通硅酸盐水泥的组分可知,普通硅酸盐水泥与硅酸盐水泥的差别仅在于普通硅酸盐水泥含有少量混合材料,而绝大部分仍是硅酸盐水泥熟料,故其特性与硅酸盐水泥基本相同;但由于掺入少量混合材料,因此与同强度等级硅酸盐水泥相比,普通硅酸盐水泥早期硬化速度稍慢、3d强度稍低、抗冻性稍差、水化热稍小、耐腐蚀性稍好。

普通硅酸盐水泥的终凝时间不大于100min,其余技术性质要求同硅酸盐水泥。

《通用硅酸盐水泥》(GB 175—2007)规定,普通硅酸盐水泥分四个强度等级:42.5、42.5R、52.5、52.5R。其中代号R属于早强型水泥。普通硅酸盐水泥不同强度等级的各龄期强度应不低于表3-3中的数值。

表3-3 普通硅酸盐水泥各龄期的强度要求(GB 175—2007)

| 强度等级 | 抗压强度(MPa) | | 抗折强度(MPa) | |
| --- | --- | --- | --- | --- |
| | 3d | 28d | 3d | 28d |
| 42.5 | ≥17.0 | ≥42.5 | ≥3.5 | ≥6.5 |
| 42.5R | ≥22.0 | ≥42.5 | ≥4.0 | ≥6.5 |
| 52.5 | ≥23.0 | ≥52.5 | ≥4.0 | ≥7.0 |
| 52.5R | ≥27.0 | ≥52.5 | ≥5.0 | ≥7.0 |

### 七、水泥的贮运

水泥在运输与贮存时不得受潮或混入杂物,不同品种和强度等级的水泥在贮运中避免混杂。一般水泥的储存期为三个月,使用存放三个月以上的水泥,必须重新检验其强度,否则不得使用。

## 第三节 大掺量混合材料的硅酸盐水泥

### 一、混合材料

在磨制水泥时加入的天然或人工矿物材料称为混合材料,包括非活性混合材料和活性混合材料两大类。

(一)活性混合材料

凡常温下与石灰、石膏或硅酸盐水泥一起拌和后能发生水化反应,生成具有水硬性的胶凝水化产物的混合材料称为活性混合材料。常用的活性混合材料有粒化高炉矿渣、火山灰质混合材料和粉煤灰等。活性混合材料作用是改善水泥的某些性能、调节水泥强度等级、降低水化热、降低生成成本、增加水泥产量、扩大水泥品种等。

**1. 粒化高炉矿渣**

粒化高炉矿渣是高炉冶炼生铁时的熔融矿渣,经水淬冷后得到的松散颗粒。粒化高炉矿渣的主要化学成分是 CaO、$SiO_2$、$Al_2O_3$ 和少量 MgO、$Fe_2O_3$。急冷的矿渣结构为不稳定的玻璃体，具有较大的化学潜能，其主要活性成分是活性 $SiO_2$ 和活性 $Al_2O_3$。

**2. 火山灰质混合材料**

火山灰质混合材料是指具有火山灰性质的天然或人工的矿物材料，其品种很多，天然的有火山灰、浮石、浮石岩、沸石、硅藻土等；人工的有烧页岩、烧黏土、煤渣、烧煤矸石或自燃煤矸石、硅灰等。火山灰的主要活性成分是活性 $SiO_2$ 和活性 $Al_2O_3$。

**3. 粉煤灰**

粉煤灰是燃煤电厂所排放的工业废气经排烟通道收集的粉末，又称飞灰。粉煤灰是由煤粉悬浮态燃烧后急冷形成的，因此，粉煤灰大多为直径 0.001～0.05mm 实心或空心玻璃态球粒。粉煤灰化学活性的高低取决于活性 $SiO_2$ 和活性 $Al_2O_3$ 的含量和玻璃体的含量。

活性混合材的作用主要体现在如下方面：

(1) 发生火山灰反应。因活性混合材活性较低，该反应较慢；同时此反应具有温度敏感性，适合蒸养。

(2) 填充效应。小粒径的混合材颗粒可填充在未水化水泥和水化产物之间的微小空隙中，提高水泥石的密实度。

(3) 形态效应。某些混合材的颗粒圆形度比水泥大，加之早期不易与水发生化学反应，起到类似滚珠的作用，改善工作性。

(二) 非活性材料

凡是不具有活性或活性很低的天然或人工矿物质材料统称为非活性混合材料，它与水泥的水化产物基本不发生化学反应。非活性混合材料的作用是调整水泥强度等级、增加产量、减少生产成本、降低水化热等。常用的非活性混合材料有石英砂、石灰石和慢冷矿渣。

## 二、矿渣硅酸盐水泥、火山灰质硅酸盐水泥及粉煤灰硅酸盐水泥

(一) 材料组分

**1. 矿渣硅酸盐水泥**

矿渣硅酸盐水泥由硅酸盐水泥熟料和粒化高炉矿渣、适量石膏磨细制成，代号 P·S。粒化高炉矿渣的掺量按质量百分比为 >20% 且 ≤70%，并分为 A 型和 B 型。A 型矿渣掺量 >20% 且 ≤50%，代号 P·S·A；B 型矿渣掺量 >50% 且 ≤70%，代号 P·S·B。

**2. 火山灰质硅酸盐水泥**

火山灰质硅酸盐水泥由硅酸盐水泥熟料和火山灰质混合材料、适量石膏磨细制成的，代号 P·P。火山灰质混合材料掺量按质量百分比为 >20% 且 ≤40%。

**3. 粉煤灰硅酸盐水泥**

粉煤灰硅酸盐水泥由硅酸盐水泥熟料和粉煤灰、适量石膏磨细制成，代号为 P·F。粉煤灰掺量按质量分数计为 >20% 且 ≤40%。

## (二)技术要求

《通用硅酸盐水泥》(GB 175—2007)规定的技术要求如下:

(1)细度:80μm方孔筛筛余不大于10%或45μm方孔筛筛余不大于30%。

(2)凝结时间:初凝不得早于45min,终凝不得迟于600min。

(3)氧化镁:水泥中氧化镁的含量不超过5.0%,如果水泥经压蒸安定性试验合格,则氧化镁的含量允许放宽到6.0%。其中要求P·S·A型、P·P型、P·F型、P·C型水泥中的氧化镁含量不大于6.0%,如果水泥压蒸试验合格,则水泥中氧化镁的含量(质量分数)允许放宽至6.0%。如果水泥中氧化镁的含量(质量分数)大于6.0%,需进行水泥压蒸安定性试验并试验合格。当有更低要求时,该指标由买卖双方协商确定。

(4)三氧化硫:矿渣水泥中不得超过4.0%,火山灰水泥、粉煤灰水泥中不得超过3.5%。

(5)安定性:用沸煮法检验合格。

(6)强度:水泥强度等级按规定龄期的抗压强度和抗折强度划分。三种水泥各强度等级、各龄期强度不得低于表3-4所规定数值。

表3-4  矿渣水泥、火山灰水泥及粉煤灰水泥各龄期的强度要求(GB 175—2007)

| 强度等级 | 抗压强度(MPa) | | 抗折强度(MPa) | |
| --- | --- | --- | --- | --- |
|  | 3d | 28d | 3d | 28d |
| 32.5 | ≥10.0 | ≥32.5 | ≥2.5 | ≥5.5 |
| 32.5R | ≥15.0 | ≥32.5 | ≥3.5 | ≥5.5 |
| 42.5 | ≥15.0 | ≥42.5 | ≥3.5 | ≥6.5 |
| 42.5R | ≥19.0 | ≥42.5 | ≥4.0 | ≥6.5 |
| 52.5 | ≥21.0 | ≥52.5 | ≥4.0 | ≥7.0 |
| 52.5R | ≥23.0 | ≥52.5 | ≥4.5 | ≥7.0 |

(7)碱含量:水泥中碱含量按$Na_2O+0.658K_2O$计算值表示。若使用活性骨料,用户要求提供低碱水泥时,水泥中的碱含量应不大于0.60%或由买卖双方协商确定。

## (三)特性与应用

### 1. 三种水泥的共性

(1)早期强度低、后期强度发展高。其原因是这些活性混合材料的水化慢,故早期(3d、7d)强度低。后期由于二次水化反应水化产物不断增多,强度可赶上或超过同强度等级的硅酸盐水泥或普通硅酸盐水泥(图3-4)。这三种水泥不适用于早期强度要求高的混凝土工程,如冬期施工现浇混凝土工程等。

(2)对温度敏感,适合高温养护。这三种水泥在低温下水化明显减慢,强度较低。采用高温养护可大大加速活性混合材料的水化,并可加速熟料的水化,大大提高早期强度,并且不影响常温下后期强度的发展。

(3)耐腐蚀性好。这三种水泥的熟料数量相对较少,水化产物氢氧化钙和水化铝酸钙数量少,且二次水化反应使氢氧化钙进一步降低,因此耐腐蚀性好,适用于有硫酸盐、镁盐、软水侵蚀作用的环境,如水工、海港、码头等混凝土工程。

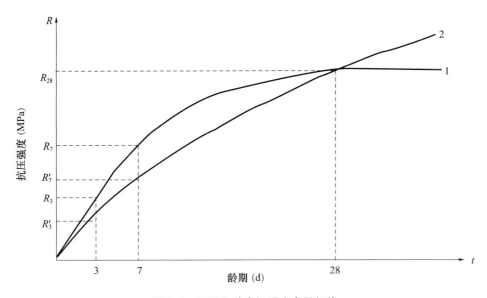

**图 3-4 不同品种水泥强度发展规律**
1—硅酸盐水泥；2—掺混合材料硅酸盐水泥

（4）水化热小。三种水泥中的熟料含量少，因而水化放热量少，尤其是早期放热速度慢，放热量少，适用于大体积混凝土过程。

（5）抗冻性较差。矿渣和粉煤灰易泌水形成连通孔隙，火山灰一般需水量较大，会增加内部的孔隙含量，故这三种水泥的抗冻性均较差。

（6）抗碳化性较差。由于这三种水泥在水化硬化后，水泥石中的氢氧化钙数量少，故抵抗碳化的能力比较差，所以不适用于二氧化碳浓度含量高的工业厂房。

**2. 三种水泥的特性**

（1）矿渣硅酸盐水泥。矿渣硅酸盐水泥的抗渗性差，且干缩较大，但矿渣硅酸盐水泥的耐热性较好，因此适用于有耐热要求的混凝土工程，但不适用于有抗渗要求的混凝土工程。

（2）火山灰质硅酸盐水泥。火山灰质硅酸盐水泥的保水性好，抗渗性较好；但火山灰质硅酸盐水泥的干缩大，水泥石的表面易产生起粉现象，并且其耐磨性也较差。火山灰质硅酸盐水泥适用于有抗渗要求的混凝土工程，不宜用于干燥环境中的地上混凝土工程及有耐磨性要求的混凝土工程。

（3）粉煤灰硅酸盐水泥。粉煤灰硅酸盐水泥抗渗性差，干缩较小，耐磨性较差。粉煤灰硅酸盐水泥适用于承载较晚的混凝土工程，不宜用于有抗渗性要求的混凝土工程及耐磨性要求高的混凝土工程。

## 三、复合硅酸盐水泥

《通用硅酸盐水泥》（GB 175—2007）规定：复合硅酸盐水泥由硅酸盐水泥熟料、两种或两种以上规定的混合材料、适量石膏磨细制成，代号 P·C。水泥中混合材料总掺量按质量分数 >20% 且 ≤50%。

复合硅酸盐水泥有 32.5、32.5R、42.5、42.5R、52.5 和 52.5R 六个强度等级，各强度等级水泥的龄期强度要求同矿渣水泥、火山灰质水泥、粉煤灰水泥,其余技术要求与火山灰质硅酸盐水泥相同。

复合水泥由于掺入了两种或两种以上规定的混合材料,其效果不只是各类混合材料的简单混合,而是互相取长补短,产生单一混合材料不能起到的优良效果。因此,复合水泥的性能介于普通硅酸盐水泥和三种混合材料硅酸盐水泥之间。

以上六种常用水泥的组成、特性及选用见表3-5~表3-7。

表3-5 六种常用水泥的组成及性质

| 项目 | 硅酸盐水泥 | 普通硅酸盐水泥 | 矿渣硅酸盐水泥 | 火山灰质硅酸盐水泥 | 粉煤灰硅酸盐水泥 | 复合硅酸盐水泥 |
|---|---|---|---|---|---|---|
| 组成 | 硅酸盐水泥熟料、很少量(0~5%)混合材料、适量石膏 | 硅酸盐水泥熟料、少量(5%~20%)混合材料、适量石膏 | 硅酸盐水泥熟料、大量(>20%且≤70%)粒化高炉矿渣、适量石膏 | 硅酸盐水泥熟料、大量(>20%且≤40%)火山灰质混合材料、适量石膏 | 硅酸盐水泥熟料、大量(>20%且≤40%)粉煤灰、适量石膏 | 硅酸盐水泥熟料、大量(>20%且≤50%)两种或两种以上的混合材料、适量石膏 |
| 共同点 | 硅酸盐水泥熟料、适量石膏 | | | | | |
| 不同点 | 无或很少的混合材料 | 少量混合材料 | 大量混合材料(化学组成或化学活性基本相同) | | | 大量活性或非活性混合材料 |
| | | | 粒化高炉矿渣 | 火山灰质混合材料 | 粉煤灰 | 两种以上活性或非活性混合材料 |
| 性质 | 1. 早期、后期强度都较高;<br>2. 耐腐蚀性差;<br>3. 水化热大;<br>4. 抗碳化性好;<br>5. 抗冻性好;<br>6. 耐腐蚀性好;<br>7. 耐热性差 | 1. 早期强度稍低,后期强度高;<br>2. 耐腐蚀性稍好;<br>3. 水化热稍小;<br>4. 抗碳化性好;<br>5. 抗冻性好;<br>6. 耐磨性较好;<br>7. 耐热性稍好;<br>8. 抗渗性较好 | 早期强度低,后期强度高 | | | 早期强度较高 |
| | | | 对温度敏感;适合高温养护;耐腐蚀性好;水化热小;抗冻性较差;抗碳化性较差 | | | |
| | | | 1. 泌水性大;<br>2. 耐热性较好;<br>3. 干缩较大 | 1. 保水性好、抗渗性差;<br>2. 干缩大;<br>3. 耐磨性差 | 1. 泌水性大、易产生失水裂纹、抗渗性差;<br>2. 干缩小、抗裂性好;<br>3. 耐磨性差 | 干缩较大 |

表3-6 通用硅酸盐水泥技术性质标准

| 项目 细度 | 硅酸盐水泥 | | 普通水泥 P·O | 矿渣水泥 P·S 火山灰水泥 P·P 粉煤灰水泥 P·F | 复合水泥 P·C |
|---|---|---|---|---|---|
| | P·Ⅰ | P·Ⅱ | | | |
| 细度 | 比表面积>300m²/kg | | | 80μm方孔筛筛余不大于10.0%或45μm方孔筛筛余不大于30% | |
| 凝结时间 | 初凝 | ≥45min | | | | |
| | 终凝 | ≤390min | | ≤600min | | |
| 体积安定性 | 安定性 | 沸煮法检验必须合格(若试饼法和雷氏法两种有争议,以雷氏法为准) | | | | |
| | MgO | 含量≤5.0% | | | | |
| | $SO_3$ | 含量≤3.5%(矿渣水泥中含量≤4.0%) | | | | |

续表

| 项目<br>细度 | | 硅酸盐水泥 | | 普通水泥<br>P·O | | 矿渣水泥 P·S<br>火山灰水泥 P·P<br>粉煤灰水泥 P·F | | 复合水泥 P·C | |
|---|---|---|---|---|---|---|---|---|---|
| | | P·Ⅰ | P·Ⅱ | | | | | | |
| 强度等级 | 龄期(d) | 抗压强度(MPa) | 抗折强度(MPa) | 抗压强度(MPa) | 抗折强度(MPa) | 抗压强度(MPa) | 抗折强度(MPa) | 抗压强度(MPa) | 抗折强度(MPa) |
| 32.5 | 3 | — | — | — | — | 10.0 | 2.5 | 11.0 | 2.5 |
| | 28 | — | — | — | — | 32.5 | 5.5 | 32.5 | 5.5 |
| 32.5R | 3 | — | — | — | — | 15.0 | 3.5 | 16.0 | 3.5 |
| | 28 | — | — | — | — | 32.5 | 5.5 | 32.5 | 5.5 |
| 42.5 | 3 | 17.0 | 3.5 | 17.0 | 3.5 | 15.0 | 3.5 | 16.0 | 3.5 |
| | 28 | 42.5 | 6.5 | 42.5 | 6.5 | 42.5 | 6.5 | 42.5 | 6.5 |
| 42.5R | 3 | 22.0 | 4.0 | 22.0 | 4.0 | 19.0 | 4.0 | 21.0 | 4.0 |
| | 28 | 42.5 | 6.5 | 42.5 | 6.5 | 42.5 | 6.5 | 42.5 | 6.5 |
| 52.5 | 3 | 23.0 | 4.0 | 23.0 | 4.0 | 21.0 | 4.0 | 22.0 | 4.0 |
| | 28 | 52.5 | 7.0 | 52.5 | 7.0 | 52.5 | 7.0 | 52.5 | 7.0 |
| 52.5R | 3 | 27.0 | 5.0 | 27.0 | 5.0 | 23.0 | 4.5 | 26.0 | 5.0 |
| | 28 | 52.5 | 7.0 | 52.5 | 7.0 | 52.5 | 7.0 | 52.5 | 7.0 |
| 62.5 | 3 | 28.0 | 5.0 | — | — | — | — | — | — |
| | 28 | 62.5 | 8.0 | — | — | — | — | — | — |
| 62.5R | 3 | 32.0 | 5.5 | — | — | — | — | — | — |
| | 28 | 62.5 | 8.0 | — | — | — | — | — | — |
| 碱含量 | 用户要求低碱水泥时,按 $Na_2O+0.658K_2O$ 计算的碱含量不大于0.60%,或由供需双方商定 | | | | | | | | |

**表 3-7 常用水泥的选用**

| 混凝土工程特点或所处环境条件 | | 优先选用 | 可以使用 | 不宜使用 |
|---|---|---|---|---|
| 普通混凝土 | 普通气候环境中的混凝土 | 普通硅酸盐水泥 | 矿渣硅酸盐水泥、火山灰质硅酸盐水泥、粉煤灰硅酸盐水泥、复合硅酸盐水泥 | — |
| | 在干燥环境中的混凝土 | 普通混硅酸水泥 | 矿渣硅酸盐水泥 | 火山灰质硅酸盐水泥、粉煤灰硅酸盐水泥 |
| | 在高湿度环境中或永远处在水下的混凝土 | 矿渣硅酸盐水泥 | 普通硅酸盐水泥、火山灰质硅酸盐水泥、粉煤灰硅酸盐水泥、复合硅酸盐水泥 | — |
| | 厚大体积的混凝土 | 矿渣硅酸盐水泥、火山灰质硅酸盐水泥、粉煤灰硅酸盐水泥、复合硅酸盐水泥 | 普通硅酸盐水泥 | 硅酸盐水泥、快硬硅酸盐水泥 |

续表

| 混凝土工程特点或所处环境条件 | | 优先选用 | 可以使用 | 不宜使用 |
|---|---|---|---|---|
| 有特殊要求的混凝土 | 要求快硬的混凝土 | 快硬硅酸盐水泥、硅酸盐水泥 | 普通硅酸盐水泥 | 矿渣硅酸盐水泥、火山灰质硅酸盐水泥、粉煤灰硅酸盐水泥、复合硅酸盐水泥 |
| | 高强混凝土 | 硅酸盐水泥 | 普通硅酸盐水泥、矿渣硅酸盐水泥 | 火山灰质硅酸盐水泥、粉煤灰硅酸盐水泥 |
| | 严寒地区的露天混凝土,寒冷地区的处在水位升降范围内的混凝土 | 普通硅酸盐水泥 | 矿渣硅酸盐水泥 | 火山灰质硅酸盐水泥、粉煤灰硅酸盐水泥 |
| | 严寒地区处在水位升降范围内的混凝土 | 普通硅酸盐水泥 | | 矿渣硅酸盐水泥、火山灰质硅酸盐水泥、粉煤灰硅酸盐水泥、复合硅酸盐水泥 |
| | 有抗渗性要求的混凝土 | 普通硅酸盐水泥、火山灰质硅酸盐水泥 | | 矿渣硅酸盐水泥 |
| | 有耐磨性要求的混凝土 | 硅酸盐水泥、普通硅酸盐水泥 | 矿渣硅酸盐水泥 | 火山灰质硅酸盐水泥、粉煤灰硅酸盐水泥 |

# 第四节 其他品种水泥

在土木工程中,除常用的硅酸盐系列水泥外,还可能使用其他品种的水泥,在此对其中一些品种做简要介绍。

## 一、道路硅酸盐水泥

(一)定义

由道路硅酸盐水泥熟料、0~10%活性混合材料和适量石膏磨细制成的水硬性胶凝材料,称为道路硅酸盐水泥,简称道路水泥,代号P·R。

(二)技术要求

(1)铝酸三钙含量:熟料中铝酸三钙含量不得大于5.0%。
(2)铁铝酸四钙含量:熟料中铁铝酸四钙含量不得小于16.0%。
铝酸三钙($C_3A$)和铁铝酸四钙($C_4AF$)含量按下式求得:
$$C_3A = 2.65Al_2O_3 - 0.64Fe_2O_3$$

$$C_4AF = 3.04\ Fe_2O_3$$

(3)游离氧化钙的含量:旋窑生产应不大于1.0%,立窑生产应不大于1.8%。

(三)特性与应用

道路水泥是一种强度较高(特别是抗折强度高)、耐磨性好、干缩性小、抗冲击性好、抗冻性和抗硫酸性比较好的专用水泥。它适用于道路路面、机场跑道道面、城市广场等工程。由于道路水泥只有干缩性小、耐磨、抗冲击等特性,可减少水泥混凝土路面的开裂和磨耗,减少维修,延长路面使用年限。

## 二、白色硅酸盐水泥

由氧化铁含量少的硅酸盐水泥熟料、适量石膏及水泥质量分数为0~10%的混合材料,磨细制成的水硬性胶凝材料称为白色硅酸盐水泥(简称白水泥),代号P·W。

白水泥的技术性质应满足《白色硅酸盐水泥》(GB/T 2015—2017)的规定,水泥白度值应不低于87。白色水泥可用于配制白色或彩色灰浆、砂浆及混凝土。

## 三、膨胀水泥及自应力水泥

膨胀水泥是在硬化过程中不产生收缩,具有一定膨胀性能的水泥。

(一)分类

**1. 按制造方法分类**

(1)氧化钙型膨胀水泥。在水泥中掺入一定量的适当温度烧制得到的氧化钙,氧化钙水化产生体积膨胀。

(2)氧化镁型膨胀水泥。在水泥中掺入一定量的适当温度烧制得到的氧化镁,氧化镁水化产生体积膨胀。

(3)钙矾石型膨胀水泥。在水泥石中形成钙矾石产生体积膨胀。

**2. 按膨胀值分类**

(1)收缩补偿水泥。这种水泥膨胀性能较弱,膨胀时所产生的压应力大致能抵消干缩引起的应力,可防止混凝土产生干缩裂缝。

(2)自应力水泥。这种水泥具有较强的膨胀性能,用于钢筋混凝土中时,由于其膨胀性能,使钢筋受到较大的拉应力,而混凝土则受到相应的压应力。当外界因素使混凝土结构产生拉应力时,就可被预先具有的压应力抵消或减低。这种靠自身水化产生膨胀来张拉钢筋达到的预应力称为自应力。

(二)膨胀水泥和自应力水泥的应用

膨胀水泥适用于收缩补偿混凝土,既可用作抗渗混凝土,也可用于填灌混凝土结构或构件的接缝及管道接头、结构的加固与修补、浇筑机器底座及固定脚螺丝等。自应力水泥适用于制造自应力钢筋混凝土压力管及配件。

### 四、抗硫酸盐硅酸盐水泥

凡以特定矿物组成的硅酸盐水泥熟料,加入适量石膏,磨细制成的具有抵抗硫酸根离子侵蚀的水硬性胶凝材料,称为抗硫酸盐硅酸盐水泥,简称抗硫酸盐水泥。

抗硫酸盐水泥按其抗硫酸盐侵蚀能力分为中抗硫酸盐水泥和高抗硫酸盐水泥两类。

减少熟料中的 $C_3S$ 和 $C_3A$,相应增加 $C_2S$、$C_4AF$ 的含量,是提高水泥抗硫酸盐腐蚀性能的主要措施。

抗硫酸盐水泥具有较高的抗硫酸盐侵蚀的性能,水化热较低,适用于受硫酸盐侵蚀的海港、水利、地下隧道、引水、道路与桥梁基础等工程。

### 五、快硬水泥

#### (一)快硬硅酸盐水泥

凡以硅酸盐水泥熟料和适量石膏磨细制成的,以3d抗压强度表示标号的水硬性胶凝材料,称为快硬硅酸盐水泥(简称快硬水泥)。

快硬水泥的制造方法与硅酸盐水泥基本相同,只是适当增加了熟料中硬化快的矿物,通常硅酸三钙含量为50%~60%,铝酸三钙含量为8%~14%,两种总量为60%~65%。同时为了加快硬化,适当增加了石膏的掺量(可达8%)并提高水泥的细度。

#### (二)铝酸盐水泥

由铝酸盐水泥熟料磨细制成的水硬性胶凝材料。其中,铝酸一钙(CA)含量为35%~60%,二铝酸一钙($CA_2$)含量为10%~20%,硅铝酸二钙($C_2AS$)含量为20%~40%,七铝酸十二钙($C_{12}A_7$)含量小于1%,以及其他微量矿物。

铝酸盐水泥早期强度高,水化热较大,抗硫酸盐侵蚀性能强,耐高温性好,但长期强度降低。因此铝酸盐水泥适用于紧急抢修工程和制备耐火材料。

#### (三)硫铝酸盐水泥

以适当成分的生料,在经煅烧所得的以硫铝酸钙和硅酸二钙为主要矿物成分的水泥熟料中掺加不同量的石灰石、适量石膏,磨细制成的具有水硬性胶凝材料。

硫铝酸盐水泥具有快凝快硬、早强高强、微膨胀、抗渗、抗海水侵蚀等性能,已成功应用于抢修抢建、海工水工等工程。

# 复习思考题

1. 简述硅酸盐水泥生产工艺。
2. 硅酸盐水泥的矿物组成是什么?各矿物对水泥石的性质有怎样的影响?
3. 简述硅酸盐水泥各熟料矿物的水化反应及其产物组成。

4. 影响硅酸盐水泥凝结硬化的因素有哪些?怎么影响?

5. 试说明下述各条"必须"的原因:

(1) 生产硅酸盐水泥时必须加入适量石膏;

(2) 水泥粉磨必须具有一定的细度;

(3) 水泥体积安定性必须合格。

6. 简述硅酸盐水泥腐蚀类型、机理及防治。

7. 简述硅酸盐水泥的技术性质及实用意义。

8. 简述引起安定性不良的原因及检验方法。

9. 简述活性和非活性混合材定义、品种和作用,及对应的通用硅酸盐水泥品种及定义。

10. 简述各类硅酸盐水泥的特性。

# 第四章　普通混凝土与建筑砂浆

## 第一节　混凝土概述

### 一、混凝土的定义

凡由胶凝材料把散粒状材料（骨料）胶结到一起，并形成具有一定强度的人造石材，统称为混凝土。在工程建设领域，混凝土常简写为"砼"。顾名思义，"砼"形象地表明混凝土是一种人工石。

### 二、混凝土的分类

混凝土可按以下几种方法分类：

**1. 按所用胶凝材料分类**

混凝土按其所用胶凝材料可分为水泥混凝土、沥青混凝土、聚合物混凝土、树脂混凝土、石膏混凝土、水玻璃混凝土、硅酸盐混凝土等。目前使用最多的混凝土为水泥混凝土，它以水泥为胶凝材料，将按一定粒径分布的砂石骨料胶结在一起。水泥混凝土是当今世界上用途最广、用量最大的人造土木工程材料，同时也是重要的工程结构材料。本章讲述的混凝土，如无特别说明，均指水泥混凝土。

**2. 按表观密度分类**

按照表观密度，混凝土可分为重混凝土、普通混凝土和轻混凝土。

（1）重混凝土。重混凝土的表观密度大于 $2600kg/m^3$。它以密度很大的重晶石、铁矿石、钢屑等作为骨料配制而成，也可以同时采用重水泥——钡水泥、锶水泥作为胶凝材料进行配制。重混凝土具有阻挡射线穿透的能力，主要用于防辐射结构，例如核能工程的屏蔽结构、核废料容器等。

（2）普通混凝土。普通混凝土的表观密度为 $2100\sim2500kg/m^3$，一般在 $2400kg/m^3$ 左右。它以普通的天然砂、石作为骨料配制而成，通常简称为混凝土。在工程建设领域，普通混凝土大量用作各种建筑物、结构物的承重材料。

（3）轻混凝土。轻混凝土的表观密度小于 $1950kg/m^3$。它采用轻质多孔的骨料配制而成，或者不用骨料而掺入加气剂或发泡剂等形成多孔结构，包括轻骨料混凝土、多孔混凝土、大孔混凝土等。

**3. 按用途分类**

混凝土按其用途可分为结构混凝土、防水混凝土、耐热混凝土、耐酸混凝土、装饰混凝土、大体积混凝土、膨胀混凝土、防辐射混凝土、道路混凝土等多种。

**4. 按生产和施工方法分类**

混凝土按生产和施工方法可分为预拌混凝土(商品混凝土)、泵送混凝土、喷射混凝土、压力灌浆混凝土(预填骨料混凝土)、挤压混凝土、离心混凝土、真空吸水混凝土、碾压混凝土、热拌混凝土等。

### 三、混凝土的性能特点

混凝土在土木工程领域中得到广泛应用,主要是由于它具有以下优点:

(1)原材料来源丰富,造价低廉,可就地取材,价格便宜。

(2)混凝土拌和物具有良好的可塑性,可按工程结构要求浇筑成各种形状,可塑性好。

(3)改变混凝土组成材料的品种及比例,可配制出不同物理力学性能的混凝土,以满足各种工程建设的不同需要。一般工程中的混凝土强度为20~40MPa,特殊工程中的混凝土强度也可高达80~100MPa,适应性好。

(4)混凝土与钢筋之间有牢固的黏结力,且混凝土与钢筋的线膨胀系数基本相同。二者复合成钢筋混凝土后,能保证共同工作,从而大大扩展了混凝土的应用范围。

(5)耐久性良好。混凝土在一般服役环境中不需要维护保养,维修费用少。

(6)耐火性好。混凝土的耐火性远比木材、钢材和塑料好,可耐数小时的高温破坏而保持其力学性能不出现明显下降,为发生火灾时提供足够的救援时间。

(7)生产能耗较低。混凝土生产的能源消耗远低于土制品的烧制及金属材料的生产过程中的能源消耗。

混凝土的不足之处主要有:

(1)自重大,比强度小。每1$m^3$普通混凝土达2400kg左右,致使在土木工程建设领域中为了支撑混凝土结构而采用肥梁、胖柱、厚基础,这对高层、大跨度建筑不利。

(2)抗拉强度低。一般混凝土的抗拉强度为其抗压强度的1/20~1/10,因此受拉时混凝土中易产生裂缝,发生脆裂破坏。

(3)导热系数大。普通混凝土的导热系数为1.40W/(m·K),为烧结砖的两倍,故混凝土易导热、保温隔热性能较差。

(4)硬化较慢,生产周期长。混凝土的硬化时间取决于其中胶凝材料的硬化时间,一般情况下,混凝土的初始硬化时间在几个小时左右。

综上所述,混凝土材料具有许多优点,但也存在着一些难以克服的缺点。随着现代混凝土技术的发展,混凝土的不足之处已经得到很大改善。例如,采用轻骨料可使混凝土的自重和导热系数显著降低;在混凝土中掺入纤维或聚合物,可大大降低混凝土的脆性;混凝土采用快硬水泥或掺入早强剂、减水剂等,可明显缩短其硬化时间。由于混凝土具有以上优点,同时其不足之处也容易得到适当控制,从而使得许多其他强度大、效益高的结构材料无法与之竞争。混凝土早已成为当代主要的土木工程材料,广泛应用于工业与民用建筑工程、水利工程、地下工程、公路、铁路、桥梁及国防建设中。

# 第二节　普通混凝土的组成材料

普通混凝土是由胶凝材料、水、细骨料和粗骨料组成,另外还常加入适量的外加剂。其中,胶凝材料包括水泥和矿物掺合料。混凝土是一种多相、多组分、不均匀、不连续的非匀质材料。

## 一、混凝土组成材料的作用

普通混凝土的基本组成材料中,水泥和水占总体积的20%~30%,粗骨料和细骨料占总体积的70%~80%。混凝土中的水泥和水混合搅拌后叫作水泥浆,具有润滑作用,同时可以将骨料黏结,赋予混凝土拌和物流动性和可塑性,方便混凝土的浇筑和施工。水泥浆硬化后称为水泥石,具有胶结作用,能将散粒状的骨料联结起来,形成坚固的整体。虽然水泥和水在混凝土中所占比例较小,但它们所起的作用至关重要,水泥浆是混凝土拌和物整体流动性、可塑性的来源,同时水泥石也是混凝土硬化后整体强度的重要来源。

混凝土中的骨料首先起到骨架和填充作用;其次,由于骨料颗粒坚硬,体积稳定性好,相互搭接可以形成坚实的骨架,起到抵抗外力的作用,同时,骨料还可以限制水泥石的收缩,确保混凝土的体积稳定性;最后,在混凝土中大部分体积的骨料成本远低于水泥,使得混凝土整体的生产成本大大降低。

## 二、胶凝材料

(一)水泥

水泥是混凝土中最重要的组分,合理选用水泥需注意以下两方面的问题。

**1. 水泥品种的选择**

配制混凝土用的水泥品种,应根据混凝土所应用于工程的性质与特点、工程所处的环境和施工条件,以及各种水泥的特性进行合理选择。

**2. 水泥强度等级的选择**

水泥强度等级的选择应当与混凝土的设计强度等级相适应,原则上是配制高强度等级的混凝土选用高强度等级的水泥,低强度等级的混凝土选用低强度等级的水泥。普通强度等级的混凝土宜采用混凝土强度等级1.5倍左右的水泥;对高强混凝土而言,这个强度等级倍数可取1左右。

若用低强度等级的水泥来配制高强度等级的混凝土,为满足强度要求就要使用大量水泥,这样不仅不经济,还会使混凝土收缩和水化热增大,同时为保证高强度要求采用低水灰比而造成混凝土在搅拌时太干,施工困难,不易捣实,不能保证混凝土的施工质量。反之,如果用高强度等级的水泥来配制低强度等级的混凝土,单从强度的角度考虑只需用少量水泥就可满足要求,但为了满足混凝土拌和物的和易性及混凝土硬化后的耐久性要求,势必增加水泥的用量,这样往往产生超强现象,也不经济。

(二)掺合料

混凝土掺合料是在混凝土搅拌前或在搅拌过程中,与其他组分一起加入的一种外掺料。用于混凝土的掺合料多是具有一定活性的工业固体废渣,这种活性来源于其中的矿物成分,所以掺合料往往称为矿物掺合料。掺合料取代部分水泥掺加在混凝土中,不仅可以将废料资源化,还可以降低混凝土的生产成本,同时改善混凝土拌和物的和易性和硬化后混凝土的耐久性能。因此,混凝土中掺加掺合料的技术、经济和环境效益十分显著。

混凝土的常用掺合料有粉煤灰、粒化高炉矿渣粉和硅灰等,其中粉煤灰和矿渣粉是目前用量最大、使用范围最广的掺合料。

**1. 粉煤灰**

粉煤灰是指火力发电厂中的煤在经过燃烧后排放出来的烟道灰,属于火山灰质活性掺合料,其主要成分是硅、铝和铁的氧化物,具有潜在的化学活性。由于煤粉微细,且在高温燃烧过程中易形成玻璃微珠,因此粉煤灰颗粒多数呈球状,其粒径多在 $45\mu m$ 以下,是可以不用粉磨直接用作混凝土的掺合料。

粉煤灰按照燃煤来源和氧化钙含量分为 F 类粉煤灰和 C 类粉煤灰两类。F 类粉煤灰是由无烟煤或烟煤煅烧收集的,其 CaO 含量不大于 10%,又称为低钙粉煤灰;C 类粉煤灰是由褐煤或次烟煤煅烧收集的,其 CaO 含量大于或等于 10%,又称高钙粉煤灰。与 F 类粉煤灰相比,C 类粉煤灰一般具有活性高和自硬性好等特征,但由于 C 类粉煤灰中往往含有游离氧化钙,所以在用作混凝土掺合料时,必须对其体积安定性进行检验。

粉煤灰又根据其理化性能要求分为Ⅰ级、Ⅱ级和Ⅲ级三个等级,《用于水泥和混凝土中的粉煤灰》(GB/T 1596—2017)规定的粉煤灰的理化性能要求列于表 4-1。扫码获取《用于水泥和混凝土中的粉煤灰》(GB/T 1596—2017)的详细内容。

表 4-1 不同等级粉煤灰理化性能要求

| 项　目 | | 技术要求 | | |
| --- | --- | --- | --- | --- |
| | | Ⅰ级 | Ⅱ级 | Ⅲ级 |
| 细度($45\mu m$ 方孔筛筛余)(%) | F 类粉煤灰 | ≤12.0 | ≤30.0 | ≤45.0 |
| | C 类粉煤灰 | | | |
| 需水量比(%) | F 类粉煤灰 | ≤95 | ≤105 | ≤115 |
| | C 类粉煤灰 | | | |
| 烧失量(%) | F 类粉煤灰 | ≤5.0 | ≤8.0 | ≤10.0 |
| | C 类粉煤灰 | | | |
| 含水量(%) | F 类粉煤灰 | ≤1.0 | | |
| | C 类粉煤灰 | | | |

续表

| 项 目 | | 技术要求 | | |
|---|---|---|---|---|
| | | Ⅰ级 | Ⅱ级 | Ⅲ级 |
| 三氧化硫(%) | F类粉煤灰 | ≤3.0 | | |
| | C类粉煤灰 | | | |
| 游离氧化钙(%) | F类粉煤灰 | ≤1.0 | | |
| | C类粉煤灰 | ≤4.0 | | |
| 二氧化硅($SiO_2$)、三氧化二铝($Al_2O_3$)和三氧化二铁($Fe_2O_3$)总质量分数(%) | F类粉煤灰 | ≥70.0 | | |
| | C类粉煤灰 | ≥50.0 | | |
| 密度($g/cm^3$) | F类粉煤灰 | ≤2.6 | | |
| | C类粉煤灰 | | | |
| 安定性 雷氏夹沸煮后增加距离(mm) | C类粉煤灰 | ≤5.0 | | |

### 2. 粒化高炉矿渣粉

炼铁厂在高炉冶炼生铁时得到的以硅铝酸钙为主要成分的熔融物经水淬成粒后,得到工业固体废渣——粒化高炉矿渣,其大部分为玻璃质。粒化高炉矿渣经干燥、粉磨可制得粒化高炉矿渣粉,粉磨时可添加适量的石膏和助磨剂。粒化高炉矿渣粉简称矿渣粉,又称矿渣微粉。

按照《用于水泥、砂浆和混凝土中的粒化高炉矿渣粉》(GB/T 18046—2017)的规定,矿渣粉应符合表4-2的技术要求。扫码获取《用于水泥、砂浆和混凝土中的粒化高炉矿渣粉》(GB/T 18046—2017)的详细内容。

表4-2 矿渣粉技术要求

| 项 目 | | 级 别 | | |
|---|---|---|---|---|
| | | S105 | S95 | S75 |
| 密度($g/cm^3$) | | ≥2.8 | | |
| 比表面积($m^2/kg$) | | ≥500 | ≥400 | ≥300 |
| 活性指数(%) | 7d | ≥95 | ≥70 | ≥55 |
| | 28d | ≥105 | ≥95 | ≥75 |
| 流动度比(%) | | ≥95 | | |
| 含水量(%,质量分数) | | ≤1.0 | | |
| 三氧化硫(%,质量分数) | | ≤4.0 | | |
| 氧化镁(%,质量分数) | | ≤13.5 | | |
| 氯离子(%,质量分数) | | ≤0.06 | | |
| 烧失量(%,质量分数) | | ≤1.0 | | |

表4-2中,矿渣粉的活性指数是指以矿渣粉取代50%水泥后的试验砂浆强度与对比的水泥砂浆强度之比。矿渣粉按其活性指数分为三个等级:S105、S95和S75。

## 三、骨料

石子和砂在混凝土中起到骨架作用,因此称为骨料。普通混凝土所用骨料按粒径大小分为两种,粒径大于4.75mm的称为粗骨料,粒径在0.15~4.75mm的称为细骨料。

普通混凝土所用的粗骨料通常有碎石和卵石两种。碎石是指天然岩石、卵石或矿山废石(经机械破碎、筛分制成的,粒径大于4.75mm的岩石颗粒);卵石是指由自然风化、水流搬运和分选、堆积形成的,粒径大于4.75mm的岩石颗粒。普通混凝土中所用细骨料有天然砂和机制砂两种。天然砂是指自然生成的,经人工开采和筛分的,粒径小于4.75mm的岩石颗粒,包括河砂、湖砂、山砂和淡化海砂,但不包括软质、风化的岩石颗粒;机制砂与碎石制备方法相同,是指经除土处理,由机械破碎、筛分制成的,粒径小于4.75mm的岩石、矿山尾矿或工业废渣颗粒,但不包括软质、风化的颗粒,机制砂又称人工砂。

粗、细骨料的总体积一般占混凝土体积的70%~80%,所以骨料质量的优劣,将直接影响到混凝土各项性质的好坏。为此,我国在《建设用砂》(GB/T 14684—2022)和《建设用卵石、碎石》(GB/T 14685—2022)这两个国家标准中,对砂、石提出了明确的质量要求,砂、石按照其技术要求均分为Ⅰ、Ⅱ、Ⅲ三类。Ⅰ类宜用于强度等级大于C60的混凝土;Ⅱ类宜用于强度等级C30~C60及满足抗冻、抗渗或其他要求的混凝土;Ⅲ类宜用于强度等级小于C30的混凝土。下面简要介绍用于普通混凝土的骨料的技术要求。扫码获取《建设用砂》(GB/T 14684—2022)和《建设用卵石、碎石》(GB/T 14685—2022)的详细内容。

**1. 表观密度、堆积密度、空隙率**

骨料的表观密度、堆积密度及空隙率是进行混凝土配合比计算所必需的原始数据,骨料的这些性质会影响混凝土内部骨架的坚硬程度和颗粒级配,从而影响混凝土的性质。规范要求普通混凝土用砂的表观密度大于$2500kg/m^3$,松散堆积密度大于$1400kg/m^3$,空隙率要小于44%。普通混凝土用卵石、碎石的表观密度应大于$2600kg/m^3$,Ⅰ类、Ⅱ类、Ⅲ类石子在连续级配松散堆积状态下的空隙率应分别小于43%、45%、47%。

**2. 颗粒形状和表面特征**

骨料的颗粒形状是不规则的,有的颗粒三维尺寸比较接近;有的呈薄片状或细长针状;有的带棱角,有的呈圆滑形状。石子中细长或薄片状的骨料颗粒分别称为针状或片状颗粒。这两种颗粒形状的骨料在受力时容易折断,从而影响骨架整体强度;并且这两种颗粒在拌和时不易搅拌,对混凝土的力学性能和施工性能均不利。因此,要求混凝土用粗骨料中的针状和片状颗粒总含量不能超过一定的限值,见表4-3。

表4-3 石子的针片状含量

| 项 目 | 指标 | | |
|---|---|---|---|
| | Ⅰ类 | Ⅱ类 | Ⅲ类 |
| 针片状含量(%,质量分数) | ≤5 | ≤10 | ≤15 |

骨料的表面特征是指颗粒表面的粗糙或光滑程度,不同骨料的表面粗糙程度也不相同。天然卵石、河砂等由于长期受水流的冲刷和磨蚀,表面光滑,有利于增强混凝土拌和

物的和易性；但是它们与水泥浆体的黏结性差，相互搭接形成的骨架性能也差，因此不利于维持强度。机械破碎的碎石和山砂，表面粗糙，相互之间搭接能形成坚固的骨架，与水泥浆体的黏结性强，有利于提高强度；但它们表面需要用较多的水泥浆包裹，颗粒之间需要较厚的水泥浆层来润滑，以达到需要的混凝土和易性。在实际工程中要根据工程所要求的强度和工作性指标综合考虑选取骨料。

**3. 含泥量、泥块含量、石粉含量**

含泥量指粗骨料和天然砂中粒径小于 $75\mu m$ 的颗粒含量。石粉含量指人工砂中粒径小于 $75\mu m$ 的颗粒含量。泥块含量是指在细骨料中粒径大于 1.18mm，经水浸洗、手捏后小于 0.6mm 的颗粒含量；在粗骨料中粒径大于 4.75mm，经水浸洗、手捏后小于 2.36mm 的颗粒含量。

骨料中的泥颗粒和石粉颗粒极细，会黏附在骨料表面，影响水泥石与骨料之间的胶结能力，降低混凝土的强度及耐久性，增加混凝土的干缩性。骨料中的泥块强度很低，经浸水溃散、干燥收缩后，会在混凝土中形成薄弱部分，对混凝土的质量影响更大。所以国家标准对各类骨料的含泥量、泥块含量及石粉含量均有一定要求，不符合要求的骨料要进行冲洗等处理。表 4-4 列出了骨料的含泥量和泥块含量要求。对于人工砂来说，主要采用亚甲蓝（MB）值判定其中粒径小于 $75\mu m$ 的颗粒是泥还是与被加工母岩化学成分相同的石粉，小于 1.4 则说明人工砂的颗粒主要为石粉，越大于 1.4，泥的含量逐渐增多。表 4-5 和表 4-6 分别列出了 MB 值小于和大于 1.4 时人工砂中石粉含量的限值。

表 4-4　骨料的含泥量和泥块含量

| 项目 | | 说明 | 指标 | | |
|---|---|---|---|---|---|
| | | | Ⅰ类 | Ⅱ类 | Ⅲ类 |
| 砂 | 含泥量（%，质量分数） | 天然砂 | ≤1.0 | ≤3.0 | ≤5.0 |
| | 泥块含量（%，质量分数） | 天然砂、人工砂 | 0 | ≤1.0 | ≤2.0 |
| 石子 | 含泥量（%，质量分数） | | ≤0.5 | ≤1.0 | ≤1.5 |
| | 泥块含量（%，质量分数） | | 0 | ≤0.2 | ≤0.5 |

表 4-5　亚甲蓝值小于 1.4 时人工砂中石粉含量限值

| 类别 | 指标 | | |
|---|---|---|---|
| | Ⅰ类 | Ⅱ类 | Ⅲ类 |
| MB 值 | ≤0.5 | ≤1.0 | ≤1.4 |
| 石粉含量（%，质量分数） | ≤10.0 | | |

表 4-6　亚甲蓝值大于 1.4 时人工砂中石粉含量限值

| 类别 | 指标 | | |
|---|---|---|---|
| | Ⅰ类 | Ⅱ类 | Ⅲ类 |
| 石粉含量（%，质量分数） | ≤1.0 | ≤3.0 | ≤5.0 |

**4. 有害物质含量**

普通混凝土使用的粗细骨料中不应混有草根、树叶、树枝、塑料、炉渣、煤块等杂物。

细骨料中的云母、轻物质(密度小于2000kg/m³的物质)等,与水泥浆黏性差,会影响混凝土的强度及耐久性。骨料中的硫化物及硫酸盐、氯化物和有机物等可以阻碍水泥水化,对混凝土中的钢筋具有锈蚀作用或者与水泥水化产物反应生成有害膨胀产物。因此,这些物质含量要符合表4-7的规定。

表4-7 骨料中有害物质含量

| | 项 目 | 指 标 | | |
|---|---|---|---|---|
| | | Ⅰ类 | Ⅱ类 | Ⅲ类 |
| 砂 | 云母含量(%,质量分数) | ≤1.0 | ≤2.0 | ≤2.0 |
| | 轻物质含量(%,质量分数) | ≤1.0 | ≤1.0 | ≤1.0 |
| | 有机物(比色法,%) | 合格 | 合格 | 合格 |
| | 硫化物及硫酸盐含量(%,按$SO_3$质量分数) | ≤0.5 | ≤0.5 | ≤0.5 |
| | 氯化物含量(%,按氯离子质量分数) | ≤0.01 | ≤0.02 | ≤0.06 |
| 石子 | 有机物(比色法,%) | 合格 | 合格 | 合格 |
| | 硫化物及硫酸盐含量(%,按$SO_3$质量分数) | ≤0.5 | ≤1.0 | ≤1.0 |

**5. 坚固性**

骨料的坚固性指骨料在气候、环境变化或其他物理因素作用下抵抗破坏的能力,可用硫酸钠溶液进行检验,试样经5次循环后其质量损失应符合表4-8的规定。

表4-8 骨料坚固性指标

| 项 目 | | 指 标 | | |
|---|---|---|---|---|
| | | Ⅰ类 | Ⅱ类 | Ⅲ类 |
| 石子 | %,质量损失 | ≤5 | ≤8 | ≤12 |
| 砂 | | ≤8 | ≤8 | ≤10 |

**6. 碱活性**

骨料中若含有活性氧化硅,就会与水泥中的碱($Na_2O$或$K_2O$)发生碱-骨料反应,生成膨胀性产物。如果膨胀应力大于混凝土的强度,就会导致混凝土的开裂。因此,当骨料用于重要工程混凝土或对骨料有怀疑时,须按标准规定,采用化学法或长度法对骨料进行碱活性检验。标准规定经碱-骨料反应试验后由骨料制备的试件应无裂缝、酥裂、胶体外溢等现象,在规定的试验龄期膨胀率应小于0.10%。

**7. 骨料的强度**

粗骨料在混凝土中起到整体骨架的作用,粗骨料本身的强度会直接影响到混凝土的整体强度,因此,标准对粗骨料的强度有一定的要求。粗骨料的强度测量方法有两种,即检测母体岩石的抗压强度和压碎指标值。

所谓岩石抗压强度,是将其母岩制成边长为50mm的立方体(或直径与高均为50mm的圆柱体)试件,将其浸没在水中48h,在水饱和状态下测定的抗压强度值。通常要求岩石抗压强度与混凝土强度等级之比不小于1.5。

压碎指标值是直接测定堆积后的卵石或碎石承受压力而不破碎的能力,能更直接地反映骨料在混凝土中的受力状态,因此是衡量骨料坚硬程度的重要力学性能指标。试验时采用 9.50～19.0mm 粒级、气干状态的石子,并去除针片状颗粒;按标准规定方法将 3000g 试样装入受压试模内,按规定的加荷速率对试样施加 200kN 的压力,并稳荷 5s,然后卸载;倒出试样,用孔径为 2.36mm 的标准筛筛除被压碎的细粒,称出留在筛上的试样质量。按下式计算压碎指标值:

$$Q_a = \frac{G_1 - G_2}{G_1} \times 100 \qquad (4-1)$$

式中　$Q_a$——压碎指标值(%);
　　　$G_1$——试样总质量(g);
　　　$G_2$——压碎试验后孔径 2.36mm 筛上筛余的试样质量(g)。

压碎指标值越小,表明石子越坚硬,抗压能力越强,各类别石子其压碎指标必须满足表 4-9 所规定的数值。

表 4-9　粗骨料压碎指标

| 项　目 | 指　标 | | |
|---|---|---|---|
| | Ⅰ类 | Ⅱ类 | Ⅲ类 |
| 碎石压碎指标(%) | ≤10 | ≤20 | ≤30 |
| 卵石压碎指标(%) | ≤12 | ≤14 | ≤16 |

**8. 骨料的含水状态**

骨料的含水状态不同,在配制混凝土时会导致混凝土用水量和骨料用量变化很大,进而影响混凝土性能。饱和面干状态的骨料既不从混凝土中吸取水分,也不向混凝土拌和物中释放水分,因此可以准确控制混凝土的用水量。

骨料在饱和面干状态时的含水率,称为饱和面干吸水率。饱和面干吸水率越小,骨料颗粒越密实,质量越好。一般坚固骨料的饱和面干吸水率在 1% 左右,而气干状态密实骨料的含水率在 1% 以下,与饱和面干吸水率相差无几,故工程中常以气干状态骨料的含水率为基准进行混凝土配合比设计。不过在工程施工中,必须经常测定骨料的含水率,以及时调整混凝土组成材料实际用量的比例,从而保证混凝土质量的稳定性。

砂的体积和堆积密度与其含水状态紧密相关。气干状态的砂随着含水率的增大,砂颗粒表面形成一层吸附水膜,推挤砂粒分开并引起砂的体积的增大,这种现象称为砂的湿胀。砂的湿胀与其粒径有关,细砂的湿胀要大于粗砂。砂的含水率从零开始增大时,其体积由于湿胀而逐渐增加。砂的含水率增大至 5%～8% 时,其体积可增加 20%～30%。若砂的含水率继续增大,砂表面水膜增厚,水的自重超过砂粒表面对水的吸附而导致砂粒表面的水膜破裂消失,砂的体积减小。当砂的含水率增大至 20% 左右时,湿砂的体积降低至与干砂相近。

**9. 骨料的级配**

骨料的级配,是指骨料中不同粒径颗粒的分布情况。骨料的粒径分布全在同一尺寸范围内,会产生很大的空隙率,如图 4-1(a)所示;骨料的粒径分布在两种尺寸范围内时,空隙率会减小,如图 4-1(b)所示;若骨料的粒径分布在更多的尺寸范围内,空隙率会进一步减小,如图 4-1(c)所示。由此可见,只有适宜的骨料粒径分布,才能达到良好级配的要求。良好的级配能使骨料的空隙率和总表面积减少,从而降低所需水泥浆量,同时还可以提高混凝土的密实度、强度及其他性能;在考虑降低空隙率的同时,也应兼顾混凝土流动性的要求,细骨料应该有一定程度的盈余使其能够包裹粗骨料流动。

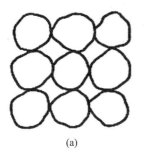 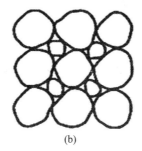 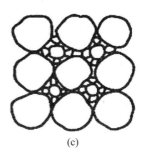

(a)          (b)          (c)

**图 4-1 骨料的颗粒级配**

骨料的级配可分为宏观级配和细观级配,骨料宏观级配指的是砂率,骨料细观级配包括砂、石各自的颗粒级配。

(1)砂率

砂率($S_p$)指的是混凝土中砂的质量($S$)占砂、石($G$)总质量的百分数,即:

$$S_p = \frac{S}{S+G} \times 100\% \tag{4-2}$$

(2)砂的颗粒级配和粗细程度

砂的颗粒级配和粗细程度是用筛分析方法测定的。砂的筛分析方法是用一套孔径分别为 4.75mm、2.36mm、1.18mm、0.60mm、0.30mm 和 0.15mm 的标准筛,将抽样所得 500g 干砂,由粗到细依次过筛,然后称得留在各筛上砂的质量,并计算出各筛上的分计筛余百分率 $a_1$、$a_2$、$a_3$、$a_4$、$a_5$、$a_6$(各筛上的筛余量占砂样总质量的百分率),及计筛余百分率 $A_1$、$A_2$、$A_3$、$A_4$、$A_5$、$A_6$(各筛与比该筛粗的所有筛的分计筛余百分率之和)。累计筛余与分计筛余的关系见表 4-10,任意一组累计筛余($A_1 \sim A_6$)表征了一个级配。

**表 4-10 累计筛余与分计筛余的关系**

| 筛孔尺寸(mm) | 分计筛余(%) | 累计筛余(%) |
| --- | --- | --- |
| 4.75 | $a_1$ | $A_1 = a_1$ |
| 2.36 | $a_2$ | $A_2 = a_1 + a_2$ |
| 1.18 | $a_3$ | $A_3 = a_1 + a_2 + a_3$ |
| 0.60 | $a_4$ | $A_4 = a_1 + a_2 + a_3 + a_4$ |
| 0.30 | $a_5$ | $A_5 = a_1 + a_2 + a_3 + a_4 + a_5$ |
| 0.15 | $a_6$ | $A_6 = a_1 + a_2 + a_3 + a_4 + a_5 + a_6$ |

砂的粗细程度用细度模数表示,细度模数($M_x$)按下式计算:

$$M_x = \frac{(A_2 + A_3 + A_4 + A_5 + A_6) - 5A_1}{100 - A_1} \tag{4-3}$$

细度模数越大,表示砂越粗。普通混凝土用砂的细度模数范围一般为 1.6~3.7。其中,$M_x$ 在 3.1~3.7 为粗砂,$M_x$ 在 2.3~3.0 为中砂,$M_x$ 在 1.6~2.2 为细砂,配制混凝土时宜优先选用中砂。

应当注意,细度模数相同的砂,级配可以不同。砂的细度模数并不能反映其级配的优劣,因此,配制混凝土时必须同时考虑砂的级配曲线。标准规定,按照 0.60mm 筛孔的累计筛余百分率,将砂分成三个级配区,见表 4-11。以累计筛余百分率为纵坐标,以筛孔尺寸为横坐标,根据表 4-11 的规定数值可以画出砂的 1、2、3 三个级配区上下限的筛分曲线(图 4-2,以天然砂为例)。

表 4-11 砂的颗粒级配区范围

| 方筛孔尺寸 (mm) | 累计筛余(%) | | | | | |
|---|---|---|---|---|---|---|
| | 天然砂 | | | 机制砂 | | |
| | 1 区 | 2 区 | 3 区 | 1 区 | 2 区 | 3 区 |
| 9.50 | 0 | 0 | 0 | 0 | 0 | 0 |
| 4.75 | 10 ~ 0 | 10 ~ 0 | 10 ~ 0 | 10 ~ 0 | 10 ~ 0 | 10 ~ 0 |
| 2.36 | 35 ~ 5 | 25 ~ 0 | 15 ~ 0 | 35 ~ 5 | 25 ~ 0 | 15 ~ 0 |
| 1.18 | 65 ~ 35 | 50 ~ 10 | 25 ~ 0 | 65 ~ 35 | 50 ~ 10 | 25 ~ 0 |
| 0.60 | 85 ~ 71 | 70 ~ 41 | 40 ~ 16 | 85 ~ 71 | 70 ~ 41 | 40 ~ 16 |
| 0.30 | 95 ~ 80 | 92 ~ 70 | 85 ~ 55 | 95 ~ 80 | 92 ~ 70 | 85 ~ 55 |
| 0.15 | 100 ~ 90 | 100 ~ 90 | 100 ~ 90 | 97 ~ 85 | 94 ~ 80 | 94 ~ 75 |

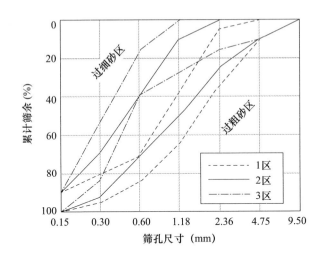

图 4-2 砂的级配区曲线

试验时,将砂样筛分析试验得到的各筛累计筛余标注在级配区图中,连线就可观察到此筛分曲线落在哪个级配区。判定砂级配是否合格的方法如下:

(1)先看 0.60mm 的累计筛余落在哪个级配区,其余各筛上的累计筛余百分率原则上应完全处于该级配区;

(2)4.75mm 和 0.60mm 筛号上不允许有任何超出;

(3)其余筛号允许有少量超出,但各级累计筛余超出值总和应小于 5%。

配制混凝土时宜优先选用 2 区砂;当采用 1 区砂时,应提高砂率,并保持足够的水泥用量,以满足混凝土的和易性;当采用 3 区砂时,宜适当降低砂率,以保证混凝土强度。

(3)石子的颗粒级配和最大粒径

石子的颗粒级配分为连续粒级和单粒级两种,石子的级配通过筛分试验确定,一套标准筛由孔径为 2.36mm、4.75mm、9.50mm、16.0mm、19.0mm、26.5mm、31.5mm、37.5mm、53.0mm、63.0mm、75.0mm 和 90.0mm 的 12 个筛子组成,可按需选用筛号进行筛分,然后计算得出每个筛号的分计筛余百分率和累计筛余百分率(计算方法与砂相同)。碎石和卵石的级配范围要求是相同的,应符合表 4-12 的规定。

**表 4-12 碎石或卵石的颗粒级配范围**

| 公称粒级 (mm) | | 累计筛余(%,质量分数) | | | | | | | | | | |
|---|---|---|---|---|---|---|---|---|---|---|---|---|
| | | 筛孔尺寸(方筛孔,mm) | | | | | | | | | | |
| | | 2.36 | 4.75 | 9.50 | 16.0 | 19.0 | 26.5 | 31.5 | 37.5 | 53.0 | 63.0 | 75.0 | 90.0 |
| 连续粒级 | 5~16 | 95~100 | 85~100 | 30~60 | 0~10 | 0 | | | | | | | |
| | 5~20 | 95~100 | 90~100 | 40~80 | — | 0~10 | 0 | | | | | | |
| | 5~25 | 95~100 | 90~100 | — | 30~70 | — | 0~5 | 0 | | | | | |
| | 5~31.5 | 95~100 | 90~100 | 70~90 | — | 15~45 | — | 0~5 | 0 | | | | |
| | 5~40 | — | 95~100 | 70~90 | — | 30~65 | — | — | 0~5 | 0 | | | |
| 单粒粒级 | 5~10 | 95~100 | 80~100 | 0~15 | 0 | | | | | | | | |
| | 10~16 | | 95~100 | 80~100 | 0~15 | | | | | | | | |
| | 10~20 | | 95~100 | 85~100 | — | 0~15 | 0 | | | | | | |
| | 16~25 | | | 95~100 | 55~70 | 25~40 | 0~10 | | | | | | |
| | 16~31.5 | | 95~100 | | 85~100 | | | 0~10 | 0 | | | | |
| | 20~40 | | | 95~100 | | 80~100 | | | 0~10 | 0 | | | |
| | 40~80 | | | | | 95~100 | | | 70~100 | | 30~60 | 0~10 | 0 |

粗骨料中,公称粒级的上限为该骨料的最大粒径,当骨料粒径增大时,其比表面积减小。与小粒径的骨料相比,包裹同等质量的大粒径骨料所需的水泥浆用量要少,可节约水泥。因此,在条件许可的情况下,应尽量选用最大粒径较大的粗骨料。但是,最大粒径受

一些条件制约,例如要考虑构件截面的最小尺寸、钢筋的间距及板材的厚度等。通常骨料的最大粒径不能超过结构截面最小尺寸的1/4,不超过钢筋间净距的3/4;浇筑实心混凝土板时,骨料的最大粒径不能超过板厚的1/2且不能超过50mm。对于泵送施工的混凝土,还要考虑泵送过程中的管道堵塞问题,通常骨料的最大粒径不能大于管道内径的1/3,以避免发生堵泵现象。

扫码可获取混凝土用骨料的相关试验内容。

### 四、混凝土用水

水是混凝土的重要组成成分之一,水质的好坏不仅影响混凝土的凝结和硬化,也影响混凝土的强度和耐久性。混凝土预制和养护用水不得含有影响水泥正常凝结硬化的有害物质,凡是能饮用的自来水及清洁的天然水都能用来拌制和养护混凝土。

### 五、混凝土外加剂

#### (一)外加剂的定义及分类

外加剂是指在混凝土拌和前或拌和时掺入的能显著改善混凝土某项或多项性能的一类材料,其掺量一般不大于胶凝材料质量的5%。外加剂在混凝土中并不能代替水泥,但能显著改善混凝土的和易性、强度、耐久性、调节凝结时间以及节约水泥等。外加剂的使用促进了混凝土的飞速发展,使得高强、高性能混凝土的生产和应用成为现实,并且解决了许多实际工程中的技术难题。目前,外加剂已成为除胶凝材料、水、砂子、石子以外的第五种重要的组成材料(称为第五组分),应用日益广泛。

混凝土外加剂按其主要作用可分为四类:

(1)改善混凝土拌和物工作性的外加剂,如减水剂、引气剂和泵送剂等。

(2)调节混凝土凝结时间和硬化性能的外加剂,如缓凝剂、速凝剂和早强剂等。

(3)改善混凝土耐久性的外加剂,如引气剂、防水剂和阻锈剂等。

(4)提供混凝土特殊性能的外加剂,如膨胀剂、防冻剂、减缩剂等。

#### (二)几种常用的混凝土外加剂

**1. 减水剂**

减水剂是指在混凝土流动性基本相同的条件下能减少拌和用水量,或者在混凝土配合比和材料不变的情况下能增加混凝土流动性的外加剂。减水剂是一种表面活性剂,其分子由亲水基和憎水基两个部分组成。减水剂加入水泥浆体中后,其分子中的亲水基指向水,憎水基定向吸附于水泥颗粒表面,使水泥颗粒表面带有同一种电荷,形成静电排斥作用,水泥颗粒互相分散;水泥颗粒表面的减水剂吸附膜能与水分子形成一层稳定的溶剂化水膜,这层水膜具有很好的润滑作用,可改善混凝土的流动性,如图4-3所示。减水剂品种繁多,根据减水效果可分为普通型和高效型;根据化学成分可分为木质素磺酸盐类、多元醇类、多环芳香族磺酸盐类、萘系、聚羧酸盐等。

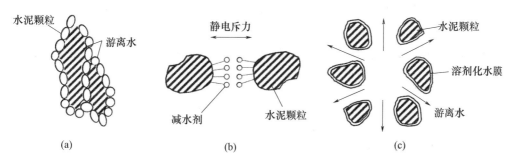

图 4-3 减水剂减水机理示意图

在混凝土中加入减水剂后,可取得以下效果:
(1)在拌和物用水量不变时可显著提高混凝土流动性,有利于浇筑成型;
(2)保持混凝土拌和物流动性和水泥用量不变,减少用水量,降低水灰比,提高混凝土强度和混凝土耐久性;
(3)保持混凝土流动性和强度不变时,节约水泥用量,降低成本;
(4)使混凝土更加匀质性,改善混凝土的孔结构。

**2. 引气剂**

在搅拌混凝土过程中能引入大量均匀分布、稳定而封闭的微小气泡,从而改善混凝土和易性和耐久性(主要为抗冻性)的外加剂,称为引气剂。引气剂引入的气泡直径为 0.02~1mm,大多在 0.2mm 以下。常用的引气剂有松香类、皂甙类、烷基磺酸盐等,在混凝土中的掺量一般为胶凝材料质量的 0.005%~0.01%。

**3. 缓凝剂**

缓凝剂是指能延长混凝土凝结时间的外加剂。常用的缓凝剂有糖类多羟基化合物(葡萄糖、蔗糖、糖蜜、糖钙等)、羟基羧酸类(柠檬酸、酒石酸、葡萄糖酸、水杨酸等)和多元醇类(山梨醇、聚乙烯醇、麦芽糖醇、木糖醇等)。缓凝剂的掺量一般为胶凝材料的 0.01%~0.1%。

**4. 泵送剂**

泵送剂指能改善混凝土拌和物泵送性能的外加剂。泵送剂一般由减水组分与缓凝组分、引气组分复合而成,减水组分能提高混凝土拌和物的流动性,引气组分具备润滑作用,缓凝组分可以抑制拌和物的坍落度损失。

**5. 早强剂**

早强剂是指能加速混凝土早期强度发展的外加剂。常用的早强剂种类有无机盐类(氯盐、硫酸盐、硫酸复盐、硝酸盐、亚硝酸盐等)和水溶性有机化合物类(三乙醇胺、甲酸盐、乙酸盐、丙酸盐等),其中三乙醇胺的掺量小于 0.05%,其他种类早强剂的掺量一般小于 2%。值得注意的是无机盐类早强剂的使用会影响混凝土后期的强度发展;另外,钢筋混凝土中禁止使用氯盐类早强剂。

**6. 防冻剂**

防冻剂指能使混凝土在负温下硬化,并产生足够防冻强度的外加剂。常用的防冻剂通常由减水组分、防冻组分和引气组分复合而成。减水组分的作用是减少混凝土的用水量,从而降低混凝土中的冰胀应力;同时,水灰比的降低,细化了孔结构,减少了混凝土中的固有缺陷。防冻组分的作用是降低冰点,保证混凝土中的液相在规定的负温条件下不冻结或少冻结。引气组分则可以提高混凝土后期的抗冻性。

# 第三节　新拌混凝土的工作性能

## 一、混凝土工作性的概念与含义

混凝土从出搅拌机到浇筑振捣完毕这一阶段称为新拌混凝土。工作性指新拌混凝土能保持其组分均匀,易于运输、浇筑、捣实、成型等施工作业,而不发生离析、分层、泌水的性能,反映新拌混凝土施工的难易程度。工作性也称为和易性,包括三方面的含义,即流动性、黏聚性和保水性。

流动性是指混凝土拌和物在自重或外力作用下,能产生流动并均匀、密实地充满模型的能力。

黏聚性也叫抗离析性,是指混凝土拌和物在运输、浇筑和振捣过程中,能保持组分均匀,不发生分层、离析现象的性能。

保水性是指混凝土拌和物具有一定的保持其内部水分的能力,在施工过程中不致产生严重泌水的性能。

混凝土拌和物的流动性、黏聚性及保水性之间互相关联又互相矛盾,当流动性很大时,往往混凝土拌和物的黏聚性和保水性较差,反之亦然。因此,所谓拌和物和易性良好,指这三方面的性质在某种具体条件下全部达到良好。

## 二、混凝土工作性的评价方法

混凝土拌和物工作性的评价通常是采用一定的实验方法测定混凝土拌和物的流动性,再辅以直观经验目测评定混凝土拌和物的黏聚性和保水性。按照《混凝土质量控制标准》(GB 50164—2011)的规定,混凝土拌和物的流动性以坍落度或维勃稠度作为指标。坍落度适用于流动性较大的混凝土拌和物,维勃稠度适用于干硬性混凝土拌和物。扫码获取《混凝土质量控制标准》(GB 50164—2011)的详细内容。

**1. 坍落度**

将坍落度筒(标准截头圆锥筒,无底,上口直径、下口直径和高度分别为100mm、200mm和300mm)放在水平的、不吸水的刚性底板上并固定,将刚刚拌和的混凝土混合料分三层装入筒内,每装完一层之后,用弹头型金属捣棒由边缘向中心沿螺旋线方向插捣至上一层拌和物上表面,每一层都插捣25次,最后将上口抹平,垂直提起坍落度筒,筒内的拌和物在自重作用下向下坍落,测量坍落后试样的最高点与坍落度筒之间的高度之差,即为坍落度值(mm)(图4-4)。坍落度越大,表示混凝土拌和物的流动性越好。由于此法简便,目前世界各国普遍采用。在测完坍落度后,应用捣棒敲击已坍落的混凝土拌和物试体,观察其受击后下沉、坍落及四周泌水情况,然后再凭目测判定混凝土拌和物黏聚性和保水性的优劣。

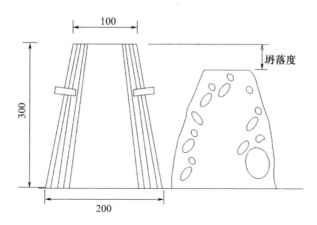

图4-4 混凝土拌和物坍落度测定

坍落度试验只适用于骨料最大粒径不大于40mm、坍落度值大于10mm的非干硬性混凝土。根据坍落度大小,将混凝土拌和物分为5级,见表4-13。

**2. 维勃稠度**

坍落度值小于10mm的混凝土叫作干硬性混凝土。干硬性混凝土难以用坍落度值来反映其流动性的大小,故常采用维勃稠度(VB稠度值)来反映其干硬程度。该方法所用仪器叫作维勃稠度仪,如图4-5所示。在振动台上安装圆筒形容器,在筒内按坍落度试验方法装料,提起坍落度筒后在混凝土试料上面放置透明的压板,然后启动振动台,从开始振动至混凝土试样与压板全面接触的时间为维勃稠度值,用来定量地评价干硬性混凝土的稠度。维勃稠度值越大,表明混凝土拌和物越干硬,流动性越差。混凝土拌和物根据其维勃稠度值的大小,可分为5级,见表4-13。

表4-13 新拌混凝土根据坍落度和维勃稠度的分级

| 级 别 | 名 称 | 坍落度(mm) | 级 别 | 名 称 | 维勃稠度(s) |
|---|---|---|---|---|---|
| S1 | 低塑性混凝土 | 10~40 | V0 | 超干硬性混凝土 | ≥31 |
| S2 | 塑性混凝土 | 50~90 | V1 | 特干硬性混凝土 | 21~30 |
| S3 | 流动性混凝土 | 100~150 | V2 | 干硬性混凝土 | 11~20 |
| S4 | 泵送混凝土 | 160~210 | V3 | 半干硬性混凝土 | 6~10 |
| S5 | 流态混凝土 | ≥210 | V4 | 低干硬性混凝土 | 3~5 |

**3. 坍落扩展度**

对于坍落度大于210mm的流态混凝土,除需测试坍落度之外,还要测试坍落扩展度。

坍落扩展度试验在传统的坍落度试验基础上,同时测定拌和物的水平扩展度和扩展到一定直径(一般为50cm)时所用的时间,以此反映拌和物的变形能力和变形速度。

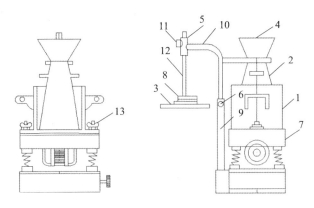

图 4-5 维勃稠度仪

1—容器；2—坍落度筒；3—透明圆盘；4—喂料斗；5—套筒；6—定位螺钉；7—振动台；
8—荷重；9—支柱；10—螺旋架；11—测杆螺钉；12—测杆；13—固定螺钉

## 三、混凝土工作性的影响因素

**1. 单位用水量**

在配制混凝土时，当所用粗、细骨料的品种及数量一定时，如果每 $1m^3$ 混凝土用水量一定，即使水泥用量稍有波动（水泥用量增减 $50\sim100kg/m^3$），混凝土的流动性保持不变，这一规律称为固定需水量法则。混凝土中用水量增大，流动性随之变好，但用水量增大带来的不利影响是混凝土拌和物的黏聚性和保水性变差，易产生泌水和分层离析。

**2. 水灰比**

水灰比的大小反映水泥浆的稀稠程度。在水泥浆用量一定时，降低水灰比，水泥浆变稠，水泥浆的黏聚力增大，混凝土的黏聚性和保水性变好，而流动性变差；增加水灰比则情况相反。

**3. 水泥浆用量**

混凝土拌和物在自重或外界振动力的作用下要产生流动，必须克服其内部的阻力。拌和物内的阻力主要来自两个方面，一为骨料间的摩阻力，一为水泥浆的黏聚力。骨料间摩阻力的大小主要取决于骨料颗粒表面水泥浆层的厚度，即水泥浆的数量；水泥浆的黏聚力大小主要取决于水泥浆的干稀程度，即水泥浆的稠度。

混凝土拌和物在保持水灰比不变的情况下，水泥浆用量越多，包裹在骨料颗粒表面的浆层越厚，润滑作用越好，使骨料间摩擦阻力减小，混凝土拌和物易于流动，流动性越好；反之流动性越差。

**4. 水泥品种**

在水泥用量和用水量一定的情况下，采用矿渣硅酸盐水泥或火山灰硅酸盐水泥的混凝土拌和物的流动性比采用硅酸盐水泥或普通硅酸盐水泥的要差，这是因为前者的密度较小，在相同水泥用量的情况下绝对体积较大，进而在相同用水量情况下混凝土显得更为干稠。另外，矿渣硅酸盐水泥拌制的混凝土拌和物保水性也较差。由于粉煤灰的形态效应,掺加粉煤灰的硅酸盐水泥的混凝土拌和物体现出较好的流动性,同时保水性和黏聚性均较好。

**5. 骨料性质**

骨料性质指混凝土所用骨料的品种、级配、颗粒粗细及表面性状等。在混凝土骨料用量一定的情况下,采用卵石和河砂拌制的混凝土拌和物,其流动性比采用碎石和山砂拌制的好,这是因为前者骨料表面光滑,摩阻力小。在水泥浆量一定的情况下,级配良好的骨料空隙较少,用于填充空隙的水泥浆较少,相对来说包裹骨料颗粒表面的水泥浆层较厚一些,从而提高了混凝土拌和物的和易性。

**6. 砂率**

砂率表示混凝土中砂与石子二者的组合关系,砂率的变动,会使骨料的总表面积和空隙率发生很大的变化,因此会对混凝土拌和物的工作性产生显著的影响。砂率增大,混凝土拌和物的流动性变好,黏聚性和保水性也会提高;但当砂率过大时,骨料的总表面积和空隙率均增大,当混凝土拌和物中水泥浆量一定的情况下,骨料颗粒表面的水泥浆层相对减薄,混凝土拌和物变得干稠,流动性变差,同时黏聚性也变差。反之,若砂率过小,则混凝土拌和物中石子过多而砂过少,形成的砂浆量不足以包裹石子表面,混凝土拌和物的流动性降低,黏聚性也会变差。

由上可知,在配制混凝土时,砂率不能过大,也不能太小,应该选用合理砂率。合理砂率指在用水量及水泥用量一定的情况下,能使混凝土拌和物获得最大的流动性,且能保持黏聚性及保水性良好的砂率,如图4-6所示。当采用合理砂率时,混凝土拌和物能在获得所要求的流动性及良好的黏聚性与保水性条件下,水泥用量最少,如图4-7所示。一般来说,合理砂率应使砂填满石子空隙并有一定的富余量。

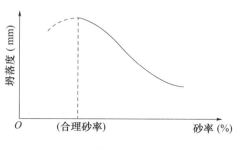

图4-6 坍落度与砂率的关系
(水和水泥用量一定)

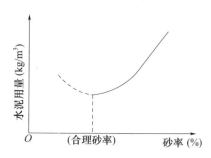

图4-7 水泥用量与砂率的关系
(达到相同坍落度)

**7. 化学外加剂**

混凝土拌和物中掺入减水剂或引气剂,其流动性明显提高,引气剂还可有效地改善混凝土拌和物的黏聚性和保水性。

**8. 存放时间及环境温度**

混凝土拌和物会随着时间的延长逐渐变得干稠,坍落度逐渐减小,这一现象称为坍落度经时损失。这是由于混凝土拌和物中的一些水分会逐渐被骨料吸收、一部分水被蒸发、水泥的水化与凝聚结构的逐渐形成,以及混凝土拌和物中减水剂被水泥颗粒吸附等。

混凝土拌和物的和易性还将受到环境温度的影响。随着环境温度的升高,混凝土的坍落度损失得更快,因为此时混凝土的水分蒸发及水泥的化学反应进行得更快。

### 四、新拌混凝土浇筑后的性能

从混凝土浇筑完成至凝结的时长约几个小时,混凝土拌和物在此时间段内将呈现塑性和半流动状态,这个状态下混凝土的性能称为新拌混凝土浇筑后的性能。

**1. 离析**

新拌混凝土中的各组分由于密度不同,在重力作用下产生相对运动,骨料下沉,而浆体上浮,造成混凝土拌和物不均匀和失去连续性的现象称为离析。新拌混凝土离析的发生主要与其中固体颗粒的表面积与拌和用水的质量有关联,这里的固体颗粒包括骨料与胶凝材料。新拌混凝土的抗离析指数为混凝土中固体颗粒表面积与拌和水的质量之比,混凝土中固体颗粒表面积越大,拌和水的质量越小,混凝土越不容易发生离析。

**2. 泌水**

新拌混凝土在浇筑成型后到初凝期间,混凝土中的固体颗粒下沉,水分上升并在表面析出的现象称为泌水。混凝土泌出的水或者向外蒸发,或者因为水泥水化被吸回,二者都会使混凝土体积减小。泌水对混凝土性能有两方面影响:一方面,顶部或靠近顶部的混凝土因含水多,形成疏松的水化物结构,对其耐磨性等十分有害;另一方面,部分上升的水积存在骨料下方形成水囊,削弱水泥浆与骨料间的过渡区,影响硬化混凝土的强度和耐久性。

**3. 塑性沉降**

混凝土拌和物由于泌水产生整体沉降的现象称为塑性沉降。新拌混凝土的塑性沉降受到阻碍(例如钢筋)则会产生塑性沉降裂缝,裂缝从表面向下直至钢筋的上方。

**4. 凝结时间**

混凝土产生凝结的根本原因是水泥的水化反应,但混凝土拌和物的凝结时间与其所用水泥的凝结时间不一定完全相同。水泥的凝结时间是水泥净浆在规定的温度和稠度条件下测得的,混凝土拌和物的存在条件与水泥凝结时间测定条件不一定相同。混凝土的水灰比、环境温度和外加剂的性能等均对其凝结时间产生很大影响。

混凝土拌和物的凝结时间通常用贯入阻力法进行测定,所使用的仪器为贯入阻力仪。先用 4.75mm 的方孔筛从混凝土拌和物中筛取砂浆,并将其按一定的方法装入规定的容器,然后每隔 0.5h 测定贯入砂浆 25mm 的贯入阻力。绘制贯入阻力与时间的关系曲线,以贯入阻力 3.5MPa 和 28.0MPa 画两条平行于时间坐标的直线,直线与曲线交点的时间分别为混凝土拌和物的初凝时间和终凝时间。初凝时间表示此时混凝土拌和物不能再正常地浇筑和捣实,即施工的时间极限,终凝时间表示此时混凝土强度开始以相当的速度发展。

扫码可获取混凝拌和物的相关试验内容。

# 第四节 硬化混凝土的结构

## 一、结构特点

混凝土凝结硬化产生强度后称为硬化混凝土。从结构特点来说,硬化混凝土是颗粒状的粗细骨料均匀地分散在水泥石中形成的分散体系,如图4-8所示。具体来说,水泥净浆包裹砂,并填充砂的空隙,形成水泥砂浆;水泥砂浆包裹石子,并填充石子的空隙,形成混凝土。

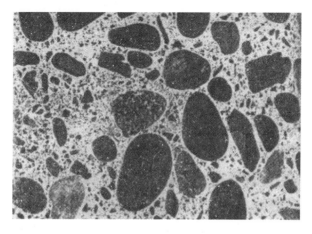

图4-8 硬化混凝土的结构

从组成来说,混凝土是一个三相组成的复合结构,包括骨料相、水泥石相和过渡区相。骨料相中的砂和石是混凝土的骨架,分散在水泥石中,因此又称为分散相。水泥石相是混凝土的基体,因此又称为连续相。过渡区相指骨料与水泥石之间的界面过渡区,因此又称为界面相。

## 二、界面过渡区

界面过渡区是骨料周围存在的一层水泥石薄层,厚度大约为20~50μm。混凝土中的骨料颗粒数量众多,界面过渡区的体积可达到硬化水泥浆体的1/3~1/2,其对混凝土的性能影响不可忽略。

对于界面过渡区在混凝土中的形成,一般认为是因为混凝土在凝结硬化之前,骨料颗粒受重力作用向下沉降,水由于密度小而向上迁移,它们之间的相对运动使水在骨料颗粒的下方富集,骨料周围逐渐形成一层高水灰比的水泥浆膜,待混凝土硬化后,这里就形成了界面过渡区。

与水泥石相相比,界面过渡区内的水灰比较高。同时界面过渡区内水化硅酸钙凝胶的数量较少,密实度较差,孔隙率较大,尤其是大孔较多。除此之外,氢氧化钙、钙矾石等晶体在界面过渡区内的尺寸较大,含量较多,且大多取向生长(垂直于骨料表面定向生长),如图4-9所示。由于骨料和水泥石的变形模量、收缩性能等存在差别,或者由于水分蒸发等原因,界面过渡区内存在大量的原生微裂缝,是混凝土中的薄弱环节。

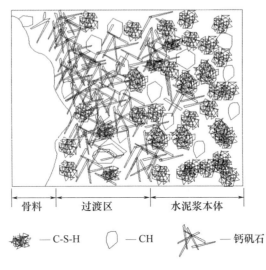

图 4-9 过渡区示意图

### 三、孔结构

水泥石由水泥的水化产物(包括结晶体和凝胶体)、未完全水化的水泥颗粒、大小不一的孔隙以及其中的孔隙水或气体组成。其中,各种孔在混凝土的力学性能、变形性能和耐久性能上都有着重要的影响。一般认为混凝土中的孔可以分为四类,分别为特征尺寸为 0.5~10nm 的凝胶孔、平均半径为 5nm~5μm 的毛细孔、尺寸为 20~200μm 的引气孔以及尺寸为 1000μm 的由于不够密实形成的大孔。

吴中伟院士提出将混凝土中孔分为四级,分别为小于 20nm 的无害孔、20~50nm 的少害孔、50~200nm 的有害孔和大于 200nm 的多害孔。其中 50nm 以上的孔对混凝土的强度和耐久性危害较大,小于 20nm 的孔对混凝土性能的影响极其微小。

一般认为,混凝土中的孔对混凝土的有害作用主要体现在如下三点:首先,孔是侵蚀物质侵入混凝土内部的通道;其次,毛细孔可以降低混凝土的强度;最后,凝胶孔失水会导致混凝土的收缩。与此同时,孔对混凝土也具有一定的积极作用,因为孔的存在为混凝土的后期强度发展提供了空间,而封闭球形的小孔可以改善混凝土的抗冻性。

## 第五节 硬化混凝土的力学性能

### 一、受压破坏过程

如前所述,硬化后的混凝土在未受外力作用之前,其界面过渡区内已存在一定的原生微裂缝,当混凝土受荷时,这些界面微裂缝会逐渐扩大、延长并联通起来,形成可见的裂缝,致使混凝土丧失连续性而遭到完全破坏。

当对混凝土立方体试件进行单轴静力受压试验时,混凝土的破坏过程和裂缝状态呈现四个不同的阶段,如图4-10和图4-11所示。各阶段具体情况如下:

Ⅰ阶段:在荷载达"比例极限"(约为极限荷载的30%)以前,荷载与变形近似直线关系(图4-10中的 OA 段),界面裂缝无明显变化

Ⅱ阶段:荷载超过"比例极限"后,变形速度大于荷载的增加速度,荷载与变形之间不再是线性关系(图4-10中的 AB 段)。在此阶段,界面裂缝的数量、长度及宽度不断增大,界面借摩阻力继续分担荷载,而砂浆内尚未出现明显的裂缝。

Ⅲ阶段:荷载超过"临界荷载"(约为极限荷载的70%~90%)以后,变形速度进一步加快,曲线明显弯向变形坐标轴(图4-10中的 BC 段)。此阶段中,界面裂缝继续发展,砂浆中开始出现裂缝,部分界面裂缝连接成连续裂缝。

Ⅳ阶段:外荷超过极限荷载以后,荷载减小而变形迅速增大,曲线下弯而终止(图4-10中的 CD 段)。此时,连续裂缝急速发展,混凝土承载能力下降,以致完全破坏。

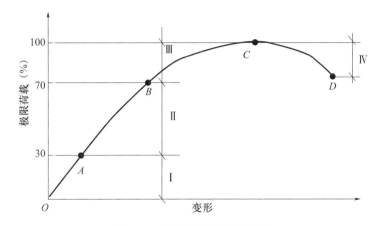

**图 4-10　混凝土受压变形曲线**

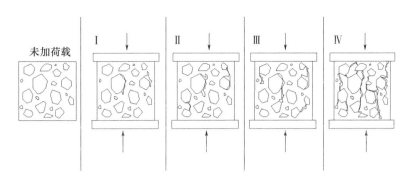

**图 4-11　不同受力阶段裂缝示意图**

Ⅰ—界面裂缝无明显变化;Ⅱ—界面裂缝明显增长;Ⅲ—出现砂浆裂缝和连续裂缝;Ⅳ—连续裂缝迅速发展

## 二、混凝土的强度

强度是硬化混凝土最重要的技术性质,也是工程施工中控制和评定混凝土质量的主

要指标。混凝土的强度有抗压、抗拉、抗折和抗剪等强度,其中以抗压强度为最大,因此在结构工程中混凝土主要用于承受压力。

**1. 立方体抗压强度**

我国采用立方体抗压强度作为混凝土的强度特征值。根据国家标准《混凝土物理力学性能试验方法标准》(GB/T 50081—2019),规定制作边长为 150mm 的立方体标准试件,在标准养护条件[温度(20±2)℃,相对湿度为95%以上]下,养护到 28d 龄期,用标准试验方法测得的抗压强度值称为混凝土立方体抗压强度,用 $f_{cc}$ 表示。混凝土立方体试件的抗压强度按下式计算:

$$f_{cc} = \frac{F}{A} \tag{4-4}$$

式中　$F$——试件破坏荷载(N);
　　　$A$——试件承压面积($mm^2$)。

立方体试件抗压强度值的确定应符合以下三条规定:

(1)取 3 个试件测值的算术平均值作为该组试件的强度值,应精确至 0.1MPa。

(2)当 3 个测值中的最大值或最小值中有一个与中间值的差值超过中间值的 15% 时,把最大值和最小值剔除,取中间值作为该组试件的抗压强度值。

(3)当最大值和最小值与中间值的差值均超过中间值的 15% 时,该组试件的试验结果无效。

混凝土采用边长为 150mm 的标准试件在标准条件下测定其抗压强度,是为了确保不同批次试验之间的可比性。在实际施工中,允许采用非标准尺寸的试件,但应将其抗压强度换算成标准试件的抗压强度,换算系数见表 4-14。

表 4-14　混凝土立方体试件边长与强度换算系数

| 试件边长(mm) | 抗压强度换算系数 |
| --- | --- |
| 100 | 0.95 |
| 150 | 1.00 |
| 200 | 1.05 |

混凝土试件尺寸越大,测得的抗压强度值越低。这是由于测试时产生的环箍效应及试件存在缺陷的概率不同所致。首先,将混凝土立方体试件置于压力机上受压时,在沿加荷方向发生纵向变形的同时,混凝土试件及上、下钢压板也由于泊松比效应产生横向自由变形,但由于压力机钢压板的弹性模量比混凝土大 10 倍左右,而泊松比仅是混凝土的 2 倍左右,所以在压力作用下,钢压板的横向变形小于混凝土的横向变形,造成上、下钢压板与混凝土试件接触的表面之间产生摩阻力,它对混凝土试件的横向膨胀起着约束作用,从而提高了混凝土的强度,称为环箍效应,如图 4-12 所示。这种约束作用随离试件端部越远而变小,所以试件抗压破坏后呈一对顶棱锥体,如图 4-13 所示。混凝土立方体试件尺寸较大时,环箍效应的相对作用较小,测得的抗压强度因而偏低;反之,测得的抗压强度偏高。再者,混凝土试件中存在的微裂缝和孔隙等缺陷,减小混凝土试件的实际受力面积以及引起应力集中,导致强度降低。大尺寸混凝土试件中存在缺陷的概率较大,所测强度要

较小尺寸混凝土试件偏低。扫码获取《混凝土物理力学性能试验方法标准》(GB/T 50081—2019)的详细内容。

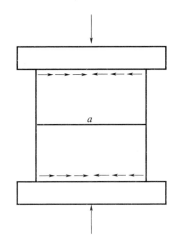

图 4-12　压力机压板对试块的约束作用

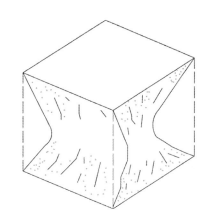

图 4-13　受压板约束试块破坏残存的棱锥体

**2. 轴心抗压强度**

混凝土轴心抗压强度又称棱柱体抗压强度。在实际结构中,钢筋混凝土受压构件大部分为棱柱体或圆柱体。为了使所测混凝土的强度能接近于混凝土结构的实际受力情况,规定在钢筋混凝土结构设计中计算轴心受压构件(如柱、桁架的腹杆等)时,均需用混凝土的轴心抗压强度作为依据。

根据《混凝土物理力学性能试验方法标准》(GB/T 50081—2019)的规定,混凝土轴心抗压强度 $f_{cp}$ 应采用 150mm × 150mm × 300mm 的棱柱体作为标准试件,如确有必要,可采用非标准尺寸的棱柱体试件,但其高宽比值应在 2~3 的范围内。标准棱柱体试件的制作条件与标准立方体试件相同,但测得的抗压强度值前者较小。试验表明,当标准立方体抗压强度 $f_{cc}$ 在 10~50MPa 范围内时,$f_{cp} = (0.7 \sim 0.8) f_{cc}$,一般取 0.76。

**3. 轴心抗拉强度**

混凝土的轴心抗拉强度 $f_t$ 很低,只有其抗压强度的 1/20~1/10(通常取 1/15),且这个比值是随着混凝土强度等级的提高而降低。所以,混凝土受拉时呈脆性断裂,破坏时无明显残余变形。为此,在钢筋混凝土结构设计中,不考虑混凝土承受拉力,而是在混凝土中配以钢筋,由钢筋来承担结构中的拉力。

**4. 劈裂抗拉强度**

混凝土轴心抗拉强度较难测定,试验时受到的外界干扰较多,例如,荷载作用线难以与试件轴线保持重合,易发生偏心。目前国内外都采用劈裂法反映混凝土的抗拉性能,测定混凝土的劈裂抗拉强度,简称劈拉强度。我国标准规定,混凝土劈拉强度采用边长为 150mm 的立方体作为标准试件。这个方法的原理是:在立方体试件上、下表面中部划定的劈裂面位置线上,作用一对均匀分布的压力,在此外力作用下的试件竖向平面内,产生均布拉伸应力(图 4-14),该拉应力可以根据弹性理论计算得出。混凝土劈裂抗拉强度计

算公式为:

$$f_{ts} = \frac{2F}{\pi A} \tag{4-5}$$

式中 $f_{ts}$——混凝土劈裂抗拉强度(MPa);

$F$——破坏荷载(N);

$A$——试件劈裂面积($mm^2$)。

#### 5. 抗折强度

混凝土的抗折强度也称抗弯拉强度。根据标准,混凝土抗折强度试验采用的标准试件为 150mm×150mm×600mm 或 150mm×150mm×550mm 的棱柱体试件。试验采用"三分点加荷方式"进行,即以两个相等的荷载同时垂直作用在试件跨度的两个三分点处,如图 4-15 所示。根据材料力学理论,混凝土的抗折强度可由试件断裂时的最大拉伸应力得到:

$$f_f = \frac{Fl}{bh^2} \tag{4-6}$$

式中 $f_f$——混凝土抗折强度(MPa);

$F$——破坏荷载(N);

$l$——支座间跨度(mm);

$b$——试件截面宽度(mm);

$h$——试件截面高度(mm)。

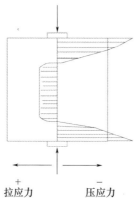

图 4-14 劈裂试验时垂直于受力面的应力分布

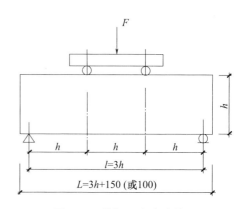

图 4-15 抗折强度试验装置

扫码获取硬化混凝土力学性能试验的相关内容。

### 三、混凝土的强度等级

按《混凝土强度检验评定标准》(GB/T 50107—2010)的相关规定,混凝土的强度等级应按其立方体抗压强度标准值确定。混凝土立方体抗压强度标准值为按照标准方法制作和养护的边长为 150mm 的立方体试件,用标准试验方法在 28d 龄期测得的混凝土抗压强度总体分布中的一个值,强度低于该值的概率为 5%,以 $f_{cc,k}$ 表示。换言之,混凝土强度总体分布中大于立方体抗压强度标准值的概率为95%,即混凝土强度总体分布中大于设计强度等级值的强度保证率为95%。混凝土强度等级采用符号"C"与立方体抗压强度标准值(以 N/mm² 计)表示。目前我国建筑工程中所用的混凝土强度划分为 14 个等级,即 C15、C20、C25、C30、C35、C40、C45、C50、C55、C60、C65、C70、C75、C80。扫码获取《混凝土强度检验评定标准》(GB/T 50107—2010)的详细内容。

**1. 抗压强度分布**

在正常生产施工条件下,影响混凝土强度的因素都是随机变化的,因此混凝土的强度也应是随机变量。在一定施工条件下,对同一种混凝土进行随机取样,制作 $n$ 组试件($n \geqslant 25$),测得其 28d 龄期的抗压强度,然后以混凝土强度为横坐标,以混凝土强度出现的概率为纵坐标,绘制出混凝土强度概率分布曲线。混凝土的强度分布曲线一般符合正态分布,如图 4-16 所示。混凝土强度正态分布曲线具有以下特点:

(1)曲线呈钟形,两边对称,对称轴在平均强度值 $\bar{f}_{cc}$ 处,曲线的最高峰出现在这里,表明混凝土强度接近其平均强度值出现的次数最多。随着距离对称轴越远,强度测定值比平均值越低或越高出现的概率越少。

(2)曲线和横坐标之间所包围的面积为概率的总和,等于100%。对称轴两侧的概率相等,各为50%。

(3)对称轴两侧的曲线上各有一个拐点,拐点至对称轴的垂直距离为强度正态分布的标准差。

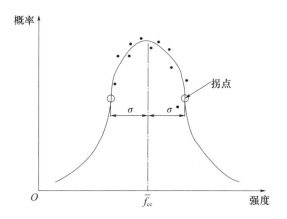

**图 4-16  混凝土强度的正态分布曲线**

**2. 强度保证率**

混凝土强度保证率 $P(\%)$ 是指混凝土强度总体中大于或等于设计强度等级值($f_{cc,k}$)的概率,在混凝土强度正态分布曲线图中以阴影面积表示,如图 4-17 所示。强度保证率 $P(\%)$ 可由正态分布曲线方程积分求得,即:

$$P(t) = \int_{t}^{+\infty} \phi(t) \, dt = \frac{1}{\sqrt{2}} \int_{t}^{+\infty} e^{\frac{t^2}{2}} dt \tag{4-7}$$

计算混凝土强度保证率的过程中,先根据混凝土的设计强度等级值 $f_{cc,k}$、强度平均值 $\bar{f}_{cc}$ 和标准差计算出概率密度系数:

$$t = \frac{\bar{f}_{cc} - f_{cc,k}}{\sigma} \tag{4-8}$$

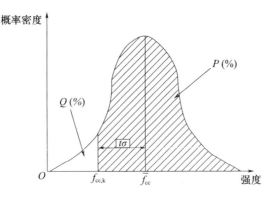

图 4-17 混凝土强度保证率

进而可以按标准正态分布曲线方程求出强度保证率 $P(\%)$ 或按表 4-15 进行查取。由表 4-15 可知,概率密度系数为 1.645 时,混凝土的强度保证率达到 95%。

表 4-15 不同 $t$ 值的强度保证率值

| $t$ | 0.00 | 0.50 | 0.80 | 0.84 | 1.00 | 1.04 | 1.20 | 1.28 | 1.40 | 1.50 | 1.60 |
|---|---|---|---|---|---|---|---|---|---|---|---|
| $P(\%)$ | 50.0 | 69.2 | 78.8 | 80.0 | 84.1 | 85.1 | 88.5 | 90.0 | 91.9 | 93.5 | 94.5 |
| $t$ | 1.645 | 1.70 | 1.75 | 1.81 | 1.88 | 1.96 | 2.00 | 2.05 | 2.33 | 2.50 | 3.00 |
| $P(\%)$ | 95.0 | 95.5 | 96.0 | 96.5 | 97.0 | 97.5 | 97.7 | 98.0 | 99.0 | 99.4 | 99.87 |

### 四、混凝土强度的影响因素

#### 1. 水灰比

理论上,水泥水化时所需的水一般只占水泥质量的 23% 左右,但在制备混凝土拌和物时,为了获得施工要求的和易性,常常需要多加一些水。混凝土中这些多余的水分将会形成水泡或水道,随着混凝土的硬化而蒸发,最后留下大量孔隙,孔隙的存在减少混凝土实际的受力面积;同时在混凝土受力时,易在孔隙周围产生应力集中。在充分密实的情况下,水灰比越大,多余水分越多,留下的孔隙也越多,混凝土强度也就越低;反之则混凝土强度越高,如图 4-18 所示。在不能充分密实的情况下,若水灰比过小,混凝土拌和物和易性太差,反而导致混凝土强度严重下降。

试验证明,在材料相同的情况下,混凝土的强度 $f_{cc}$ 与其水灰比($W/C$)的关系,呈近似双曲线形状(图 4-18 中的实线),而混凝土强度与灰水比的关系呈直线关系,如图 4-19 所示。综合考虑水泥强度并应用数理统计方法,可建立混凝土强度与水泥强度及灰水比之间的经验公式,即混凝土强度经验公式(又称鲍罗米公式):

$$f_{cc} = \alpha_a f_{ce}\left(\frac{C}{W} - \alpha_b\right) \tag{4-9}$$

式中 $f_{cc}$——混凝土 28d 龄期的抗压强度(MPa);

$C$——$1m^3$ 混凝土中的水泥用量(kg);

$W$——$1m^3$ 混凝土中的用水量(kg);

$\dfrac{C}{W}$——混凝土的灰水比;

$f_{ce}$——水泥 28d 抗压强度实测值(MPa);

$\alpha_a$、$\alpha_b$——与骨料种类有关的回归系数。

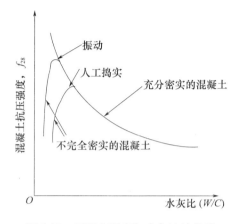

图 4-18　混凝土强度与水灰比的关系　　　　图 4-19　混凝土与灰水比的关系

$\alpha_a$ 和 $\alpha_b$ 应按工程所使用的水泥和骨料,通过试验建立的灰水比与混凝土强度关系式来确定。当不具备上述试验统计资料时,则可按表 4-16 的值取用。

表 4-16　回归系数 $\alpha_a$ 和 $\alpha_b$ 的取值

| 粗骨料种类 | $\alpha_a$ | $\alpha_b$ |
| --- | --- | --- |
| 碎石 | 0.53 | 0.20 |
| 卵石 | 0.49 | 0.13 |

### 2. 骨料

混凝土中,骨料质量与水泥质量之比称为骨灰比。骨灰比对 35MPa 以上的混凝土强度影响较大,在相同水灰比和坍落度下,骨料增多后表面积增大,吸水量也增加,从而降低了有效水灰比,因此混凝土强度随骨灰比的增大而提高。

骨料的表面特征对混凝土的强度也有影响。碎石表面粗糙富有棱角,与水泥石胶结性好,且骨料颗粒间有嵌固作用,所以在原材料及坍落度相同的情况下,用碎石拌制的混凝土较用卵石拌制的混凝土强度高。当水灰比小于 0.40 时,碎石混凝土强度可比卵石混凝土高约三分之一。但随着水灰比的增大,二者之间的强度差值逐渐减小,当水灰比达到 0.65 后,二者的强度无明显差异。这是因为当水灰比较小时,混凝土强度的主要矛盾是界面过渡区的强度,而当水灰比较大时,水泥石强度逐渐成为了主要矛盾。

### 3. 矿物掺合料

活性矿物掺合料在混凝土中与氢氧化钙发生火山灰反应,生成额外的水化硅酸钙凝胶,可以提高混凝土的后期强度。另外,矿物掺合料可以干扰水化产物的结晶过程,使得水化产物结晶尺寸变小,富集程度和取向程度都下降,界面过渡区的孔隙率下降,降低了界面过渡区内缺陷的数量。除此之外,细小的矿物掺合料可填充水泥颗粒之间的空隙,使混凝土基体更加密实。

**4. 外加剂**

与混凝土强度最为密切的两种外加剂为减水剂和早强剂。减水剂可以降低混凝土的用水量,进而降低水灰比,提高混凝土的强度。早强剂通过促进水泥水化反应的进程可以提高混凝土的早期强度,值得注意的是,掺加早强剂的混凝土其后期强度有倒缩风险。

**5. 施工方法**

制备混凝土时采用机械搅拌比人工拌和更为均匀,实践证明,在相同配合比和成型密实条件下,机械搅拌的混凝土强度一般要比人工搅拌时的混凝土强度提高10%左右。

浇筑混凝土时,采用机械振动成型的混凝土比人工捣实更加密实。振动作用暂时破坏了水泥浆的凝聚结构,降低了水泥浆的黏度,同时骨料间的摩阻力也大大减小,有利于混凝土密实度和强度的提高。

**6. 养护条件**

混凝土硬化过程中,人为变化混凝土周围环境的温度与湿度条件,使其微观结构和性能达到所需要的结果,称为对混凝土的养护。

温度是决定水泥水化作用速度快慢的重要条件,养护温度高,水泥早期水化速度快,混凝土的早期强度就高。值得注意的是,混凝土硬化初期的温度对其后期强度有影响,混凝土初始养护温度越高,其后期强度增长率越低。这是因为较高初始温度(40℃以上)下水泥水化速率加快,使正在水化的水泥颗粒周围聚集了高浓度的水化产物,减缓了水泥进一步水化的速度,水化产物来不及扩散,从而形成分布不均匀的多孔结构,成为水泥浆体中的薄弱区,最终对混凝土长期强度产生了不利影响。相反,在较低养护温度(如5~20℃)下,水泥水化缓慢,水化产物生成速率低,但有充分的扩散时间以形成均匀的结构,从而获得较高的最终强度。当温度降至0℃以下时,水泥水化反应停止,混凝土强度停止发展;同时混凝土中的水结冰产生体积膨胀(约9%),对孔壁产生压应力(可达100MPa),致使硬化中的混凝土结构遭到破坏。因此在冬期施工时,要特别注意对混凝土的保温养护,以免混凝土早期受冻破坏。

湿度是决定水泥能否正常进行水化作用的必要条件。若浇筑后的混凝土所处环境湿度适宜,水泥水化反应可顺利进行,混凝土强度能得以充分发展。若环境湿度较低,水泥不能正常进行水化作用,甚至会停止水化,严重降低混凝土的强度。混凝土强度与潮湿养护期的关系如图4-20所示。由图可知,混凝土受干燥日期越早,其强度损失越大。混凝土硬化期间缺水,还将导致其结构疏松,易形成干缩裂缝,增大渗水而影响混凝土的耐久性。为此,施工规范《混凝土结构工程施工质量验收规范》(GB 50204—2015)规定,在混凝土浇筑完毕后,应在12h内进行覆盖并开始浇水,在夏季施工混凝土进行自然养护时,更要特别注意浇水养护。扫码获取《混凝土结构工程施工质量验收规范》(GB 50204—2015)的详细内容。

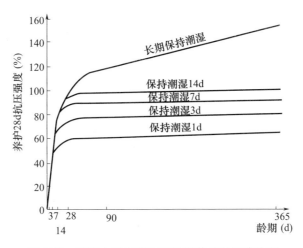

图 4-20　混凝土抗压强度与潮湿养护时间的关系

### 7. 养护龄期

在正常养护条件下,混凝土的抗压强度随龄期的增加而不断增大,最初在 7～14d 以内发展较快,然后逐渐缓慢,28d 后趋向于稳定。但只要具有一定的温度和湿度条件,混凝土的抗压强度增长可延续数十年之久。混凝土抗压强度与龄期的关系从图 4-20 中可以看出。

实践证明,由中等强度等级的普通水泥配制的混凝土,在标准养护条件下,其抗压强度发展大致与其龄期的常用对数成正比关系,其经验估算公式如下:

$$\frac{f_n}{f_{28}} = \frac{\lg n}{\lg 28} \tag{4-10}$$

式中　$f_n$——混凝土 $n$ 龄期的抗压强度(MPa);

　　　$f_{28}$——混凝土 28d 龄期的抗压强度(MPa);

　　　$n$——养护龄期(d),$n \geqslant 3d$。

## 五、提高混凝土强度的措施

在实际工程中,为了满足混凝土施工或工程结构的要求,常需提高混凝土的强度。根据影响混凝土强度的因素,提高混凝土强度通常可采取以下措施:

### 1. 采用高强度等级水泥或早强型水泥

在混凝土配合比不变的情况下,采用高强度等级水泥可提高混凝土 28d 龄期的强度;采用早强型水泥可提高混凝土的早期强度,有利于加快工程进度。

### 2. 采用低水灰比

降低水灰比是提高混凝土强度最有效的途径。在低水灰比的混凝土拌和物中游离水少,硬化后留下的孔隙少,混凝土密实度高,强度可显著提高。若水灰比减小过多,将影响拌和物流动性,造成施工困难,为此一般采取同时掺加减水剂的办法,可使混凝土在低水灰比的情况下,仍然具有良好的和易性。

**3. 掺加混凝土外加剂和矿物掺合料**

混凝土掺加外加剂是使其获得早强、高强的重要手段之一。混凝土中掺入早强剂，可显著提高其早期强度；当掺入减水剂尤其是高效减水剂时，可大幅度减少拌和用水量，提高混凝土的强度。对于高强混凝土和高性能混凝土，除了必须掺入高效减水剂外，还要同时掺加粉煤灰、磨细矿渣或硅灰等矿物掺合料，以适应混凝土高强和高性能的需要。

**4. 采用机拌机振**

当施工采用低水灰比混凝土时，必须同时采用机械搅拌和机械振捣，否则混凝土难以达到密实成型和较高强度的要求。

**5. 采用湿热养护**

蒸汽养护是将混凝土放在90℃以上的常压蒸汽中进行养护，以加速水泥的水化作用，经16h左右，其强度可达正常条件下养护28d强度的70%~80%。蒸汽养护混凝土的目的在于获得足够高的早期强度，加快拆模，提高模板及场地的周转率，有效提高生产效率和降低成本。

蒸压养护是将混凝土放在173℃及8个大气压的蒸压釜中进行养护，在此高温高压下水泥的水化与硬化加快，同时水化时析出的氢氧化钙与二氧化硅反应，生成结晶较好的水化硅酸钙，可有效地提高混凝土的强度。这种方法对掺有矿物掺合料的混凝土或者使用掺有活性混合材水泥的混凝土更为有效。

## 第六节 混凝土的质量控制与评定

在生产与施工过程中，混凝土的质量受到不同因素的影响：

(1) 原材料的影响。如水泥品种与强度等级的改变；水泥强度的波动；砂石杂质含量、级配、粒径、粒形的变化；骨料含水率的变化等。

(2) 施工操作的影响。如组成材料计量的误差；水灰比的波动；搅拌时间控制不同；浇捣条件的变化；养护时温、湿度的变化等。

(3) 试验条件的影响。如取样方法的不同；试件成型、养护条件的差异；试验时加荷速度的快慢；试验操作者本身的误差等。

混凝土质量的波动是客观存在的，要使混凝土质量在一定的范围内波动，以达到质量稳定的目的，必须要对其进行质量管理。由于混凝土的抗压强度与混凝土其他性能有着紧密的相关性，能较好地反映混凝土的全面质量，因此工程中常以混凝土抗压强度作为重要的质量控制指标，并以此作为评定混凝土生产质量水平的依据。

### 一、衡量指标

衡量混凝土施工质量水平的指标主要包括正常生产控制条件下混凝土强度的平均值、标准差和变异系数等。

**1. 强度平均值**

混凝土的强度平均值表示混凝土强度总体的平均水平,计算式为:

$$\bar{f}_{cc} = \frac{1}{n}\sum_{i=1}^{n} f_{cc,i} \tag{4-11}$$

式中 $\bar{f}_{cc}$——$n$ 组混凝土试件抗压强度的算术平均值(MPa);
 $f_{cc,i}$——第 $i$ 组混凝土试件抗压强度(MPa);
 $n$——试件的组数,$n \geqslant 25$。

**2. 强度标准差**

混凝土的强度平均值并不能反映混凝土强度的波动情况,采用强度标准差表征混凝土强度的离散性,强度标准差值越小说明强度离散性小,混凝土质量控制较稳定;而强度标准差越大,表明强度值离散性大,质量控制差。强度标准差($\sigma$)的计算式为:

$$\sigma = \sqrt{\frac{\sum_{i=1}^{n}(f_{cc,i} - \bar{f}_{cc})^2}{n-1}} \tag{4-12}$$

或

$$\sigma = \sqrt{\frac{\sum_{i=1}^{n} f_{cc,i}^2 - n\bar{f}_{cc}^2}{n-1}} \tag{4-13}$$

**3. 变异系数**

变异系数是评定混凝土质量均匀性的指标。变异系数越小,表明混凝土质量越稳定;变异系数越大,表明混凝土质量稳定性越差。变异系数($C_v$)的计算式如下:

$$C_v = \sigma/\bar{f}_{cc} \tag{4-14}$$

## 二、强度的检验评定

混凝土强度评定分为统计法和非统计法两种。当检验结果满足统计法或非统计法的规定时,则该批混凝土强度评定为合格。

**1. 统计法**

当连续生产的混凝土的生产条件在较长时间内能保持一致,且同一品种、同一强度等级混凝土的强度变异性保持稳定时,应由连续的三组试件组成一个检验批,其强度应同时满足下列要求:

$$\bar{f}_{cc} \geqslant f_{cc,k} + 0.7\sigma \tag{4-15}$$

$$f_{cc,\min} \geqslant f_{cc,k} - 0.7\sigma \tag{4-16}$$

式中 $\bar{f}_{cc}$——同一检验批混凝土立方体抗压强度的平均值(MPa);
 $f_{cc,\min}$——同一检验批混凝土立方体试件抗压强度的最小值(MPa);
 $f_{cc,k}$——混凝土立方体试件抗压强度标准值(MPa);
 $\sigma$——检验批混凝土立方体抗压强度的标准差(MPa)。

当混凝土强度等级不高于 C20 时,其强度的最小值还应满足下列要求:

$$f_{cc,\min} \geqslant 0.85 f_{cc,k} \tag{4-17}$$

当混凝土强度等级高于 C20 时,其强度的最小值应满足下列要求:

$$f_{cc,min} \geq 0.90 f_{cc,k} \qquad (4-18)$$

**2. 非统计法**

对试件数量有限,不具备按统计法评定混凝土强度的条件,可采用非统计法评定。按非统计法评定混凝土强度时,其最小值应满足下列要求:

$$f_{cc,min} \geq 0.95 f_{cc,k} \qquad (4-19)$$

当混凝土强度等级不高于 C60 时,其强度的平均值还应满足下列要求:

$$\bar{f}_{cc} \geq 1.15 f_{cc,k} \qquad (4-20)$$

当混凝土强度等级高于 C60 时,其强度的平均值应满足下列要求:

$$\bar{f}_{cc} \geq 1.10 f_{cc,k} \qquad (4-21)$$

# 第七节　混凝土的变形性能

混凝土在硬化和使用过程中,由于受物理、化学及力学等因素的影响,常会发生各种变形,这些变形是导致混凝土产生裂缝的主要原因之一,从而影响混凝土的强度及耐久性。混凝土的变形通常有以下几种。

## 一、化学收缩

混凝土在硬化过程中,水泥水化生成物的固相体积小于水化前反应物的总体积,导致混凝土产生的体积收缩,称为化学收缩。混凝土的化学收缩不可恢复,其收缩量随混凝土硬化龄期的延长而增加,一般在 40d 内渐趋稳定。混凝土的化学收缩值很小(小于 1%),对混凝土结构物没有破坏作用,但在混凝土内部可能产生微裂缝。

## 二、塑性收缩

塑性阶段,混凝土由于表面失水速率大于内部水分迁出速率而产生的收缩称为塑性收缩。混凝土塑性收缩一般发生在养护条件差、环境恶劣或大面积施工的条件下,时间集中在浇筑后的 3~12h。混凝土的塑性收缩过大会使混凝土表面产生微小裂缝。

## 三、干燥收缩

干燥环境下,混凝土中因水分散失而导致的体积收缩现象称为混凝土的干燥收缩。

**1. 干燥收缩机理**

混凝土内部存在着许多孔径大小不一、形状不同的孔隙,孔隙中通常有水存在,当环境湿度下降时,孔隙中的水会逐步失去,从而导致混凝土的干燥收缩。

(1)自由水。存在于较大的气孔中或凝胶体及晶体表面,易蒸发,对混凝土的收缩基本没有影响。

(2)毛细孔水。存在于毛细孔中,当环境的相对湿度为 40%~50% 时即可蒸发,同时产生毛细孔负压力,混凝土开始干燥收缩。混凝土中的毛细孔负压力由下式计算:

$$\Delta P = -\frac{2\gamma}{r} = \frac{RT}{M}\ln\frac{P_v}{P_0} \tag{4-22}$$

式中 $M$——水溶液的摩尔质量(kg/mol);

$\gamma$——水溶液的表面张力(N/m);

$R$——气体常数[8.314J/(K·mol)];

$T$——体系绝对温度(K);

$r$——孔隙水弯液面的曲率半径(m);

$P_v$——外界蒸气压;

$P_0$——饱和蒸气压。

由上式可知,外界湿度越大,蒸气压越大,在接近饱和蒸气压时,毛细孔负压力接近于零;当外界越干燥时,毛细孔负压力越大。

(3)凝胶吸附水。在分子引力作用下,吸附于水泥凝胶粒子表面,当相对湿度下降至30%时凝胶粒子表面的吸附水开始减少,凝胶颗粒之间紧缩,混凝土干燥收缩变大。

(4)凝胶水。在水泥凝胶体粒子之间通过氢键牢固地与凝胶粒子键合,又叫作层间水。只有在环境非常干燥时(相对湿度小于11%)才会失去,可使混凝土明显产生收缩。

当混凝土处于水中或潮湿环境时,气孔和毛细孔中充满水。当外部环境开始干燥时,首先是气孔中的自由水蒸发,然后是毛细孔水蒸发,这时将使毛细孔负压增大而产生收缩力,使毛细孔被压缩,从而使混凝土体积发生收缩。如果再继续失水,将使凝胶粒子表面的吸附水膜减薄,胶粒之间紧缩,混凝土干燥收缩进一步增大。干燥收缩后的混凝土如果再吸收水分,孔隙内充水,体积膨胀,可恢复大部分收缩变形,但其中30%~50%不可恢复。

**2. 干燥收缩的危害**

混凝土的干燥收缩主要发生在水泥浆体中,水泥浆的干缩值通常可达到400~1000微应变($10^{-6}$mm/mm)。硬化后的混凝土属脆性材料,变形能力极差,抗拉强度低。干燥收缩产生的收缩应力大于混凝土的抗拉强度时,混凝土中就会产生裂缝。收缩开裂不但会降低混凝土的强度,也会为侵蚀介质的侵入提供通道,降低混凝土的抗冻、抗渗、抗侵蚀等耐久性能。由于混凝土在后期的强度会逐渐增长,混凝土更不易开裂,所以混凝土早期干燥收缩的危害要大于后期干燥收缩的危害。

### 四、温度变形

因水泥水化放热,形成混凝土内外温差。混凝土的温度膨胀系数为(0.6~1.3)×$10^{-5}$/℃,一般取$1.0\times10^{-5}$/℃,即温度每改变1℃,1m长的混凝土将产生0.01mm的膨胀或收缩变形,混凝土内外温差导致混凝土内部膨胀变形超过外部变形。同时混凝土内外温差造成混凝土内产生温度梯度,并产生应力。二者共同作用下,导致混凝土的温度变形开裂。

为了减少温度变形对结构物的影响,在混凝土配合比方面应尽量少用水泥,多采用低热水泥或掺入磨细矿渣、粉煤灰等矿物质掺合料,从而有效地降低混凝土的水化热。夏季施工时气温很高,故应对原材料进行降温处理。大体积混凝土(指最小边尺寸在1m以上的混凝土结构)应实行分层浇筑,待浇筑的混凝土热量大致放出后,再浇筑下一层。对于较长的混凝土路面、面积较大的地面等,为防止由于温度变形引起的龟裂,还可以采取每隔一段距离设置一道伸缩缝,或者在结构中设置钢筋等,增加抗拉能力。

### 五、短期荷载作用下的弹塑性变形

**1. 混凝土的弹塑性变形**

混凝土是一种多相复合材料,它是一种弹塑性体。混凝土在静力受压时,其应力 $\sigma$ 与应变 $\varepsilon$ 之间的关系如图 4-21 所示。由图可知,当在 $A$ 点卸荷时,应力-应变曲线为 $AC$ 弧线,卸荷后弹性变形 $\varepsilon_{弹}$ 恢复了,留下塑性变形 $\varepsilon_{塑}$。

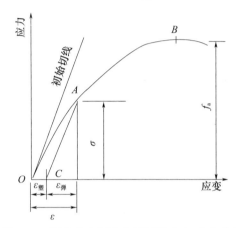

图 4-21 混凝土在压力作用下的应力-应变曲线

**2. 弹性模量的测定**

由于混凝土是弹塑性体,故要准确测定其弹性模量并非易事,但可间接地求其近似值。即在低应力(轴心抗压强度 $f_{cp}$ 的 30%~50%)下,随着荷载重复次数(3~5 次)的增加,混凝土的塑性变形的增加量逐渐减少,最后得到一条曲率很小的应力-应变曲线,几乎与初始切线(混凝土最初受压时的应力-应变曲线在原点的切线)相平行,如图 4-22 中的 $A'C'$ 弧线。由 $A'C'$ 弧线对应的应力与应变的比值可得到混凝土的割线弹性模量,近似为混凝土的静力受压弹性模量。

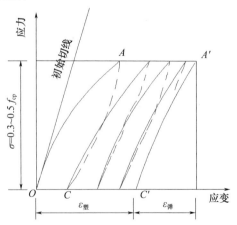

图 4-22 混凝土在低应力重复荷载下的应力-应变曲线

按我国标准的规定,混凝土弹性模量的测定采用 150mm × 150mm × 300mm 的棱柱体试件,取其轴心抗压强度值 $f_{cp}$ 的三分之一作为试验控制应力荷载值,经 3 次以上反复加荷和卸荷后,测得应力与应变的比值,即为混凝土的弹性模量。

混凝土的弹性模量与混凝土的抗压强度正相关,二者之间的经验公式见式(4-23)。通常当混凝土强度等级在 C10 ~ C60 时,其弹性模量约在 17.5 ~ 36.0GPa 之间。混凝土的弹性模量增加,则混凝土的变形降低。

$$E_c = \frac{10^5}{2.2 + \frac{34.74}{f_{cc}}} \quad (4-23)$$

## 六、长期荷载作用下的徐变

混凝土在长期荷载作用下会发生徐变现象。混凝土在长期恒载作用下,随着时间的延长,沿着作用力的方向变形不断增大,一般要延续 2 ~ 3 年才逐渐趋向稳定。混凝土这种在长期荷载作用下发生的变形会随时间发展的性质,称为混凝土的徐变。混凝土无论是受压、受拉或受弯时,均会产生徐变现象。混凝土在长期荷载作用下,其变形与持荷时间的关系如图 4-23 所示。

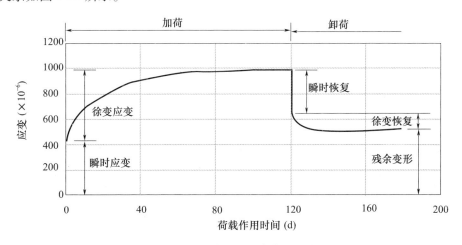

**图 4-23 混凝土的应变与持荷时间的关系**

由图 4-23 可知,当混凝土受荷后会立即产生瞬时变形,这时主要为弹性变形,随后则随受荷时间的延长而产生徐变变形,此时以塑性变形为主。混凝土的徐变变形为瞬时弹性变形的 1 ~ 3 倍。当作用应力不超过一定值时,这种徐变变形在加荷初期发展较快,以后逐渐减慢,最后渐行停止。持荷一定时间后卸除荷载,则部分变形可瞬时恢复,称为瞬时弹性恢复,值得注意的是瞬时弹性恢复要小于瞬时弹性变形。还有少部分变形将在若干天内逐渐恢复,即徐变恢复,徐变恢复要远远小于徐变变形。最后,大部分不能恢复的变形称为残余变形。

混凝土产生徐变的原因,一般认为是由于在长期荷载作用下,水泥石中的凝胶层状结构内部产生切向滑移,同时凝胶中的吸附水或结晶水向毛细管内迁移渗透所致。徐变变形与受力方向一致,在结构设计中徐变对结构物的影响如下:

(1)增加结构物的变形量

设计桥梁、建筑物的梁等受弯构件时,不仅要考虑承载能力,而且对跨中最大挠度有一定要求。徐变使挠度随荷载作用时间延长而增大,对结构安全和正常使用不利,因此在设计时要充分考虑到徐变对结构物变形的影响。

(2)引起预应力钢筋混凝土结构的预应力损失

利用钢筋抗拉强度高的特性,先对钢筋混凝土中的钢筋施加预拉应力,并待其与混凝土黏结或锚固为一体后,再卸掉荷载,利用钢筋试图恢复弹性变形的特性,对混凝土施加压应力,使混凝土在未受外力作用时,内部已经产生预加的压应力。当预应力钢筋混凝土受拉力作用时,混凝土内部的预压应力可以抵消一部分拉力,从而提高混凝土的抗拉和抗裂性能。但是由于混凝土具有徐变的性质,随时间延长,混凝土将在受压方向上增大变形,使钢筋的拉伸变形量得以部分恢复,造成预应力损失。

(3)降低温度应力,减少微裂缝

由于混凝土的水化热,大体积混凝土内部往往存在较大的温度应力,导致温度变形裂缝。徐变能够使混凝土在应力方向上缓慢地产生变形,从而降低温度应力,减轻温度变形裂缝形成的风险。

(4)产生应力松弛,缓解应力集中

徐变能够产生应力松弛,缓解混凝土构件内部的裂缝或其他有缺陷部位产生的应力集中,对结构是有利的。

# 第八节　混凝土的耐久性

用于建筑物和构筑物的混凝土,不仅应具有设计要求的强度,以保证其能安全承受荷载作用,还应具有耐久性能,满足在所处环境及使用条件下经久耐用的要求。

## 一、耐久性的概念及意义

混凝土的耐久性是指混凝土结构物在环境因素作用下,能长期保持原有性能、抵抗劣化变质和破坏的性能。环境因素包括物理作用、化学作用和生物作用等方面。物理作用包括温度变化与冻融循环、湿度变化与干湿循环等;化学作用包括酸、碱、盐类物质的水溶液或其他有害物质的侵蚀作用,以及日光、紫外线等对材料的作用;生物作用包括菌类、昆虫等的侵蚀等。

谈到混凝土结构物的安全性,人们往往首先想到承载能力,即强度。但工程实践表明,仅仅由承载力不够导致混凝土结构物破坏的事例并不多,许多混凝土结构物是由于水的侵蚀、冻融循环、化学物质的腐蚀等造成破坏,使用寿命缩短,造成重大浪费。因此,混凝土结构设计不仅要考虑强度,还要考虑耐久性。提高混凝土的耐久性,对于结构物的安全性和经济性能均具有重要意义。

## 二、常见的耐久性问题

混凝土材料的耐久性能包括抗渗性、抗冻性、抗侵蚀、抗碳化作用、碱-骨料反应等方面。要直接考察材料的这些性能需要进行长期观察和测试,实际工程中通常根据这些侵蚀性因素的基本原理,模拟实际使用条件或强化试验条件,进行加速试验,以评估材料的相关耐久性能。

**1. 抗渗性**

混凝土的抗渗性是指混凝土抵抗压力液体(水、油、溶液等)渗透作用的能力。抗渗性是决定混凝土耐久性最主要的因素,若混凝土的抗渗性差,环境中水等液体物质容易渗入混凝土内部,当遭遇负温条件或水中含有侵蚀性介质时,混凝土就易遭受冰冻或侵蚀作用。抗渗性不良还会引起钢筋混凝土内部钢筋锈蚀并导致表面混凝土保护层开裂与剥落。因此,对于受压力水(或油)作用的工程,如地下建筑、水池、水塔、压力水管、水坝、油罐以及港口工程、海洋工程等,必须要求混凝土具有一定的抗渗能力。

混凝土的抗渗性用抗渗等级 $P$ 表示。按《普通混凝土长期性能和耐久性能试验方法标准》(GB/T 50082—2009)中规定的逐级加压法以 6 个试件进行抗渗性试验,采用下式进行计算:

$$P = 10H - 1 \tag{4-24}$$

式中 $H$——6 个试件中已有 3 个试件端面渗水,而第 4 个尚未渗水时的水压力(MPa)。

混凝土按照抗渗性分为 P4、P6、P8、P10 和 P12 五个等级,分别表示混凝土在 0.4MPa、0.6MPa、0.8MPa、1.0MPa 和 1.2MPa 的水压力作用下不渗水。

混凝土渗水的主要原因是其内部存在有连通的渗水孔道,这些孔道主要来源于水泥浆中多余水分蒸发和泌水后留下的毛细管道,以及粗骨料下缘聚积的水隙;另外也可产生于混凝土浇捣不密实及硬化后因干缩、热胀等变形造成的裂缝。混凝土的抗渗性主要受到以下因素影响。首先,水泥浆产生的渗水孔道的多少,主要与混凝土的水灰比有关,水灰比越小,混凝土中毛细孔隙率和孔径越小,混凝土抗渗性越好,反之则越差;其次,采用矿物掺合料的混凝土由于矿物掺合料的二次水化作用填充了部分孔隙,提高了混凝土抗渗性;此外,混凝土中骨料的最大粒径增加,混凝土的抗渗性降低;最后,加强养护对提高混凝土的抗渗性也有一定的作用。

**2. 抗冻性**

混凝土的抗冻性是指硬化混凝土在水饱和状态下,能经受多次冻融循环作用而不破坏,同时也不严重降低强度的性能。混凝土受冻融破坏,是由于其内部空隙和毛细孔隙中的水结冰时产生体积膨胀和冷水迁移所致。当膨胀压和渗透压的综合作用应力超过混凝土的抗拉强度时,混凝土发生微细裂缝,在反复冻融作用下,混凝土内部的微细裂缝逐渐增多和扩展,导致混凝土表面产生酥松剥落,直至完全破坏。

按《普通混凝土长期性能和耐久性能试验方法标准》(GB/T 50082—2009)的规定,混凝土的抗冻性可用慢冻法的抗冻标号或快冻法的抗冻等级来确定。慢冻法以标准养护 28d 龄期、边长为 100mm 的立方体试件,在水饱和后,于 -18℃ 和 20℃ 的情况下进行反复冻融循环(气冻和水融时间各自大于 4h),最后以抗压强度损失率达到 25% 或质量损失率达到 5% 时,混凝土所能承受的最大冻融循环次数来表示。混凝土的抗冻标号分为 D25、D50、D100、D150、D200、D250、D300 和 D300 以上,其中数字即表示混凝土能承受的最大冻融循环次数。

对于抗冻性要求高的混凝土,可采用快冻法进行。快冻法以标准养护 28d 龄期、尺寸为 100mm×100mm×400mm 的棱柱体试件,在水饱和后,于 -18℃和 5℃的情况下进行单次循环时间为 2~4h 的反复冻融循环,最后以相对动弹性模量降低到 60% 或质量损失率达到 5% 时的最大循环次数表示混凝土的抗冻等级,分为 F25、F50、F100、F150、F200、F250 和 F300 等 7 个等级。

混凝土的抗冻性与混凝土内部的孔隙数量、孔隙特征、孔隙内充水程度、环境温度降低的程度及反复冻融的次数等有关。首先,混凝土的水灰比降低,混凝土的密实度提高,对其抗冻性有利;其次,混凝土中封闭小孔多或开口孔中水未充满时,混凝土抗冻性好;此外,掺加减水剂可以细化混凝土的孔结构,提高其抗冻性;最后,引气剂的使用可显著提高混凝土的抗冻性。

**3. 抗侵蚀性**

混凝土的抗侵蚀性指混凝土抵抗所处环境中的盐、酸、强碱等侵蚀性介质破坏的性能。环境中混凝土遭受的侵蚀主要包括硫酸盐侵蚀、氯离子侵蚀和酸性侵蚀。混凝土的抗侵蚀性主要取决于其所用水泥的品种及混凝土的密实度。所以,提高混凝土抗侵蚀性的措施,主要是合理选用水泥品种、降低水灰比、提高混凝土的密实度,以及尽量减少混凝土中的开口孔隙。

**4. 抗碳化**

混凝土的碳化指混凝土内水泥石中的氢氧化钙与空气中的二氧化碳,在湿度相宜时发生化学反应,生成碳酸钙和水。

碳化对混凝土的不利影响首先是减弱了对钢筋的保护作用。混凝土中水泥水化生成大量氢氧化钙,使得混凝土内部碱度较高,钢筋处在碱性环境中会在表面生成一层钝化膜,保护钢筋不受锈蚀。当碳化穿透混凝土保护层达到钢筋表面时,降低了钢筋所处环境的碱度,破坏了钢筋钝化膜,进而诱发钢筋锈蚀,产生体积膨胀,致使混凝土保护层产生开裂。一旦开裂,混凝土的碳化会进一步加快,钢筋锈蚀也更加剧烈,最后导致混凝土产生顺筋开裂而破坏。另外,碳化增加了混凝土的收缩,引起混凝土表面产生拉应力而出现微小裂缝,从而降低混凝土的抗渗性。

碳化对混凝土有一些有利影响。首先碳化产生的碳酸钙填充了水泥石的孔隙,可以提高混凝土碳化层的密实度;另外,碳化放出的水分有助于水泥水化,对提高混凝土的抗压强度有利。

**5. 碱-骨料反应**

碱-骨料反应指混凝土内水泥中的碱性氧化物(氧化钠和氧化钾)与骨料中的活性二氧化硅发生化学反应,生成碱-硅酸凝胶,其吸水后会产生很大的体积膨胀(体积增大可达 3 倍以上),从而导致混凝土产生膨胀开裂而破坏。

混凝土发生碱-骨料反应必须具备以下三个条件:

(1) 水泥中碱含量高。水泥中的 $Na_2O$ 含量(以 $Na_2O + 0.658K_2O$ 计)大于 0.6%;

(2) 骨料中含有活性二氧化硅成分。含活性二氧化硅成分的矿物有蛋白石、玉髓、鳞石英等,它们常存在于流纹岩、安山岩、凝灰岩等天然岩石中;

(3) 有水存在。没有水存在的情况下,混凝土不可能发生碱-骨料反应。

混凝土中碱-骨料反应进行缓慢,通常要经若干年后才会出现,但问题一经出现难以修复,故必须防患于未然。为应对碱-骨料反应,应采取下列措施:

(1)选用低碱水泥(碱含量小于0.6%);

(2)采用非碱活性骨料;

(3)掺加矿物掺合料,如粉煤灰、矿渣和硅灰等。矿物掺合料中含有的二氧化硅活性更高,会先于骨料与水泥中的碱发生反应,降低了混凝土中的碱含量;

(4)掺加碱-骨料反应抑制剂,如碳酸锂和氯化锂等锂盐。

扫码获取《普通混凝土长期性能和耐久性能试验方法标准》(GB/T 50082—2009)的详细内容。

### 三、混凝土结构耐久性设计

所谓混凝土的耐久性设计,即混凝土结构应根据设计使用年限和环境类别进行耐久性设计。混凝土结构的耐久性设计主要包括以下内容:确定结构所处的环境类别;提出对混凝土材料的耐久性基本要求;确定构件中钢筋的混凝土保护层厚度;不同环境条件下的耐久性技术措施;提出结构使用阶段的检测与维护要求等。

混凝土结构的环境类别按表4-17的要求划分。

表4-17 混凝土结构的环境类别

| 环境类别 | 条件 |
| --- | --- |
| 一 | 室内干燥环境;<br>无侵蚀性静水浸没环境 |
| 二a | 室内潮湿环境;<br>非严寒和非寒冷地区的露天环境;<br>非严寒和非寒冷地区与无侵蚀性的水或土壤直接接触的环境;<br>严寒和寒冷地区的冰冻线以下与无侵蚀性的水或土壤直接接触的环境 |
| 二b | 干湿交替环境;<br>水位频繁变动环境;<br>严寒和寒冷地区的露天环境;<br>严寒和寒冷地区的冰冻线以上与无侵蚀性的水或土壤直接接触的环境 |
| 三a | 严寒和寒冷地区冬季水位变动区环境;<br>受除冰盐影响环境;<br>海风环境 |
| 三b | 盐渍土环境;<br>受除冰盐作用环境;<br>海岸环境 |
| 四 | 海水环境 |
| 五 | 受人为或自然的侵蚀性物质影响的环境 |

处于一、二、三类环境中的混凝土结构设计使用年限为50年时,混凝土材料的基本要求应符合表4-18的规定。此三类环境中设计使用年限100年的混凝土结构应采取专门

的有效措施。耐久性环境类别为四、五类的混凝土结构,其耐久性要求应符合有关标准的规定。

表4-18 结构混凝土材料的耐久性基本要求

| 环境等级 | 最大水胶比 | 最低强度等级 |
| --- | --- | --- |
| 一 | 0.60 | C20 |
| 二 a | 0.55 | C25 |
| 二 b | 0.50(0.55) | C30(C25) |
| 三 a | 0.45(0.50) | C35(C30) |
| 三 b | 0.40 | C40 |

注:混凝土使用引气剂时可采用括号中的有关参数。

# 第九节 混凝土的配合比设计

混凝土的配合比指单位体积的混凝土中各组成材料的质量比例关系。混凝土配合比有两种表示方法,一种是直接以每 $1m^3$ 混凝土中各种材料的用量来表示;另一种是以混凝土各组成材料间的质量比例关系来表示,其中以胶凝材料质量为1,如胶凝材料:水:砂:石 $=1:W/B:x:y$。

采用工程所用原材料,确定混凝土中各原材料的比例用量,以获得具有特定性能混凝土的过程就称为混凝土配合比设计。

## 一、混凝土配合比设计的基本要求、依据及方法

按照《普通混凝土配合比设计规程》(JGJ 55—2011)规定,普通混凝土的配合比应根据原材料性能及对混凝土的技术要求进行计算,并经实验室试配、调整后确定。

**1. 配合比设计的基本要求**

(1)满足结构设计的强度等级要求;
(2)满足混凝土施工所要求的和易性;
(3)满足工程所处环境对混凝土耐久性的要求;
(4)符合经济原则,即节约水泥以降低混凝土成本。

**2. 配合比设计的依据**

(1)混凝土配合比设计基本参数确定的原则

水灰比、单位用水量和砂率是混凝土配合比设计的三个基本参数,它们与混凝土各项性能之间有着十分密切的关系,因此,混凝土配合比设计主要是正确地确定出这三个参数,才能保证制备出满足要求的混凝土。

混凝土配合比设计中确定三个参数的原则是:在满足混凝土强度和耐久性的基础上,确定混凝土的水灰比;在满足混凝土和易性的基础上,根据粗骨料的种类和规格确定混凝土的单位用水量;砂率应以砂填充石子空隙后略有富余的原则来确定。

(2)配合比设计的算料基准

混凝土配合比设计以计算 $1m^3$ 混凝土中各材料用量为基准,计算时其中的骨料以干燥状态为准。所谓干燥状态的骨料指细骨料的含水率小于 0.5%,粗骨料的含水率小于 0.2%。如需以饱和面干骨料为基准进行计算时,则应作相应的修改。

混凝土外加剂的掺量一般较小,在计算混凝土体积时,外加剂的体积可忽略不计,在计算混凝土表观密度时,外加剂的质量也可忽略不计。

**3. 配合比设计的方法和原理**

普通混凝土配合比设计的方法有体积法和质量法两种。

(1)体积法

混凝土配合比设计体积法的基本原理是:假定刚浇捣完毕的混凝土拌和物的体积,等于其各组成材料的绝对体积及其所含少量空气体积之和。若以 $V_h$、$V_c$、$V_f$、$V_s$、$V_g$、$V_w$ 和 $V_k$ 分别表示混凝土、水泥、矿物掺合料、砂、石、水和空气的体积,则体积法原理可用公式表达为

$$V_h = V_c + V_f + V_s + V_g + V_w + V_k \quad (4-25)$$

若在每 $1m^3$ 混凝土中,以 $m_{c0}$、$m_{f0}$、$m_{s0}$、$m_{g0}$、$m_{w0}$ 分别表示混凝土中的水泥、水、砂、石的用量,并以 $\rho_c$、$\rho_f$、$\rho_s$、$\rho_g$、$\rho_w$ 分别表示水泥密度、矿物掺合料密度、砂的表观密度、石的表观密度、水的密度,假设混凝土拌和物中含空气体积百分数为 $\alpha$,则上式可改写为

$$\frac{m_{c0}}{\rho_c} + \frac{m_{f0}}{\rho_f} + \frac{m_{s0}}{\rho_s} + \frac{m_{g0}}{\rho_g} + \frac{m_{w0}}{\rho_w} + 0.01\alpha = 1 \quad (4-26)$$

式中 $\alpha$——混凝土含气量的百分数(%),在不使用引气型外加剂时,$\alpha$ 可取 1。

(2)质量法

混凝土配合比设计质量法的基本原理是:当混凝土所用原材料比较稳定时,配制的混凝土的表观密度接近一个恒值。预先假定出每 $1m^3$ 新拌混凝土的质量,就可建立下列关系式:

$$m_{c0} + m_{f0} + m_{s0} + m_{g0} + m_{w0} = m_{cp} \quad (4-27)$$

每 $1m^3$ 混凝土的假定质量($m_{cp}$)可在 $2350 \sim 2450 kg/m^3$ 范围内选定。

扫码获取《普通混凝土配合比设计规程》(JGJ 55—2011)的详细内容。

## 二、混凝土配合比设计计算

**1. 确定混凝土配制强度($f_{cc,0}$)**

根据混凝土的强度保证率概念可知,如果配制的混凝土平均强度($\bar{f}_{cc}$)等于设计强度等级值($f_{cc,k}$),则其强度保证率仅有 50%。如果要达到高于 50% 的强度保证率,混凝土的配制强度必须高于设计强度等级值。当混凝土的设计强度等级小于 C60 时,若令混凝土的配制强度等于平均强度,根据式(4-28),则有

$$f_{cc,0} = \bar{f}_{cc} = f_{cc,k} + t\sigma \quad (4-28)$$

我国目前要求混凝土的强度保证率为 95%,查表 4-15,得 $t = 1.645$,代入上式得到配制强度为:

$$f_{cc,0} = f_{cc,k} + 1.645\sigma \tag{4-29}$$

式中，$\sigma$ 值可根据历史统计的混凝土强度资料，由式(4-12)或式(4-13)计算得到。若无混凝土强度资料时，可参考表 4-19。

表 4-19　标准差 $\sigma$ 值（MPa）

| 混凝土强度等级 | 低于 C20 | C20 ~ C35 | 高于 C35 |
|---|---|---|---|
| $\sigma$ | 4.0 | 5.0 | 6.0 |

当混凝土的设计强度等级不小于 C60 时，混凝土的配制应按下式确定：

$$f_{cc,0} \geq 1.15 f_{cc,k} \tag{4-30}$$

**2. 确定水胶比（W/B）**

根据已知的混凝土配制强度（$f_{cc,0}$）及所用胶凝材料 28d 胶砂抗压强度（$f_b$），则可由混凝土强度经验公式求得水胶比，即：

$$\frac{W}{B} = \frac{\alpha_a f_b}{f_{cc,0} + \alpha_a \alpha_b f_b} \tag{4-31}$$

式中，两个回归系数 $\alpha_a$ 和 $\alpha_b$ 可按工程所使用的原材料，通过试验建立的水胶比与混凝土强度关系式来确定。当不具备上述试验统计资料时，可按表 4-16 的值取用。

当胶凝材料 28d 胶砂抗压强度值（$f_b$）无实测值时，可根据水泥 28d 胶砂抗压强度（$f_{ce}$）按下式计算：

$$f_b = \gamma_f \gamma_s f_{ce} \tag{4-32}$$

式中　$\gamma_f$ 和 $\gamma_s$——粉煤灰影响系数和粒化高炉矿渣粉影响系数，可按表 4-20 选用。

表 4-20　粉煤灰影响系数（$\gamma_f$）和粒化高炉矿渣粉影响系数（$\gamma_s$）

| 掺量（%） | 粉煤灰影响系数，$\gamma_f$ | 粒化高炉矿渣粉影响系数，$\gamma_s$ |
|---|---|---|
| 0 | 1.00 | 1.00 |
| 10 | 0.85 ~ 0.95 | 1.00 |
| 20 | 0.75 ~ 0.85 | 0.95 ~ 1.00 |
| 30 | 0.65 ~ 0.75 | 0.90 ~ 1.00 |
| 40 | 0.55 ~ 0.65 | 0.80 ~ 0.90 |
| 50 | — | 0.75 ~ 0.85 |

当水泥 28d 胶砂抗压强度值（$f_{ce}$）无实测值时，可根据水泥强度等级值（$f_{ce,g}$）按下式计算：

$$f_{ce} = \gamma_c f_{ce,g} \tag{4-33}$$

式中　$\gamma_c$——水泥强度等级值的富余系数，可按实际统计资料确定；当缺乏实际统计资料时，可按表 4-21 选用。

表 4-21　水泥强度等级值的富余系数，$\gamma_c$

| 水泥强度等级值 | 32.5 | 42.5 | 52.5 |
|---|---|---|---|
| 富余系数，$\gamma_c$ | 1.12 | 1.16 | 1.10 |

如计算所得的水胶比大于表 4-18 规定的最大水胶比时,应取表中规定的最大水胶比。

**3. 选定混凝土拌和用水量($m_{w0}$)**

混凝土拌和物的单位用水量($m_{w0}$),可根据所用粗骨料的种类、最大粒径及施工要求的坍落度值,按表 4-22 规定的值选用。

表 4-22 塑性和干硬性混凝土的用水量($kg/m^3$)

| 项目 | 指标 | 卵石最大粒径(mm) | | | | 碎石最大粒径(mm) | | | |
|---|---|---|---|---|---|---|---|---|---|
| | | 10 | 20 | 31.5 | 40 | 16 | 20 | 31.5 | 40 |
| 坍落度<br>(mm) | 10~30 | 190 | 170 | 160 | 150 | 200 | 185 | 175 | 165 |
| | 35~50 | 200 | 180 | 170 | 160 | 210 | 195 | 185 | 175 |
| | 55~70 | 210 | 190 | 180 | 170 | 220 | 205 | 195 | 185 |
| | 75~90 | 215 | 195 | 185 | 175 | 230 | 215 | 205 | 195 |
| 维勃稠度<br>(s) | 16~20 | 175 | 160 | — | 145 | 180 | 170 | — | 155 |
| | 11~15 | 180 | 165 | — | 150 | 185 | 175 | — | 160 |
| | 5~10 | 185 | 170 | — | 155 | 190 | 180 | — | 165 |

注:1. 本表用水量系采用中砂时的取值,如采用细砂或粗砂,则每 $1m^3$ 混凝土用水量应相应增加或减少 5~10kg;
2. 掺用各种外加剂或矿物掺合料时,用水量应相应调整。

对于坍落度大于 100mm 的流动性较大的混凝土,需掺用减水剂。此时混凝土用水量的确定,应先以表 4-22 中坍落度 90mm 的用水量为基础,按坍落度每增大 20mm,用水量增加 5kg 计算出不掺外加剂时 $1m^3$ 混凝土的用水量 $m'_{w0}$,然后再按下式计算出掺减水剂时每 $1m^3$ 混凝土的用水量 $m_{w0}$:

$$m_{w0} = m'_{w0} \times (1-\beta) \quad (4-34)$$

式中 $\beta$——外加剂的减水率(%),由试验确定。

**4. 计算胶凝用量($m_{b0}$)**

每 $1m^3$ 混凝土中的胶凝材料用量($m_{b0}$)按下式计算:

$$m_{b0} = \frac{m_{w0}}{W/B} \quad (4-35)$$

除配制 C15 及其以下强度等级的混凝土外,混凝土的最小胶凝材料用量应符合表 4-23 的规定。若通过式(4-35)计算得到的胶凝材料用量小于表 4-23 规定的最小胶凝材料用量时,应取表中规定的最小胶凝材料用量。

表 4-23 混凝土的最小胶凝材料用量($kg/m^3$)

| 最大水胶比 | 最小胶凝材料用量 | | |
|---|---|---|---|
| | 素混凝土 | 钢筋混凝土 | 预应力混凝土 |
| 0.60 | 250 | 280 | 300 |
| 0.55 | 280 | 300 | 300 |
| 0.50 | 320 | | |
| ≤0.45 | 330 | | |

每 $1m^3$ 混凝土的矿物掺合料用量($m_{f0}$)按下式计算:
$$m_{f0} = m_{b0}\beta_f \tag{4-36}$$
式中 $\beta_f$——矿物掺合料掺量(%),由试验确定。

每 $1m^3$ 混凝土的水泥用量($m_{c0}$)按下式计算:
$$m_{c0} = m_{b0} - m_{f0} \tag{4-37}$$

**5. 计算外加剂用量($m_{b0}$)**

每 $1m^3$ 混凝土中外加剂用量($m_{a0}$)按下式计算:
$$m_{a0} = m_{b0}\beta_a \tag{4-38}$$
式中 $\beta_a$——外加剂掺量(%),由混凝土试验确定。

**6. 确定砂率($\beta_s$)**

混凝土砂率的确定应根据骨料的技术指标、混凝土拌和物性能和施工要求,参考既有历史资料确定。当缺乏资料时,混凝土砂率的确定应符合下列规定:坍落度小于 10mm 的混凝土,其砂率应经试验确定;坍落度为 10~60mm 的混凝土,其砂率可根据粗骨料品种、最大公称粒径及水胶比按表 4-24 选取;坍落度大于 60mm 的混凝土,其砂率可经试验确定,也可在表 4-24 的基础上,按坍落度每增大 20mm、砂率增大 1% 的幅度予以调整。

表 4-24 混凝土的砂率(%)

| 水胶比 | 卵石最大公称粒径(mm) | | | 碎石最大公称粒径(mm) | | |
|---|---|---|---|---|---|---|
| | 10 | 20 | 40 | 16 | 20 | 40 |
| 0.40 | 26~32 | 25~31 | 24~30 | 30~35 | 29~34 | 27~32 |
| 0.50 | 30~35 | 29~34 | 28~33 | 33~38 | 32~37 | 30~35 |
| 0.60 | 33~38 | 32~37 | 31~36 | 36~41 | 35~40 | 33~38 |
| 0.70 | 36~41 | 35~40 | 34~39 | 39~44 | 38~43 | 36~41 |

注:1. 表中数值系中砂的选用砂率,对细砂或粗砂,可相应地减小或增大砂率;
  2. 只用一个单粒级粗骨料配制混凝土时,砂率应适当增大;
  3. 采用人工砂配制混凝土时,砂率可适当增大。

**7. 计算砂、石用量($m_{s0}$、$m_{g0}$)**

当采用体积法计算混凝土配合比时,粗、细骨料的用量按式(4-26)计算,砂率按式(4-39)计算:
$$\frac{m_{s0}}{m_{s0}+m_{g0}} \times 100\% = \beta_s \tag{4-39}$$

联立以上两方程,即可求粗、细骨料的用量。

当采用质量法计算混凝土配合比时,粗、细骨料的用量按式(4-27)计算,砂率按式(4-39)计算,同样联立两方程,可得粗、细骨料的用量。

**8. 写出混凝土计算配合比**

至此,已得到混凝土的计算配合比。

### 三、混凝土配合比的试配、调整与确定

在实际施工时,应采用工程实际使用的材料在已计算配合比的基础上进行试配,经调整配合比参数使混凝土拌和物性能符合设计和施工要求,修正计算配合比,提出试拌配合比。在试拌配合比的基础上进行混凝土强度试验和表观密度调整,得到实验室配合比。最后根据实际工程中使用骨料的含水率,对实验室配合比进行调整,得到施工配合比。

**1. 试拌配合比**

混凝土试配时应采用工程中实际使用的原材料,混凝土的搅拌方法也应与施工时采用的方法相同。试配时,每盘混凝土试配的最小搅拌量应符合表 4-25 的规定,并不应小于搅拌机额定搅拌量的四分之一。

**表 4-25　混凝土试配用最小拌和量**

| 粗骨料最大公称粒径(mm) | 拌和物数量(L) |
| --- | --- |
| 31.5 及以下 | 20 |
| 40 | 25 |

在计算配合比的基础上进行试拌,用以检验混凝土拌和物的性能。如试拌拌和物的坍落度大于目标坍落度,则应保持水胶比不变,同时降低胶凝材料用量和用水量;若坍落度小于目标坍落度,则应保持水胶比不变,同时增加胶凝材料用量和用水量;若黏聚性和保水性不合格,应调整砂率,直到符合要求为止。修正计算配合比,提出供混凝土强度试验用的试拌配合比。

**2. 实验室配合比**

在试拌配合比的基础上进行混凝土强度试验,采用三个不同的配合比,其中一个为试拌配合比,另外两个配合比的水胶比应较试拌配合比分别增加和减少 0.05,用水量与试拌配合比相同,砂率可分别增加或减少 1%。进行混凝土强度试验时,拌和物和易性应符合设计和施工要求。

根据混凝土强度试验结果,绘制强度和胶水比的线性关系图,确定出与混凝土配制强度($f_{cc,0}$)对应的胶水比。在试拌配合比的基础上,用水量($m_w$)和外加剂用量($m_a$)选取试拌配合比中的用量,并根据制作强度试件时测得的坍落度进行适当调整;胶凝材料用量($m_b$)以用水量乘以确定的胶水比计算得出;粗骨料和细骨料用量($m_g$ 和 $m_s$)选取试拌配合比中的粗骨料和细骨料用量,并根据用水量和胶凝材料用量进行调整。

根据强度试验对试拌配合比调整后,还应进行混凝土表观密度校正,配合比调整后的混凝土拌和物的表观密度($\rho_{c,c}$)计算如下:

$$\rho_{c,c} = m_b + m_w + m_s + m_g \tag{4-40}$$

试拌配合比调整后进行试拌混凝土,测得其表观密度实测值($\rho_{c,t}$),然后按下式算出校正系数 $\delta$:

$$\delta = \frac{\rho_{c,t}}{\rho_{c,c}} \tag{4-41}$$

当混凝土表观密度实测值与计算值之差的绝对值不超过计算值的2%时,则上述经过调整的配合比即可确定为混凝土的实验室配合比;若二者之差超过2%时,应将配合比中每项材料用量均乘以校正系数,得到实验室配合比。

**3. 施工配合比**

混凝土实验室配合比计算用料是以干燥骨料为基准的,但实际工地使用的骨料常含有一定的水分,因此须将实验室配合比换算成工地实际施工用的配合比。

设施工配合比每 $1m^3$ 混凝土中胶凝材料、水、砂、石、外加剂的用量分别为 $m'_b$、$m'_w$、$m'_s$、$m'_g$、$m'_a$,工地砂子含水率 $a\%$,石子含水率为 $b\%$,则施工配合比每 $1m^3$ 混凝土中各材料用量应为

$$m'_b = m_b \tag{4-42}$$

$$m'_s = m_s(1 + a\%) \tag{4-43}$$

$$m'_g = m_g(1 + b\%) \tag{4-44}$$

$$m'_w = m_w - m_s \cdot a\% - m_g \cdot b\% \tag{4-45}$$

$$m'_a = m_a \tag{4-46}$$

至此,得到可用于现场施工使用的混凝土施工配合比。

### 四、混凝土配合比设计实例

**例** 处于室内干燥环境的某框架结构工程现浇钢筋混凝土梁,混凝土设计强度等级为C30,施工采用机拌机振,混凝土坍落度要求为35~50mm,并根据施工单位历史资料统计,混凝土强度标准差 $\sigma$ 为5MPa。所用原材料情况如下:

水泥:42.5级普通硅酸盐水泥,水泥密度为 $3.00g/cm^3$,水泥强度等级标准值的富余系数为1.16;

粉煤灰:Ⅱ级粉煤灰,密度为 $2.10g/cm^3$,掺量为胶凝材料的10%;

砂:中砂,级配合格,表观密度为 $2650kg/m^3$;

石:5.0~31.5mm碎石,级配合格,表观密度为 $2700kg/m^3$;

外加剂:聚羧酸高效减水剂(非引气型),掺量为0.2%,此掺量下减水率为8%。

试求:

1. 混凝土计算配合比;

2. 经试配混凝土的和易性和强度等均符合要求,无须作调整。又知现场砂子含水率为3%,石子含水率为1%,试计算混凝土施工配合比。

**解:1. 求混凝土计算配合比**

(1) 确定混凝土配制强度($f_{cc,0}$)

$f_{cc,0} = f_{cc,k} + 1.645\sigma = 30 + 1.645 \times 5 = 38.23(MPa)$

(2) 确定水胶比($W/B$)

由于水泥强度等级标准值的富余系数为1.16,则水泥胶砂的抗压强度值为:

$f_{ce} = \gamma_c f_{ce,g} = 1.16 \times 42.5 = 49.3(MPa)$

粉煤灰掺量为10%,查表4-20可知,此掺量下粉煤灰影响系数为0.9,则胶凝材料胶

砂的抗压强度值为：

$$f_b = \gamma_f \gamma_s f_{ce} = 0.9 \times 49.3 = 44.37(\text{MPa})$$

回归系数 $\alpha_a$ 和 $\alpha_b$ 查表 4-16 可知分别为 0.53 和 0.20，则水胶比为：

$$\frac{W}{B} = \frac{\alpha_a f_b}{f_{cc,0} + \alpha_a \alpha_b f_b} = \frac{0.53 \times 44.37}{38.23 + 0.53 \times 0.20 \times 44.37} = 0.55$$

由于框架结构混凝土梁处于室内干燥环境，查表 4-18 可知最大水胶比为 0.60，计算值符合要求，取水胶比为 0.55。

(3) 确定用水量 ($m_{w0}$)

查表 4-22，对于最大粒径为 31.5mm 的碎石混凝土，当所需坍落度为 35~50mm 时，1m³ 混凝土的用水量可选用 185kg。由于减水剂的减水率为 8%，则掺加减水剂后用水量为：

$$m_{w0} = m'_{w0} \times (1-\beta) = 185 \times (1-0.08) = 170.2(\text{kg})$$

(4) 计算胶凝材料用量 ($m_{b0}$)

每 1m³ 混凝土中胶凝材料总用量为：

$$m_{b0} = \frac{m_{w0}}{W/B} = \frac{170.2}{0.55} = 309(\text{kg})$$

由于粉煤灰掺量为 10%，则粉煤灰用量为：

$$m_{f0} = m_{b0}\beta_f = 309 \times 0.1 = 30.9(\text{kg})$$

水泥用量为：

$$m_{c0} = m_{b0} - m_{f0} = 309 - 30.9 = 278.1(\text{kg})$$

查表 4-23，水胶比为 0.55 的钢筋混凝土最小胶凝材料用量为 300kg/m³，计算值符合规定，每 1m³ 混凝土取水泥用量为 278.1kg，粉煤灰用量为 30.9kg。

(5) 计算外加剂用量 ($m_{a0}$)

减水剂的掺量为 0.2%，则其用量为：

$$m_{a0} = m_{b0}\beta_a = 309 \times 0.002 = 0.618(\text{kg})$$

(6) 确定砂率 ($\beta_s$)

查表 4-24，对于采用最大粒径为 31.5mm 的碎石配制的混凝土，当水胶比为 0.55 时，砂率可选取 35%（采用插值法确定）。

(7) 计算砂、石用量 ($m_{s0}$、$m_{g0}$)

用体积法计算，即

$$\begin{cases} \dfrac{278.1}{3000} + \dfrac{30.9}{2100} + \dfrac{170.2}{1000} + \dfrac{m_{s0}}{2650} + \dfrac{m_{g0}}{2700} + 0.01 \times 1 = 1 \\ \dfrac{m_{s0}}{m_{s0} + m_{g0}} \times 100\% = 35\% \end{cases}$$

解此联立方程，得：$m_{s0}$ 为 668.8kg，$m_{g0}$ 为 1241.9kg。

(8) 写出混凝土计算配合比

每 1m³ 混凝土中各材料用量为：水泥 278.1kg，粉煤灰 30.9kg，水 170.2kg，砂 668.8kg，碎石 1241.9kg，减水剂 0.618kg。

## 2. 换算成施工配合比

设施工配合比 1m³ 混凝土中水泥、粉煤灰、水、砂、石、减水剂的用量分别为 $m'_c$、$m'_f$、$m'_w$、$m'_s$、$m'_g$、$m'_a$，则有：

$m'_c = m_c = 278.1 (\text{kg})$

$m'_f = m_f = 30.9 (\text{kg})$

$m'_s = m_s(1 + a\%) = 668.8 \times (1 + 3\%) = 688.9 (\text{kg})$

$m'_g = m_g(1 + b\%) = 1241.9 \times (1 + 1\%) = 1254.3 (\text{kg})$

$m'_w = m_w - m_s \cdot a\% - m_g \cdot b\% = 170.2 - 668.8 \times 3\% - 1241.9 \times 1\% = 137.7 (\text{kg})$

$m'_a = m_a = 0.618 (\text{kg})$

# 第十节　建筑砂浆

建筑砂浆是在建筑工程中用量大、用途广泛的建筑材料之一。建筑砂浆是由胶凝材料、细骨料和水按照一定的比例配制而成的，建筑砂浆能够把散粒材料、块状材料、片状材料等胶结成整体结构，也可以成为装饰、保护主体的材料。在建筑工程中起黏结、衬垫和传递应力的作用，主要用于砌筑、抹面、修补、装饰工程等。

根据建筑砂浆所用胶结材料的不同，可分为水泥砂浆、石灰砂浆、水泥黏土砂浆、聚合物砂浆和混合砂浆等；根据不同的使用用途可分为砌筑砂浆、抹面砂浆、特种砂浆等。按生产砂浆方式可分为现场拌制砂浆和工厂预拌砂浆两种，后者是国内外生产砂浆的发展趋向，我国建设部门要求尽快实现全面推广应用预拌砂浆。

砌筑砂浆是将砖、石、砌块等黏结成为砌体的砂浆。砌筑砂浆主要起黏结、传递荷载、使应力的分布较为均匀、协调变形等作用，是砌体的重要组成部分，应用最为广泛。本节主要论述砌筑砂浆的相关内容。

## 一、砌筑砂浆的组成材料

### 1. 胶凝材料及掺合料

通用水泥都可以用来配制砌筑砂浆，水泥品种的选择与混凝土中水泥品种的选择道理是相同的。通常对砂浆的强度要求并不很高，一般采用中等强度等级的水泥就能够满足要求。按《砌筑砂浆配合比设计规程》（JGJ/T 98—2010）要求，选用水泥时应符合《通用硅酸盐水泥》（GB 175—2007）和《砌筑水泥》（GB/T 3183—2017）的规定。一般来说水泥砂浆采用的水泥强度等级不宜大于 32.5 级；水泥混合砂浆采用的水泥强度等级不宜大于 42.5 级。

为了改善砂浆的和易性和节约水泥，常在砂浆中掺入适量的石灰、粉煤灰等掺合料。这样配制的砂浆称水泥混合砂浆。粉煤灰、生石灰等掺合料的要求应符合相应的有关规定。

### 2. 细骨料

配制建筑砂浆最常用的细骨料是天然砂。砂首先应符合混凝土用砂的技术性质要求。由于砂浆层较薄，砂的最大粒径应有所限制，理论上不应超过砂浆层厚度的 1/5 ~ 1/4，例如砖砌体用砂浆宜选用中砂,最大粒径以不大于2.5mm为宜;石砌体用砂浆宜选用粗砂,最大粒径以不大于5.0mm为宜;光滑的抹面及勾缝的砂浆宜采用细砂,最大粒径不大于1.2mm。为保证砂浆质量,对砂的含泥量应予以限制,强度等级为M2.5以上砌筑砂浆用砂的含泥量不应超过5%;强度等级为M2.5的水泥混合砂浆用砂的含泥量应不大于10%;防水砂浆用砂的含泥量应不大于3%。

**3. 水**

砂浆拌和用水的技术要求与混凝土拌和用水的技术要求相同,均需满足《混凝土用水标准》(JGJ 63—2006)的规定。

**4. 外加剂**

为了改善或赋予新拌砂浆及硬化后砂浆的某些性能,常在砂浆中掺入适量外加剂。混凝土中使用的各种外加剂,对砂浆也具有相应的作用。按标准《砌筑砂浆增塑剂》(JG/T 164—2004)规定,对砌筑用水泥砂浆可掺入砂浆增塑剂,能明显改善其和易性。但对所选外加剂的品种(引气剂、早强剂、缓凝剂、防冻剂等)和掺量必须通过试验确定。在砂浆中掺加外加剂时,除了要考虑外加剂对砂浆本身性能的影响外,还要考虑其对砂浆使用功能的影响。

扫码获取《砌筑砂浆配合比设计规程》(JGJ/T 98—2010)的详细内容。

## 二、砌筑砂浆的主要技术性质

**1. 和易性**

(1)流动性

流动性指砂浆在自重或外力作用下易于流动的性能。砂浆流动性可以用稠度表示。流动性的大小用砂浆稠度测定仪来测定,用砂浆稠度测定仪的圆锥体沉入砂浆中深度的毫米数来表示,因此亦称为沉入度。沉入度大的砂浆流动性好。

砂浆流动性的选择应根据基底材料种类、施工条件以及天气情况等因素选择,但应符合《砌体结构工程施工质量验收规范》(GB 50203—2011)的规定。可参考表4-26来选择砂浆的流动性。

**表4-26　砂浆流动性参考表(沉入度/mm)**

| 砌体种类 | 干燥气候或多孔吸水材料 | 寒冷气候或密实材料 | 抹灰工程 | 机械施工 | 手工操作 |
|---|---|---|---|---|---|
| 烧结普通砖砌体 | 80~90 | 70~80 | 准备层 | 80~90 | 110~120 |
| 烧结多孔砖、空心砖砌体 | 70~80 | 60~70 | 底层 | 70~80 | 70~80 |
| 石砌体 | 40~50 | 30~40 | 面层 | 70~80 | 90~100 |
| 普通混凝土砌体 | 60~70 | 50~60 | 灰浆面层 | — | 90~120 |
| 轻骨料混凝土砌体 | 70~90 | 60~80 | — | — | — |

扫码获取《砌体结构工程施工质量验收规范》(GB 50203—2011)的详细内容。

(2)保水性

砂浆的保水性是指新拌砂浆保持水分的能力,也表示砂浆中各组成材料是否容易离析的性质。砂浆的保水性用分层度表示。分层度的测定是将已测定稠度的砂浆装满分层度筒(分层度筒内径为150mm,分为上、下两节,上节高度为200mm,下节高度为100mm),轻轻敲击筒周围3~5下,刮去多余的砂浆并抹平。静置30min后,去掉上部200mm砂浆,取出剩余100mm砂浆倒入搅拌锅中拌2min再测稠度,前后两次测得的稠度差值即为砂浆的分层度(以mm计)。砂浆合理的分层度应控制在10~30mm,分层度大于30mm的砂浆容易离析、泌水和分层,水分流失过快、不便于施工,分层度小于10mm的砂浆硬化后容易产生干缩裂缝。

**2. 抗压强度与强度等级**

根据《建筑砂浆基本性能试验方法标准》(JGJ/T 70—2009),砂浆强度等级是以70.7mm×70.7mm×70.7mm的立方体标准试块(每组3块),按标准条件养护至28d后,用标准试验方法测得的抗压强度平均值(MPa),用$f_2$表示。砌筑砂浆的强度等级分为M30、M25、M20、M15、M10、M7.5、M5.0共7个等级。扫码获取《建筑砂浆基本性能试验方法标准》(JGJ/T 70—2009)的详细内容。

根据《砌筑砂浆配合比设计规程》(JGJ/T 98—2010)的规定,砂浆的实际强度除了与水泥的强度和用量有关外,还与基底材料的吸水性有关,因此其强度可分为下列两种情况。

(1)不吸水基层材料(如致密的石材)

影响砂浆强度的因素与混凝土基本相同,主要取决于水泥强度和水灰比,即砂浆的强度与水泥强度和灰水比成正比关系。关系式如下:

$$f_{m,cu} = \alpha f_{ce}\left(\frac{C}{W} - \beta\right) \tag{4-47}$$

式中　$f_{m,cu}$——砂浆28d抗压强度(MPa);

　　　$f_{ce}$——水泥的实测强度值(MPa);

　　　$\frac{C}{W}$——灰水比;

　　　$\alpha$、$\beta$——系数,可根据试验资料统计确定,一般情况下,$\alpha$为0.29左右,$\beta$为0.40左右。

(2)吸水性基层材料(如砖或其他多孔材料)

当基层吸水后,砂浆中保留水分的多少取决于其自身的保水性,因而砂浆即使用水量不同,但因自身具有一定的保水性和基层的吸水性,经过基层吸水后,保留在砂浆中的水分几乎是相同的。因此,砂浆的抗压强度主要取决于水泥强度和水泥用量,而与水灰比关系不大。砂浆强度计算公式如下:

$$f_{m,cu} = Af_{ce}\frac{Q_c}{1000} + B \tag{4-48}$$

式中　$f_{m,cu}$——砂浆28d抗压强度(MPa);

　　　$f_{ce}$——水泥的实测强度值(MPa);

　　　$Q_c$——每1m³砂浆的水泥用量(kg/m³);

　　　$A$、$B$——砂浆的特征系数,$A = 3.03$,$B = -15.09$。

扫码获取砌筑砂浆试验的相关内容。

## 三、砌筑砂浆的配合比设计

与混凝土不同,进行砂浆配合比设计遵循不密实填充原则,按照《砌筑砂浆配合比设计规程》(JGJ/T 98—2010)的规定,砌筑砂浆的配合比设计过程简述如下:

### (一)水泥混合砂浆配合比设计

**1. 确定砂浆的试配强度**

砌筑砂浆应具有95%的保证率,砂浆的试配强度应按下式计算:

$$f_{m,0} = kf_2 \tag{4-49}$$

式中 $f_{m,0}$——砂浆的试配强度(MPa),应精确至0.1MPa;
$f_2$——砂浆强度等级值(MPa),应精确至0.1MPa;
$k$——系数,按表4-27取值。

表4-27 砂浆试配强度系数取值表

| 施工水平 | 强度标准差 σ(MPa) | | | | | | | $k$ |
|---|---|---|---|---|---|---|---|---|
| | M5 | M7.5 | M10 | M15 | M20 | M25 | M30 | |
| 优良 | 1.00 | 1.50 | 2.00 | 3.00 | 4.00 | 5.00 | 6.00 | 1.15 |
| 一般 | 1.25 | 1.88 | 2.50 | 3.75 | 5.00 | 6.25 | 7.50 | 1.20 |
| 较差 | 1.50 | 2.25 | 3.00 | 4.50 | 6.00 | 7.50 | 9.00 | 1.25 |

砌筑砂浆现场强度标准差的确定应符合下列规定:
(1)当有统计资料时,应按下式计算:

$$\sigma = \sqrt{\frac{\sum_{i=1}^{n} f_{m,i} - n\bar{f}_m}{n-1}}$$

式中 $f_{m,i}$——统计周期内同一品种砂浆第 $i$ 组试件的强度(MPa);
$\bar{f}_m$——统计周期内同一品种砂浆 $n$ 组试件强度的平均值(MPa);
$n$——统计周期内同一品种砂浆试件的总组数,$n \geq 25$。

(2)当不具有近期统计资料时,砂浆现场强度标准差 $\sigma$ 可按表4-28取用。

表4-28 砂浆强度标准差选用值(MPa)

| 施工水平 | 砂浆强度等级 | | | | | |
|---|---|---|---|---|---|---|
| | M2.5 | M5 | M7.5 | M10 | M15 | M20 |
| 优良 | 0.50 | 1.00 | 1.50 | 2.00 | 3.00 | 4.00 |
| 一般 | 0.62 | 1.25 | 1.88 | 2.50 | 3.75 | 5.00 |
| 较差 | 0.75 | 1.50 | 2.25 | 3.00 | 4.50 | 6.00 |

**2. 计算水泥用量**

每1m³水泥用量,应按下式计算:

$$Q_c = \frac{1000(f_{m,0} - B)}{A f_{ce}} \tag{4-50}$$

式中 $Q_c$——每1m³砂浆的水泥用量(kg/m³);

$f_{m,0}$——砂浆试配强度(MPa);

$f_{ce}$——水泥的实测强度值(MPa);

$A$、$B$——砂浆的特征系数,$A = 3.03$,$B = -15.09$。

在无法取得水泥的实测强度值$f_{ce}$时,可按下式计算:

$$f_{ce} = \gamma_c f_{ce,k} \tag{4-51}$$

式中 $f_{ce,k}$——每1m³砂浆的水泥用量(kg/m³);

$\gamma_c$——水泥强度等级值的富余系数,该值应按实际统计资料确定。无统计资料时,$\gamma_c$可取1.00。

**3. 水泥混合砂浆的掺加料用量**

水泥混合砂浆的掺加料用量应按下式计算:

$$Q_D = Q_A - Q_C \tag{4-52}$$

式中 $Q_D$——每1m³砂浆的掺加料用量(kg),精确至1kg;石灰膏、黏土膏使用时的稠度为(120±5)mm;

$Q_C$——每1m³砂浆的水泥用量(kg),精确至1kg;

$Q_A$——每1m³砂浆中水泥和掺加料的总量(kg),精确至1kg;宜在300~350kg之间。

当石灰膏为其他稠度时,按表4-29进行换算。

表4-29 石灰膏不同稠度时换算系数

| 石灰膏稠度(mm) | 120 | 110 | 100 | 90 | 80 | 70 | 60 | 50 | 40 | 30 |
|---|---|---|---|---|---|---|---|---|---|---|
| 换算系数 | 1.00 | 0.99 | 0.97 | 0.95 | 0.93 | 0.92 | 0.90 | 0.88 | 0.87 | 0.86 |

**4. 确定砂子用量**

砂浆中的水、胶凝材料和掺加料的加入,基本正好填充砂子中的空隙,所以每1m³砂浆中的砂子用量$Q_S$(kg/m³),应按干燥状态(含水率小于0.50%)的堆积密度值作为计算值。

**5. 确定用水量**

每1m³砂浆中的用水量$Q_w$(kg/m³),根据砂浆稠度等要求进行选择,混合砂浆可选用240~310kg。

注意:(1)混合砂浆中的用水量,不包括石灰膏或黏土膏中的水;(2)当采用细砂或粗砂时,用水量分别取上限或下限;(3)稠度小于70mm时,用水量可小于下限;(4)施工现场气候炎热或干燥季节,可酌量增加用水量。

**(二)水泥砂浆配合比选用**

由于水泥砂浆按配合比规程计算时,会普遍出现水泥用量偏少情况,这主要因水泥强度太高,而砂浆强度太低所致。为此规程规定,水泥砂浆配合比用料可参照美国ASTM标准和英国BS标准,采用直接查表选用,见表4-30。

表 4-30　每 $1m^3$ 水泥砂浆材料用量

| 强度等级 | 水泥用量(kg) | 砂子用量(kg) | 用水量(kg) |
| --- | --- | --- | --- |
| M5 | 200~230 | 每 $1m^3$ 砂子的堆积密度值 | 270~330 |
| M7.5 | 230~260 | | |
| M10 | 260~290 | | |
| M15 | 290~330 | | |
| M20 | 340~400 | | |
| M25 | 360~410 | | |
| M30 | 430~480 | | |

注:1. 此表水泥强度等级为 32.5 级,大于 32.5 级水泥用量宜取下限;
　　2. 根据施工水平合理选择水泥用量;
　　3. 当采用细砂或粗砂时,用水量分别取上限或下限;
　　4. 稠度小于 70mm 时,用水量可小于下限;
　　5. 施工现场气候炎热或干燥季节,可酌量增加用水量;
　　6. 试配强度应按式 $f_{m,0} = kf_2$。

水泥用量应根据水泥的强度等级和施工水平合理选择,一般当水泥的强度等级较高或施工管理水平较高时,水泥用量选低值。用水量根据砂的粗细程度、砂浆稠度和气候条件选择,当砂较粗、稠度较小或气候较潮湿时,用水量选低值;反之亦然。

(三)配合比的试配、调整与确定

砂浆在计算或试配时应采用工程中实际使用的材料。搅拌采用机械搅拌,搅拌时间自投料结束后算起,水泥砂浆和水泥混合砂浆不得少 120s,水泥粉煤灰砂浆和掺用外加剂的砂浆不得少于 180s。按计算或查表所得的配合比进行试拌时,应测定其拌和物的稠度和分层度,当不能满足要求时,应调整材料用量,直至符合要求为止。此时的配合比为试配时砂浆基准配合比。

为了测定的砂浆强度能在设计要求范围内,试配时至少采用 3 个不同的配合比,其中一个为基准配合比,另外两个配合比的水泥用量按基准配合比应分别增加或减少 10%,在保证稠度和分层度合格的条件下,可将用水量或掺加料用量作相应调整。按《建筑砂浆基本性能试验方法标准》(JGJ/T 70—2009)的规定制成试件,测定砂浆强度。选定符合试配强度要求并且水泥用量最少的配合比作为砂浆配合比。

砂浆配合比以各种材料用量的比例形式表示:

水泥:掺加料:砂:水 $= Q_C : Q_D : Q_S : Q_W$;

或水泥:掺加料:砂:水 $= 1 : \dfrac{Q_D}{Q_C} : \dfrac{Q_S}{Q_C} : \dfrac{Q_W}{Q_C}$。

# 复习思考题

1. 配制混凝土时如何选择水泥?
2. 什么是骨料级配,如何表示? 骨料级配良好的特征是什么?

3. 若两种砂子的细度模数相同,那么二者的级配是否相同?若二者的级配相同,那么它们的细度模数是否相同?
4. 磨细矿渣粉和粉煤灰的性质各有什么特点?
5. 简述减水剂的作用机理,并简述混凝土掺入减水剂可获得的技术经济效果。
6. 简述混凝土拌和物和易性的概念及测定方法。
7. 什么是混凝土坍落度经时损失?主要原因是什么?
8. 影响混凝土拌和物和易性的主要因素是什么?改善混凝土拌和物和易性的主要措施有哪些?
9. 什么是混凝土的固定需水量法则?
10. 什么是混凝土的砂率?什么是混凝土的合理砂率?采用合理砂率对混凝土有什么意义?
11. 混凝土的强度等级如何确定?
12. 影响混凝土强度的主要因素有哪些?提高混凝土强度的主要措施有哪些?
13. 混凝土的收缩主要分为几种?每一种收缩是由什么原因引起的?
14. 混凝土硬化初期发生温度变形导致开裂破坏的机理是什么?
15. 什么是混凝土的徐变?混凝土徐变在结构工程中有什么影响?
16. 什么是混凝土的抗渗性?改善混凝土抗渗性的措施有哪些?
17. 什么是混凝土的抗冻性?影响混凝土抗冻性的主要因素有哪些?哪种外加剂对混凝土抗冻性的提升最为明显?
18. 什么是混凝土的碳化?碳化对钢筋混凝土的性能有何影响?
19. 混凝土配合比设计的基本原则和依据各有哪些?
20. 若采用混凝土配合比设计中第一个步骤得到的计算配合比进行实验室试配后发现混凝土的和易性不满足要求,此时应采取什么措施进行调整?
21. 对新拌砂浆的技术要求与对混凝土混合料的技术要求有何异同?
22. 新拌砂浆的和易性包括哪些方面?如何测定?砂浆和易性不良对工程应用有何影响?
23. 砌筑砂浆对组成材料有何要求?为什么要加入掺加料或塑化剂?
24. 用于吸水基层和不吸水基层的砌筑砂浆,影响其强度的因素有何不同?如何进行计算?
25. 对抹面砂浆和砌筑砂浆的组成材料及技术性质的要求有哪些不同?为什么?
26. 某工程砌筑烧结普通黏土砖用水泥石灰砂浆,要求砂浆的强度等级为 M10。现场有强度等级为 32.5 和 42.5 的矿渣硅酸盐水泥可供选用。已知所用水泥的堆积密度为 $1280kg/m^3$;中砂的含水率为 1%~3%、堆积密度为 $1550kg/m^3$;石灰膏的表观密度为 $1350kg/m^3$。施工水平优良,试计算砂浆的配合比。
27. 装饰砂浆的主要饰面形式有哪些?
28. 试述采用预拌砂浆的重要意义。

# 第五章 沥青与沥青混合料

沥青材料是由高分子碳氢化合物及其衍生物组成的、黑色或深褐色、不溶于水而几乎全溶于二硫化碳的非晶态有机材料。沥青混合料是以沥青为胶结料,将粗骨料、细骨料、填料等黏结成一体的材料。

沥青与沥青混合料是土木工程建设中不可缺少的建筑材料,在建筑、公路、桥梁和防水工程中有着广泛的应用。采用沥青作胶结料的沥青混合料已成为高等级路面的主要材料。它具有力学性能和抗滑性良好、防水性好、平稳舒适、噪声小等优点,可分层加厚且易于修补;但它同时也存在着易老化和感温性差等缺点。

## 第一节 沥青材料

### 一、沥青的分类

对于沥青材料的命名和分类,目前世界各国尚未取得统一的认识。现就我国命名和分类简述如下:沥青按其在自然界中获得的方式,可以分为地沥青和焦油沥青两大类。地沥青是天然存在的或经石油精制加工得到的沥青材料,按其产源又可分为天然沥青和石油沥青。焦油沥青是利用各种有机物干馏加工得到的焦油经再加工得到的产品,包括煤沥青、页岩沥青等。

以上各类沥青,可归纳如下:

$$沥青\begin{cases}地沥青\begin{cases}天然沥青:由沥青湖或含有沥青的砂岩、砂等提炼而成\\石油沥青:由石油原油蒸馏后的残留物经加工而得\end{cases}\\焦油沥青\begin{cases}煤沥青:由煤焦油蒸馏后的残留物经加工而得\\页岩沥青:油页岩炼油工业的副产品\end{cases}\end{cases}$$

### 二、石油沥青

(一)组成与结构

石油沥青是由石油经蒸馏、吹氧、调和等工艺加工得到的残留物,主要为可溶于二硫化碳的碳氢化合物的半固体黏稠状物质。石油沥青是由多种碳氢化合物及其非金属(氧,硫,氮)的衍生物组成的混合物,它的分子表达通式为 $C_nH_{2n+a}O_bS_cN_d$。石油沥青的化学组成主要是碳(80%~87%)、氢(10%~15%),其次是非烃元素,如氧、硫、氮等(<3%)。此外,它还含有一些微量的金属元素,如镍、钒、铁、锰、镁、钠等,但含量都极少,约为百万分之几至百万分之几十。

由于沥青化学组成结构的复杂性，所以虽然多年来有许多化学家致力于这方面的研究，可是目前仍不能直接得到沥青元素含量与工程性能之间的关系。目前对沥青组成和结构的研究主要集中在组分理论、胶体理论和高分子溶液理论。

**1. 组分组成**

化学组分就是将沥青分离为化学性质相近，而且与其工程性能有一定联系的几个化学成分划分为若干组，这些组就称为组分。我国现行《公路工程沥青与沥青混合料试验规程》(JTG E20—2011)中规定有三组分和四组分两种分析法（扫描二维码了解更多内容）。

（1）三组分分析法

石油沥青的三组分分析法是将石油沥青分离为油分、树脂和沥青质三个组分（表 5-1）。因我国富产石蜡基和中间基沥青，在油分中往往含有蜡，故在分析时还应进行油蜡分离。由于这种组分分析法兼用了选择性溶解和选择性吸附的方法，故又称为溶解-吸附法。三组分分析法的优点是组分界限很明确，组分含量能在一定程度上说明其工程性能，而主要缺点是分析流程复杂，分析时间很长。

表 5-1 石油沥青三组分分析法的各组分性状

| 组分 | 外观特征 | 平均相对分子量 | 碳氢比 | 含量(%) | 物化特征 |
|---|---|---|---|---|---|
| 油分 | 淡黄色透明液体 | 200～700 | 0.5～0.7 | 45～60 | 几乎溶于大部分有机溶剂，具有光学活性，常发现有荧光，相对密度约 0.7～1.0 |
| 树脂 | 红褐色黏稠半固体 | 800～3000 | 0.7～0.8 | 15～30 | 温度敏感性高，熔点低于 100℃，相对密度大于 1.0～1.1 |
| 沥青质 | 深褐色固体微粒 | 1000～5000 | 0.8～1.0 | 5～30 | 加热不熔化而碳化，相对密度 1.1～1.5 |

油分赋予沥青以流动性，油分含量的多少直接影响沥青的柔软性、抗裂性及施工难度。油分在一定条件下可以转化为树脂甚至沥青质。

树脂又分为中性树脂和酸性树脂，中性树脂使沥青具有一定塑性、可流动性和黏结性，其含量增加，沥青的黏结力和延伸性也随之增加。除中性树脂外，沥青树脂中还含有少量的酸性树脂，即沥青酸和沥青酸酐，为树脂状黑褐色黏稠状物质，是油分氧化后的产物，呈固态或半固态，具有酸性，能为碱皂化，易溶于酒精、氯仿，而难溶于石油醚和苯。酸性树脂是沥青中活性最大的组分，它能改善沥青对矿质材料的浸润性，特别是提高了与碳酸盐类岩石的黏附性，增加了沥青的可乳化性。

沥青质决定着沥青的黏结力、黏度和温度稳定性，以及沥青的硬度、软化点等。沥青质含量增加时，沥青的黏度和黏结力增加，硬度和温度稳定性提高。

（2）四组分分析法

四组分分析法由科尔贝特首先提出，是将沥青分离为饱和分(saturate)、环烷-芳香分(naphthene-aromatics)、极性-芳香分(polar-aromatics)和沥青质(asphaltenes)等的色层分析方法。后来也有人将上述 4 个组分称为饱和分、芳香分(aromatics)、胶质(resin)和沥青质，故这一方法也称 SARA 法。我国现行四组分分析法是将沥青分离为沥青质($A_t$)、饱和分($S$)、芳香分($A$)和胶质($R$)。石油沥青按四组分分析法所得各组分的性状见表 5-2。

表 5-2 石油沥青四组分分析法的各组分性状

| 组分 | 外观特征 | 平均相对密度 | 平均相对分子量 | 主要化学结构 |
|---|---|---|---|---|
| 饱和分 | 无色液体 | 0.89 | 625 | 烷烃、环烷烃 |
| 芳香分 | 黄色至红色液体 | 0.99 | 730 | 芳香烃、含 S 衍生物 |
| 胶质 | 棕色黏稠液体 | 1.09 | 970 | 多环结构,含 S、O、N 衍生物 |
| 沥青质 | 深棕色至黑色固体 | 1.15 | 3400 | 缩合结构,含 S、O、N 衍生物 |

按照四组分分析法,各组分对沥青性质的影响,根据科尔贝特的研究认为:饱和分含量增加,可使沥青稠度降低,即针入度增大;树脂含量增大,可使沥青的延性增加;在有饱和分存在的条件下,沥青质含量增加,可使沥青获得低的感温性;树脂和沥青质的含量增加,可使沥青的黏度提高,即针入度降低。

**2. 胶体结构**

由于沥青的组分并不能全面地反映沥青材料的性质,沥青的性质还与其结构有着密切的联系。现代胶体理论认为,沥青的胶体结构是以固态超细微粒的沥青质为分散相,通常是若干个沥青质聚集在一起,它们吸附了极性半固态的胶质,而形成胶团。由于胶溶剂-胶质的胶溶作用,而使胶团胶溶、分散于液态的芳香分和饱和分组成的分散介质中,形成稳定的胶体。

根据沥青中各组分的化学组成和相对含量的不同体结构,可分为以下三个类型。

(1)溶胶型结构

当沥青中沥青质分子量较低,并且含量很少(例如在 10% 以下),并有一定数量的芳香度较高的胶质时,胶团能够完全胶溶而分散在芳香芬和饱和芬的介质中。在此情况下,胶团相距较远,它们之间吸引力很小(甚至没有吸引力),胶团可以在分散介质黏度许可范围之内自由运动,这种胶体结构的沥青,称为溶胶型沥青[图 5-1(a)]。溶胶型沥青的特点是流动性和塑性较好,开裂后自行愈合能力较强,对温度的敏感性强,对温度的稳定性较差,温度过高会流淌。通常情况下,大部分直馏沥青都属于溶胶型沥青。

(2)溶-凝胶型结构

当沥青中沥青质含量适当(例如在 15%~25% 之间),并有较多数量芳香度较高的胶质时,这样形成的胶团数量增多,胶体中胶团的浓度增加,胶团距离相对靠近[图 5-1(b)],它们之间有一定的吸引力。这是一种介乎溶胶与凝胶之间的结构,称为溶-凝胶结构。这种结构的沥青,称为溶-凝胶型沥青。修筑现代高等级沥青路用的沥青,都应属于这类胶体结构类型。通常,环烷基稠油的直馏沥青或半氧化沥青,以及按要求组分重(新)组(配)的溶剂沥青等,往往能符合这类胶体结构。这类沥青的工程性能是在高温时具有较低的感温性,低温时又具有较好的形变能力。

(3)凝胶型结构

当沥青中沥青质含量很高(例如 >30%),并有相当数量芳香度高的胶质来形成胶团时,沥青中胶团浓度有很大程度的增加,它们之间的相互吸引力增强,使胶团靠得很近,形成空间网络结构。此时,液态的芳香芬和饱和芬在胶团的网络中成为分散相,连续的胶团成为分散介质[图 5-1(c)]。这种胶体结构的沥青,称为凝胶型沥青,这类沥青的特点为弹性和黏性较高,温度敏感性较小,开裂后自行愈合能力较差,流动性和塑性较低。在工程性能上,其虽具有较好的温度稳定性,但低温变形能力较差。

(a) 溶胶结构　　　　　　(b) 溶-凝胶结构　　　　　　(c) 凝胶结构

图 5-1　沥青胶体结构

(二)技术性质

**1. 针入度**

黏稠石油沥青的相对黏度是用针入度仪测定的针入度来表示,如图 5-2 所示。它能反映石油沥青的黏稠性,针入度值越小,表明黏度越大。黏稠石油沥青的针入度是在规定温度 25℃条件下,以规定质量 100g 的标准针,经规定时间 5s 贯入试样中的深度,以 1/10mm 为单位表示,符号为 $P_{(25℃,100g,5s)}$。

**2. 软化点**

沥青软化点是反映沥青温度敏感性的重要指标。由于沥青材料从固态到液态有一定变化间隔,故规定其中某一状态作为从固态转到黏流态(或某一规定状态)的起点,相应温度称为沥青软化点。

《公路工程沥青及沥青混合料试验规程》(JTG E20—2011)是采用环球法软化点(图 5-3)。该法是将黏稠沥青试样注入内径为 18.9mm 的钢环中,环上置 3.5g 的钢球,在规定的加热速度(5℃/min)下进行加热,沥青试样逐渐软化,直至在钢球荷重作用下,使沥青下坠 25.4mm 时的温度称为软化点。根据已有研究认为:沥青在软化点时的黏度约为 1200Pa·s,相当于针入度值为 800(0.1mm)。据此,可以认为软化点是一种人为的"等黏温度"。

图 5-2　沥青针入度仪　　　　图 5-3　沥青软化点仪

### 3. 延展性

延展性是指石油沥青在外力作用下产生变形而不破坏（裂缝或断开），除去外力后仍保持变形后的形状不变的性质，它反映的是沥青受力时所能承受的塑性变形能力。

石油沥青的延展性与其组分有关，石油沥青中树脂含量较多，且其他组分含量又适当时，则塑性较大。影响沥青塑性的因素有温度和沥青膜层厚度，温度升高，则延展性增大，膜层越厚，则塑性越高。反之，膜层越薄，则塑性越差，当膜层很薄时，塑性近于消失，即接近于弹性。

在常温下，延展性较好的沥青在产生裂缝时，也可以依靠其特有的黏塑性而自行愈合。故延展性还反映了沥青开裂后的自愈能力。沥青之所以能用来制造出性能良好的柔性防水材料，很大程度上取决于沥青的延展性。沥青的延展性对冲击振动荷载具有一定吸收能力，并能减少摩擦时产生的噪声，故沥青是一种优良的路面材料。

通常是用延度作为延展性指标。延度试验方法是：将沥青试样制成8字形标准试件（最小断面面积为$1cm^2$），在规定拉伸速度和规定温度下拉断时的长度（以 cm 计）称为延度，如图 5-4 所示。常用的试验温度有15℃和10℃。

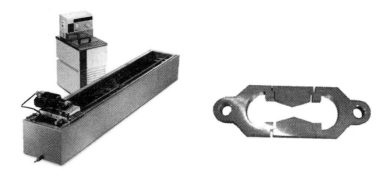

**图 5-4　延度仪和延度试模**

以上所论及的针入度、软化点和延度是评价黏稠石油沥青工程性能最常用的经验指标，所以统称为"三大指标"，扫描二维码了解石油沥青性能指标的相关实验内容。

### 4. 黏附性

黏附性是指沥青与其他材料（主要是指骨料）的界面黏结性能和抗剥落性能。沥青与骨料的黏附性直接影响沥青路面的使用质量和耐久性，所以黏附性是评价道路沥青技术性能的一个重要指标。

评价沥青与骨料的黏附性最常采用的方法是水煮法和水浸法，《公路工程沥青及沥青混合料试验规程》（JTG E20—2011）规定，沥青与粗骨料的黏附性试验，根据沥青混合料的最大粒径决定，大于13.2mm 的采用水煮法；小于（或等于）13.2mm 的采用水浸法。水煮法是选取粒径为13.2～19mm 形状接近正立方体的规则骨料5个，经沥青裹覆后，在蒸馏水中沸煮3min，按沥青膜剥落的情况分为5个等级来评价沥青与骨料的黏附性。水浸法是选取9.5～13.2mm 的骨料100g 与5.5g 的沥青在规定温度条件下拌和，配制成沥青-骨料混合料，冷却后浸入80℃的蒸馏水中保持30min，然后按剥落面积百分率来评定沥青与骨料的黏附性。

**5. 大气稳定性**

在阳光、空气和加热的综合作用下,沥青各组分会不断递变。低分子化合物将逐步转变成高分子物质,即油分和树脂逐渐减少,而沥青质逐渐增多。因此,石油沥青随着时间的进展,流动性和塑性逐渐减小,脆性逐渐增大,直至脆裂,这个过程称为石油沥青的老化。所以沥青的大气稳定性可以用抗老化性能来说明。

《公路工程沥青及沥青混合料试验规程》(JTG E20—2011)规定,石油沥青的抗老化性能是以沥青试样在加热蒸发前后的质量损失百分率、针入度比和老化后的延度来评定。其测定方法是:先测定沥青试样的质量及其针入度,然后将试样置于烘箱中,在163℃下加热蒸发5h,待冷却后再测定其质量和针入度。计算出蒸发损失质量占原质量的百分数,称为蒸发损失百分率;测得老化后针入度与原针入度的比值,称为针入度比,同时测定老化后的延度。沥青经老化后,质量损失百分率越小、针入度比和延度越大,则表示沥青的大气稳定性越好。

**6. 施工安全性**

黏稠沥青在使用时必须加热,当加热至一定温度时,沥青材料中挥发的油分蒸气与周围空气组成混合气体,此混合气体遇火焰则易发生闪火。若继续加热,油分蒸汽和饱和度增加。由于此种蒸汽与空气组成的混合气体遇火焰极易燃烧,而引发火灾,为此,必须测定沥青加热闪火和燃烧的温度,即闪点和燃点。

闪点和燃点的高低表明沥青引起火灾或爆炸可能性的大小,它关系到运输、贮存和加热使用等方面的安全性。石油沥青在熬制时,一般温度为150℃。因此通常控制沥青的闪点应大于230℃。但为安全起见,沥青加热时还应与火焰隔离。

(三)技术标准

根据石油沥青的性能不同,选择适当的技术标准,将沥青划分成不同的种类和标号(等级),以便于沥青材料的选用。目前石油沥青主要划分为三大类:道路石油沥青、建筑石油沥青和普通石油沥青,其中道路石油沥青是沥青的主要类型。

**1. 道路石油沥青分级**

道路石油沥青分为 A 级、B 级、C 级三个等级,各自的适用范围符合表5-3规定。

表5-3　道路石油沥青适用范围

| 沥青等级 | 适用范围 |
| --- | --- |
| A | 适用各个等级的公路,适用于任何场合和层次 |
| B | 1. 高速公路、一级公路沥青下面层及以下的层次,二级及二级以下公路的各个层次;<br>2. 用于改性沥青、乳化沥青、改性乳化沥青、稀释沥青的基质沥青 |
| C | 三级及三级以下公路的各个层次 |

**2. 道路石油沥青标号**

道路石油沥青按照针入度划分为160号、130号、110号、90号、70号、50号、30号七个标号,同时对各标号沥青的延度、软化点、闪点、含蜡量、薄膜加热试验等技术指标也提出相应的要求。具体要求见表5-4。

表 5-4 道路石油沥青技术要求

| 指标 | 单位 | 等级 | 160号[4] | 130号[4] | 110号 | | 90号 | | | 70号 | | | 50号 | 30号[4] | 试验方法[1] |
|---|---|---|---|---|---|---|---|---|---|---|---|---|---|---|---|
| 针入度(25℃,100g,5s) | 0.1mm | | 140~200 | 120~140 | 100~120 | | 80~100 | | | 60~80 | | | 40~60 | 20~40 | T0604 |
| 适用的气候分区[6] | | | 注[4] | 注[4] | 2-1 | 2-2 | 3-2 | 1-1 | 1-2 | 1-3 | 2-2 | 2-3 | 1-3 | 1-4 | 2-2 | 2-3 | 1-4 | 注[4] | 附录A[5] |
| 针入度PI[2] | | A | | | | | −1.5～+1.0 | | | | | | | | | T0604 |
| | | B | | | | | −1.8～+1.0 | | | | | | | | | |
| 软化点(R&B)不小于 | ℃ | A | 38 | 40 | 43 | | 45 | | | 46 | 45 | | 49 | 55 | T0606 |
| | | B | 36 | 39 | 42 | | 43 | | | 44 | 43 | | 46 | 53 | |
| | | C | 35 | 37 | 41 | | 42 | | | 43 | | | 45 | 50 | |
| 60℃动力黏度[2]不小于 | Pa·s | A | — | 60 | 120 | | 160 | | | 180 | 160 | | 200 | 260 | T0620 |
| 10℃延度[2]不小于 | cm | A | 50 | 50 | 40 | | 45 | 30 | | 20 | | | 15 | 20 | 15 | 10 | T0605 |
| | | B | 30 | 30 | 30 | | 30 | 20 | | 15 | | | 10 | 15 | 10 | 8 | |
| 15℃延度不小于 | cm | A、B | 80 | 80 | 60 | | 50 | | | 40 | | | 80 | 50 | |
| | | C | | | | | 100 | | | | | | 30 | 20 | |
| 蜡含量(蒸馏法)不大于 | % | A | | | | | 2.2 | | | | | | | | |
| | | B | | | | | 3.0 | | | | | | | | T0615 |
| | | C | | | | | 4.5 | | | | | | | | |
| 闪点不小于 | ℃ | | 230 | | | | 245 | | | | 260 | | | | T0611 |
| 溶解度不小于 | % | | | | | | 99.5 | | | | | | | | T0607 |
| 密度(15℃) | g/cm³ | | | | | | 实测记录 | | | | | | | | T0603 |

续表

| 指标 | 单位 | 等级 | 沥青标号 | | | | | | | 试验方法[1] |
|---|---|---|---|---|---|---|---|---|---|---|
| | | | 160号[4] | 130号[4] | 110号 | 90号[5] | 70号[3] | 50号 | 30号[4] | |
| | | | TFOT(或RTFOT)后[5] | | | | | | | |
| 质量变化 不大于 | % | | ±0.8 | | | | | | | T0610 或 T0609 |
| 残留针入度比 不小于 | % | A | 48 | 54 | 55 | 57 | 61 | 63 | 65 | T0604 |
| | | B | 45 | 50 | 52 | 54 | 58 | 60 | 62 | |
| | | C | 40 | 45 | 48 | 50 | 54 | 58 | 60 | |
| 残留延度(10℃) 不小于 | cm | A | 12 | 12 | 10 | 8 | 6 | 4 | — | T0605 |
| | | B | 10 | 10 | 8 | 6 | 4 | 2 | — | |
| 残留延度(15℃) 不小于 | cm | C | 40 | 35 | 30 | 20 | 15 | 10 | — | T0605 |

[1] 试验方法按照《公路工程沥青及沥青混合料试验规程》(JTG E20—2011)规定的方法执行。用于仲裁试验求取PI时的5个温度的针入度关系的相关系数不得小于0.997;
[2] 经建设单位同意,表中PI值,60℃动力黏度,10℃延度可作为选择性指标,也可不作为施工质量检验指标;
[3] 70号沥青可根据需要求供应商提供要求提供针入度范围为60~70或70~80的沥青,50号沥青可要求提供针入度范围为40~50或50~60的沥青;
[4] 30号沥青仅适用于沥青稳定基层。130号和160号沥青除寒冷地区可直接应用在中低级公路上直接应用外,通常用作乳化沥青、改性沥青、稀释沥青的基质沥青;
[5] 老化试验以TFOT为准,也可以RTFOT代替;
[6] 气候分区见《公路沥青路面施工技术规范》(JTG F 40—2004)附录A,扫描二维码了解更多内容。

### 三、其他沥青

(一)乳化沥青

乳化沥青是将黏稠沥青加热至流态,经机械力的作用而形成微滴(粒径为 $2\sim5\mu m$)分散在有乳化剂-稳定剂的水中形成均匀稳定的乳状液,亦称沥青乳液,简称乳液。它主要用于修筑道路路面,其主要组分是沥青、乳化剂、稳定剂和水等。

**1. 沥青**

沥青是乳化沥青组成的主要材料,沥青的质量直接决定乳化沥青的性能。在选择作为乳化沥青用的沥青时,首先要考虑它的易乳化性。沥青的易乳化性与其化学结构有密切关系。以工程适用为目的,可认为易乳化性与沥青中的沥青酸含量有关。通常认为沥青酸总量大于1%的沥青,采用通用乳化剂和一般工艺即易于形成乳化沥青。另外,相同油源和工艺的沥青,针入度较大者易于形成乳液,但是针入度的选择应根据乳化沥青在路面工程中的用途来决定。

**2. 乳化剂**

乳化剂是乳化沥青形成的关键材料,从化学结构上看,它是表面活性剂的一种。其分子结构中含有一种"两亲性"分子,分子的一部分具有亲水性质,而另一部分具有亲油性质。亲油部分一般由碳氢原子团,特别是由长链烷基构成,结构差别较小;亲水部分原子团则种类繁多,结构差异较大。因此乳化剂的分类,是以亲水基的结构为依据,按其亲水基在水中是否能电离而分为离子型和非离子型两大类。离子型乳化剂按其离子电性,又分为阴离子型乳化剂、阳离子型乳化剂和两性离子型乳化剂。另外,随着乳化沥青的发展,为满足各种特殊要求,衍生出了许多化学结构更为复杂的复合乳化剂。

**3. 稳定剂**

为使乳液具有良好的贮存稳定性及在施工中喷洒或拌和的机械作用下的稳定性,必要时可加入适量的稳定剂。一般稳定剂可分为有机稳定剂和无机稳定剂两类。常用的有机稳定剂包括聚乙烯醇、聚丙烯酰胺、甲基纤维素钠、糊精、MF 废液等。这类稳定剂可提高乳液的贮存稳定性和施工稳定性。常用的无机稳定剂包括氯化钙、氯化镁、氯化铵和氯化铬等。

(二)沥青的掺配

除对沥青进行改性以外,还可以通过将两三种沥青进行掺配的方法,获得所需要的特定性能(如需要的软化点条件)。需要注意的是,掺配要遵循同源原则,即同属石油沥青或同属煤沥青才可以掺配。

两种沥青的掺配比例可以使用如下公式计算:

$$Q_1 = \frac{T_2 - T_1}{T - T_1} \times 100\% \tag{5-1}$$

$$Q_2 = 100 - Q_1 \tag{5-2}$$

式中 $Q_1$——较软沥青用量(%);

$Q_2$——较硬沥青用量(%);

$T_1$——较软沥青软化点(℃);

$T_2$——较硬沥青软化点(℃);

$T$——要求配制沥青的软化点(℃)。

以估算的掺配比例和其邻近的比例(±5%～±10%)进行试配(混合熬制均匀),测定掺配后沥青的软化点,然后绘制掺配比-软化点关系曲线,即可从曲线上确定出所要求的掺配比例。同样地也可采用针入度指标按上法估算及试配。

### 四、沥青老化与改性

(一)沥青的老化

沥青在生产及使用过程中,受到周围温度、紫外线、氧气等作用,其内部化学成分及胶体结构会发生一系列变化,从而导致宏观性能的变化,这一过程称为沥青的老化。

沥青的老化会引起沥青物理化学性质的变化。通常的规律是:针入度变小、延度降低、软化点和脆点升高,具体表现为沥青变硬、变脆、延伸性降低,导致沥青产生裂缝、松散等破坏。沥青老化后物理-力学性质变化见表5-5。

表5-5 老化沥青和再生沥青的技术性质示例

| 沥青名称 | 技术性质 | | | |
| --- | --- | --- | --- | --- |
| | 针入度(1/10mm) | 延度(cm) | 软化点(℃) | 脆点(℃) |
| 原始沥青 | 106 | 73 | 48 | -6 |
| 老化沥青 | 39 | 23 | 55 | -4 |
| 再生沥青 | 80 | 78 | 49 | -10 |

在实际应用中,人们要求沥青有尽可能长的耐久性,老化的速度尽可能地小一些,因而提出了对沥青耐久性的要求。耐久性是沥青使用性能方面一个十分重要的综合性指标。由于老化带来的性能劣化,造成使用沥青材料的工程经过一定的使用年限后,都要进行大规模的翻修,所以如何提高沥青的耐久性,延长沥青材料的使用寿命,在国民经济中具有相当重要的地位,同时也是沥青科学专业研究和生产方面的一个十分迫切的课题。

与老化过程相反,利用某种工艺及材料,来改善沥青的组分组成、胶体结构以及宏观性能,使其工程性能得到一定程度的恢复,此过程称为再生。

目前沥青材料的再生,是理论界及工程界的一个热点,但由于沥青材料的复杂性,这些工作尚都处于起步阶段。

(二)改性石油沥青

沥青材料无论是用作屋面防水材料还是用作路面胶结材料,都是直接暴露于自然环境中的,而沥青的性能又受环境因素影响较大;同时现代土木工程不仅要求沥青具有较好的使用性能,还要求具有较长的使用寿命。单纯依靠自身性质,很难满足现代土木工程对沥青的多方面要求。因此,现代土木工程中,常在沥青中加入其他材料,来进一步提高沥青的性能,也称改性沥青。

目前,国内外常用的改性沥青包括以下四类:

**1. 矿物填料改性沥青**

在沥青中加入一定数量的矿物填充料,可以提高沥青的黏性和耐热性,减小沥青的温度敏感性,主要适用于生产沥青胶。

矿物填料有粉状和纤维状两种,常用的填料有滑石粉、石灰石粉、硅藻土、石棉绒、云母粉、磨细砂、粉煤灰、水泥、高岭土、白垩粉等。

**2. 树脂改性沥青**

使用树脂改性石油沥青,可以改善沥青的耐寒性、耐热性、黏结性和不透气性。在生产卷材、密封材料和防水涂料等产品时均需应用。

用作改性沥青的树脂主要是热塑性树脂,常被采用的为聚乙烯和聚丙烯。它们所组成的改性沥青性能,主要是提高沥青的黏度,改善高温抗流动性,同时可增大沥青的韧性。但其对低温性能改善有时并不明显。

**3. 橡胶改性沥青**

橡胶是石油沥青的重要改性材料,与石油沥青有很好的混溶性,能使沥青兼具橡胶的很多优点,如高温变形性小,低温柔性好,提高其强度、延伸率和耐老化性等。由于橡胶品种不同,掺入方法也有差异,故各种橡胶改性沥青的性能也不相同。

橡胶改性沥青的性能,主要取决于沥青的性能、橡胶的种类和制备、工艺等。在当前合成的橡胶改性沥青中,常用的橡胶材料有氯丁橡胶、丁基橡胶、丁苯橡胶等。

**4. 树脂和橡胶共混改性沥青**

用橡胶和树脂来改善石油沥青的性质,可使沥青兼具橡胶和树脂的特性。由于树脂比橡胶便宜,橡胶和树脂又有较好的混溶性,故树脂和橡胶共混改性沥青能取得较满意的综合效果。

橡胶、树脂和石油沥青在加热熔融状态下,沥青与高分子聚合物之间发生相互侵入和扩散,沥青分子填充在聚合物大分子的间隙内,同时聚合物分子的某些链节扩散进入沥青分子中,从而形成凝聚网状混合结构,由此而获得较优良的性能。

# 第二节 沥青混合料

沥青混合料是将粗骨料、细骨料和填料经人工合理选择级配组成的矿质混合料与适量沥青材料经拌和而成的均匀混合料,包括沥青混凝土和沥青碎石,还有开级配或间断级配沥青混合料。

由于沥青混凝土路面具有平整性好、行车平稳舒适、噪声低等特点,许多国家在建设高速公路时都优先采用沥青混凝土铺路。半刚性基层具有强度大、稳定性好及刚度大等特点,被广泛用于修建高等级公路沥青路面的基层或底基层。我国在建或已建成的高速公路路面许多采用半刚性基层沥青路面。由于沥青混合料最能适应现代交通的特点,所以它是现代高速公路的最主要的路面材料,并广泛应用于干线公路和城市道路路面。

### 一、热拌沥青混合料的分类

热拌沥青混合料是经人工组配的矿质混合料与黏稠沥青在专业设备中加热拌和而成,用保温运输工具运送至施工现场,并在热炼下进行摊铺和压实的混合料,通称热拌热铺沥青混合料,简称热拌沥青混合料。热拌沥青混合料是沥青混合料中最典型的品种,其他各种沥青混合料均由其发展而来。本节主要详述它的分类、组成结构、技术性质、组成材料和设计方法。

(1)按矿质骨料级配类型分类

① 连续级配沥青混合料。沥青混合料中的矿料是按级配原则,把从大到小的各级粒径按比例相互搭配组成的混合料,称为连续级配混合料。

② 间断级配沥青混合料。连续级配沥青混合料矿料中缺少一个或若干个档次粒径的沥青混合料称为间断级配沥青混合料。

(2)按混合料密实度分类

① 密级配沥青混凝土混合料。按密实级配原则设计的连续型密级配沥青混合料,其粒径递减系数较小,压实后剩余空隙率小于10%。密级配沥青混凝土混合料按其剩余空隙率又可分为:Ⅰ型密实式沥青混凝土混合料,其剩余空隙率为3%~6%;Ⅱ型半密实式沥青混凝土混合料,其剩余空隙率为4%~10%。

② 开级配沥青混合料。按开级配原则设计的连续型级配混合料,其粒径递减系数较大,压实后剩余空隙率大于15%。也有将剩余空隙率介于密级配和开级配之间的(即剩余空隙率为10%~15%)混合料称为半开级配沥青混合料,也称为沥青碎石混合料。

(3)按公称最大粒径分类

沥青混凝土按骨料公称最大粒径可分为下列四类:

① 粗粒式沥青混合料:骨料最大粒径为26.5mm或31.5mm的沥青混合料。

② 中粒式沥青混合料:骨料最大粒径为16mm或19mm的沥青混合料。

③ 细粒式沥青混合料:骨料最大粒径为9.5mm或13.2mm的沥青混合料。

④ 砂粒式沥青混合料:骨料最大粒径等于或小于4.75mm的沥青混合料,也称为沥青石屑或沥青砂。

沥青碎石混合料除上述四类外,尚有特粗式沥青碎石混合料,其骨料最大粒径为37.5mm以上。

### 二、沥青混合料的组成结构

**1. 沥青混合料组成结构理论**

目前就已有的研究而言,主要有两种沥青混合料组成结构理论。

表面理论认为沥青混合料是由粗骨料、细骨料和填料经人工组配成密实的级配矿质骨架,此矿质骨架由稠度较稀的沥青混合料分布其表面,而将它们胶结成为一个具有强度的整体。

胶浆理论将沥青混合料看作一种多级空间网状结构的分散体系,即以粗骨料为分散相而分散在沥青砂浆的分质中的一种粗分散系。

**2. 沥青混合料的组成结构类型**

按级配原则构成的沥青混合料,其结构通常可按下列三种方式组成,如图 5-5 所示。

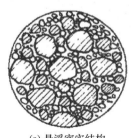 

(a) 悬浮密实结构　　　(b) 骨架空隙结构　　　(c) 骨架密实结构

图 5-5　沥青混合料矿料骨架类型

(1) 悬浮密实结构

连续级配矿质混合料组成的密实混合料,由于材料从大到小连续存在,并且各有一定数量,实际上同一档较大颗粒都被较小一档颗粒挤开,大颗粒以悬浮状态处于较小颗粒之中。这种结构通常按最佳级配原理进行设计,因此密实度与强度较高,但受沥青材料的性质和物理状态的影响较大,故稳定性较差。

(2) 骨架空隙结构

较粗石料彼此紧密相接,较细粒料的数量较少,不足以充分填充空隙。混合料的空隙较大,石料能够充分形成骨架。在这种结构中,粗骨料之间的内摩阻力起着重要的作用,其结构强度受沥青的性质和物理状态的影响较小,因而稳定性较好。

(3) 骨架密实结构

综合以上两种方式组成的结构。混合料中既有一定数量的粗骨料形成骨架,又根据粗料空隙的多少加入细料,形成较高的密实度。间断级配即是按此原理构成。

### 三、沥青混合料的技术性质

沥青混合料在路面中直接承受车辆荷载和大气因素的影响,同时沥青混合料的物理、力学性质受气候因素与时间因素影响较大,因此为了能使路面给车辆提供稳定、耐久的服务,必须要求沥青路面具有一定的稳定性和耐久性。其中包括高温稳定性、低温抗裂性、耐久性、抗滑性、施工和易性等。

**1. 高温稳定性**

沥青混合料的高温稳定性,习惯上是指沥青混合料在荷载作用下抵抗永久变形的能力。由于沥青混合料的强度与刚度(模量)随温度升高而显著下降,为了保证沥青路面在高温季节行车载荷反复作用下,不致产生诸如波浪、推移、车辙、拥包等病害,沥青路面应具有良好的高温稳定性。

《公路沥青路面施工技术规范》(JTG F40—2004)规定,采用马歇尔试验来评价沥青混合料的高温稳定性;对于高速公路、一级公路、城市快速路以及主干路用沥青混合料,还应通过动稳定度试验检验其抗车辙能力,通过浸水马歇尔试验和冻融劈裂试验检验其水稳定性。扫描二维码了解更多内容。

(1)马歇尔试验

马歇尔稳定度是指按标准方法制备的试件,在60℃的恒温水槽中保温45min,然后将试件放置于马歇尔稳定度仪上,以$(50\pm5)$mm/min的形变速度加荷载,直至试件破坏时的最大荷载(kN),即马歇尔稳定度($MS$);流值($FL$)是达到最大破坏荷重时试件的垂直变形(以0.1m计),在有$X-Y$记录仪的马歇尔稳定度仪上,可自动绘出荷载($P$)与变形($F$)的关系曲线。

马歇尔模数为稳定度除以流值的商,即:

$$T = \frac{MS \times 10}{FL} \qquad (5\text{-}3)$$

式中　$T$——马歇尔模数(kN/mm);

　　　$MS$——马歇尔稳定度(kN);

　　　$FL$——流值,0.1(mm)。

(2)车辙试验

车辙试验的方法是采用标准成型方法,首先制成300mm×300mm×50mm的沥青混合料试件,在60℃的温度条件下,以一定荷载的轮子在同一轨迹上作一定时间的反复行走,形成一定的车辙深度,然后计算试件变形1mm所需试验车轮行走次数,即为动稳定度,计算公式如下:

$$DS = \frac{N \times (t_2 - t_1)}{d_2 - d_1} \times c_1 \times c_2 \qquad (5\text{-}4)$$

式中　$DS$——沥青混合料动稳定度(次/mm);

　　　$N$——试验轮往返碾压速度,通常为42次/min;

　　　$c_1$——试验机修正系数;

　　　$c_2$——试件系数;

　　　$d_2$——试验时间$t_2$时的车辙深度(mm);

　　　$d_1$——试验时间$t_1$时的车辙深度(mm)。

对高速公路的表面层、中面层沥青混合料,其动稳定度应大于1200次/mm;对一级公路的表面层、中面层沥青混合料应大于800次/mm。如果采用改性沥青,其动稳定度标准与改性沥青的类型有关。

**2. 水稳定性**

沥青混合料的水稳定性通过浸水马歇尔试验和冻融劈裂试验进行检验。

(1)浸水马歇尔试验

浸水马歇尔稳定试验供检验沥青混合料受水损害时抵抗剥落的能力时使用,通过测试其水稳定性检验配合比设计的可行性。

浸水马歇尔试验方法与标准马歇尔试验方法的不同之处在于,试件在已达规定温度的恒温水槽中的保温时间为48h,其余均与标准马歇尔试验方法相同。

残留稳定度按下式计算:

$$MS_0 = \frac{MS_1}{MS} \times 100 \tag{5-5}$$

式中 $MS_0$——试件的浸水残留稳定度(%);

$MS_1$——试件浸水 48h 后的稳定度(kN)。

(2)冻融劈裂试验

冻融劈裂试验用于在规定条件下对沥青混合料进行冻融循环,测定混合料试件在受到水损害前后劈裂破坏的强度比,以评价沥青混合料水稳定性。标准试验温度为 25℃,加载速率为 50mm/min。

冻融劈裂试验的残留强度比按下式计算:

$$TSR = \frac{R_{T2}}{R_{T1}} \times 100 \tag{5-6}$$

式中 $TSR$——冻融劈裂试验的残留强度比(%);

$R_{T2}$——冻融循环后第二组试件的劈裂抗拉强度(MPa);

$R_{T1}$——未冻融循环的第一组试件的劈裂抗拉强度(MPa)。

**3. 低温抗裂性**

除了应具备高温的稳定性外,沥青混合料还要具有良好的低温抗裂性,以保证路面在冬季低温时不产生裂缝。

《公路沥青路面施工技术规范》(JTG F40—2004)中要求对密级配沥青混合料在温度 −10℃、加载速率 50mm/min 的条件下进行弯曲试验,测定其破坏强度、破坏应变、破坏劲度模量,并根据应力应变曲线的形状,综合评价沥青混合料的低温抗裂性能。

**4. 耐久性**

在沥青混合料的拌和、摊铺、碾压以及以后沥青路面的使用过程中,都存在老化问题。老化过程一般也分为两个阶段:即施工过程中的短期老化和路面使用过程中的长期老化。沥青混合料在拌和过程中的老化程度与拌和温度、沥青贮存温度、沥青的贮存时间等因素有关。沥青混合料在使用过程中的长期老化与沥青材料、沥青在混合料中所处的状态有关,如混合料空隙率大小,沥青用量及光、氧等自然气候条件等。

目前我国规范采用空隙率、饱和度和残留稳定度等指标来表征沥青混合料的耐久性。

**5. 抗滑性**

《沥青路面施工及验收规范》(GB 50092—1996)对抗滑层骨料提出了磨光值、道瑞磨耗值和冲击值三项指标,扫描二维码了解更多内容。

沥青混合料路面的抗滑性与矿质骨料的微表面性质、混合料的级配组成及沥青用量、含蜡量等因素有关。为保证长期高速行车的安全,配料时要特别注意粗骨料的耐磨光性。

沥青用量对抗滑性的影响非常敏感,沥青用量超过最佳用量的 0.5% 即可使抗滑系数明显降低。含蜡量对沥青混合料抗滑性有明显的影响,重交通量道路用石油沥青的含蜡量应不大于 3%。

**6. 施工和易性**

要保证室内配料在现场施工条件下顺利的实现,沥青混合料除了应具备前述的技术要求外,还应具备适宜的施工和易性。影响沥青混合料施工和易性的因素很多,诸如当地气温、施工条件及混合料性质等。

单纯从混合料材料性质而言,是影响沥青混合料施工和易性的因素,首先是混合料的级配情况。如粗细骨料的颗粒大小相距过大、缺乏中间尺寸,混合料容易分层(粗粒集中表面,细粒集中底部);如细骨料太少,沥青层就不容易均匀地分布在粗颗粒表面;细骨料过多则使拌和困难。此外当沥青用量过少,或矿粉用量过多时,混合料容易产生疏松不易压实。反之,如沥青用量过多,或矿粉质量不好,则容易使混合料黏结成团块,不易摊铺。

### 四、沥青混合料的配合比设计

沥青混合料的配合比设计目的是确定沥青混合料的材料品种及配比、矿料级配、最佳沥青用量。设计包括目标配合比设计、生产配合比设计及生产配合比验证三个阶段。我国相关规范明确规定,采用马歇尔试验配合比设计方法,配合比设计流程图如图5-6所示。如采用其他方法设计沥青混合料时,应按规范规定进行马歇尔试验及各项配合比设计检验。

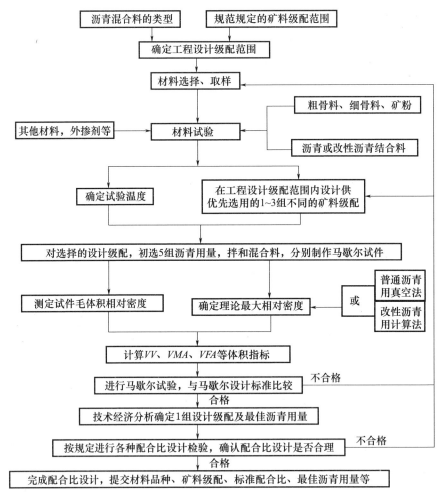

图5-6 沥青混合料配合比设计流程图

**1. 目标配合比设计**

目标配合比设计在实验室进行,分矿质混合料组成设计和沥青最佳用量确定两部分。

1)矿质混合料组合设计

矿质混合料组成设计的目的,是让各种矿料以最佳比例相混合,从而在加入沥青后,沥青混凝土既密实,又有一定空隙适应夏季沥青膨胀。

(1)选择沥青混合料种类

沥青混合料适用于各种等级公路的沥青路面。其种类应考虑骨料公称最大粒径、矿料级配、空隙率等因素选择,见表5-6。

表5-6 沥青混合料类型

| 结构层次 | 高速公路、一级公路、城市快速路、主干路 | | 其他等级公路 | | 一般城市道路及其他道路工程 | |
|---|---|---|---|---|---|---|
| | 三层式沥青混凝土路面 | 两层式沥青混凝土路面 | 沥青混凝土路面 | 沥青碎石路面 | 沥青混凝土路面 | 沥青碎石路面 |
| 上层面 | AC-13<br>AC-16<br>AC-20 | AC-13<br>AC-16 | AC-13<br>AC-16 | AC-13 | AC-5<br>AC-10<br>AC-13 | AM-5<br>AM-10 |
| 中层面 | AC-20<br>AC-25 | — | — | — | — | — |
| 下层面 | AC-25<br>AC-30 | AC-20<br>AC-30 | AC-20<br>AC-25<br>AC-35<br>AM-25<br>AM-30 | AM-25<br>AM-30 | AC-20<br>AC-25<br>AM-25<br>AM-30 | AC-25<br>AM-30<br>AM-40 |

(2)确定工程设计级配范围

沥青路面工程的混合料设计级配范围由工程设计文件或招标文件规定,不同沥青混合料的设计级配宜在规范规定的级配范围内,见表5-7。根据公路等级、工程性质、气候条件、交通条件、材料品种,通过对条件大体相当的工程使用情况进行调查研究后调整确定,必要时允许超出规范级配范围。经确定的工程设计级配范围是配合比设计的依据,不得随意变更。

2)矿质混合料配合比计算

① 矿料选择与准备。配合比设计的各种矿料必须按《公路工程集料试验规程》(JTG E42—2005)规定的方法,从工程实际使用的材料中取代表性样品。各种材料必须符合气候和交通条件的需要,其质量应符合规范规定的技术要求。记录各种材料的筛分试验结果,以供配合比设计计算使用。扫描二维码了解更多内容。

② 矿料配合比设计。高速公路和一级公路沥青路面矿料配合比设计宜借助电子计算机的电子表格用试配法进行。其他等级公路沥青路面也可参照进行。

③ 对高速公路和一级公路,宜在工程设计级配范围内计算1～3组粗细不同的配比,绘制设计级配曲线,分别位于工程设计级配范围的上方、中值及下方。设计合成级配不得有太多的锯齿形交错,且在0.3～0.6mm范围内不出现"驼峰"。当反复调整不能满意时,宜更换材料设计。

表 5-7 沥青混合料骨料级配范围

| 级配类型 | | 通过下列筛孔(mm)的质量分数(%) | | | | | | | | | | | | | | | | 供参考的沥青用量(%) |
|---|---|---|---|---|---|---|---|---|---|---|---|---|---|---|---|---|---|---|
| | | 53 | 37.5 | 31.5 | 26.5 | 19 | 16 | 13.2 | 9.5 | 4.75 | 2.36 | 1.18 | 0.6 | 0.3 | 0.15 | 0.075 | |
| 沥青混凝土 | 粗粒 AC-30 Ⅰ | | 100 | 90~100 | 79~92 | 66~82 | 59~77 | 52~72 | 49~63 | 32~52 | 25~42 | 18~32 | 13~25 | 8~18 | 5~13 | 3~7 | 4.0~6.0 |
| | Ⅱ | | 100 | 90~100 | 66~85 | 52~70 | 45~85 | 38~58 | 30~50 | 18~38 | 12~28 | 8~20 | 4~14 | 3~11 | 2~7 | 1~5 | 3.5~5.0 |
| | AC-25 Ⅰ | | | 100 | 95~100 | 75~90 | 62~80 | 53~73 | 43~63 | 32~52 | 25~42 | 18~32 | 13~25 | 8~18 | 5~13 | 3~7 | 4.0~6.0 |
| | Ⅱ | | | 100 | 90~100 | 65~85 | 52~70 | 42~62 | 32~52 | 20~40 | 13~30 | 9~23 | 6~16 | 4~12 | 3~8 | 2~5 | 3.5~5.0 |
| | 中粒 AC-20 Ⅰ | | | | 100 | 95~100 | 75~90 | 62~80 | 52~72 | 38~58 | 28~46 | 20~34 | 15~27 | 10~20 | 6~14 | 4~8 | 4.0~6.0 |
| | Ⅱ | | | | 100 | 90~100 | 65~85 | 52~70 | 40~60 | 26~45 | 16~33 | 11~25 | 7~18 | 4~13 | 3~9 | 2~5 | 4.0~5.5 |
| | AC-16 Ⅰ | | | | | 100 | 95~100 | 75~90 | 58~78 | 42~63 | 32~50 | 22~37 | 16~28 | 11~21 | 7~15 | 4~8 | 4.0~6.0 |
| | Ⅱ | | | | | 100 | 90~100 | 65~85 | 50~70 | 30~50 | 18~35 | 12~26 | 7~19 | 4~14 | 3~9 | 2~5 | 4.0~5.5 |
| | 细粒 AC-13 Ⅰ | | | | | | 100 | 95~100 | 70~88 | 48~68 | 36~53 | 24~41 | 18~30 | 12~22 | 8~16 | 4~8 | 5.0~7.0 |
| | Ⅱ | | | | | | 100 | 90~100 | 60~80 | 34~52 | 22~38 | 14~28 | 8~20 | 5~14 | 3~10 | 2~8 | 4.5~6.5 |
| | 砂粒 AC-10 Ⅰ | | | | | | | 100 | 95~100 | 55~75 | 35~58 | 26~43 | 17~33 | 10~24 | 6~16 | 4~9 | 5.0~7.0 |
| | Ⅱ | | | | | | | 100 | 90~100 | 40~60 | 24~42 | 15~30 | 9~22 | 6~15 | 4~10 | 2~6 | 4.5~6.5 |
| | AC-5 Ⅰ | | | | | | | | 100 | 95~100 | 55~75 | 35~55 | 20~40 | 12~28 | 7~18 | 5~10 | 6.0~8.0 |
| 沥青碎石 | 粗粒 AM-40 | 100 | 90~100 | 50~80 | 40~65 | 30~54 | 25~30 | 20~45 | 13~28 | 5~25 | 2~15 | 0~10 | 0~8 | 0~6 | 0~5 | 0~4 | 2.5~3.5 |
| | AM-30 | | 100 | 90~100 | 50~80 | 38~65 | 32~57 | 25~50 | 17~42 | 8~30 | 2~20 | 0~15 | 0~10 | 0~8 | 0~5 | 0~4 | 3.0~4.0 |
| | AM-25 | | | 100 | 90~100 | 50~80 | 43~73 | 38~65 | 25~55 | 10~32 | 2~20 | 0~14 | 0~10 | 0~8 | 0~6 | 0~5 | 3.0~4.5 |
| | 中粒 AM-20 | | | | 100 | 90~100 | 60~85 | 50~75 | 40~65 | 15~40 | 5~22 | 2~16 | 1~12 | 0~10 | 0~8 | 0~5 | 3.0~4.5 |
| | AM-16 | | | | | 100 | 90~100 | 60~85 | 45~68 | 18~42 | 6~25 | 3~18 | 2~14 | 0~10 | 0~8 | 0~5 | 3.0~4.5 |
| | 细粒 AM-13 | | | | | | 100 | 90~100 | 50~80 | 20~40 | 8~20 | 4~20 | 2~16 | 0~10 | 0~8 | 0~6 | 3.0~4.5 |
| | AM-10 | | | | | | | 100 | 85~100 | 35~65 | 10~35 | 5~22 | 2~16 | 0~12 | 0~9 | 0~6 | 3.0~4.5 |
| 抗滑表层 | AK-13A | | | | | | | 100 | 60~80 | 30~53 | 20~40 | 15~30 | 10~23 | 7~18 | 5~12 | 4~8 | 4.0~5.5 |
| | AK-13B | | | | | | | 85~100 | 50~70 | 18~40 | 10~30 | 8~22 | 5~15 | 3~12 | 3~9 | 2~6 | 4.0~5.5 |
| | AK-16 | | | | | | 100 | 90~100 | 45~70 | 85~45 | 15~35 | 10~25 | 8~18 | 6~13 | 4·10 | 3~7 | 4.0~5.5 |

④ 确定最佳沥青用量

现行规范采用马歇尔试验确定沥青混合料的最佳沥青用量,以 OAC 表示。沥青掺量可以采用油石比或沥青用量两种表达方式。油石比是指沥青占矿料总量的百分比;沥青用量是指沥青占沥青混合料总量的百分比。

确定最佳沥青用量,首先要根据当地的实践经验选择适宜的沥青用量,分别制作几组级配的马歇尔试件,测定试件的矿料间隙率(VMA),初选一组满足或接近设计要求的级配作为设计级配。然后再进行马歇尔试验,确定最佳沥青用量。

以预估的油石比为中值,按一定间隔(对密级配沥青混合料通常为 0.5%,对沥青碎石混合料可适当缩小间隔为 0.3~0.4%),取 5 个或 5 个以上不同的油石比分别成型马歇尔试件。通过试验测定不同油石比相关的各项马歇尔试验指标:

VV——试件的空隙率(%);

VMA——试件的矿料间隙率(%);

VFA——试件的有效沥青饱和度(即有效沥青含量占 VMA 的体积比例,%);

$\gamma_s$——试件的毛体积相对密度;

FL——流值(mm);

MS——稳定度(kN)。

以油石比或沥青用量为横坐标,以马歇尔试验的各项指标为纵坐标,将试验结果绘图,连成圆滑的曲线,如图 5-7 所示。确定符合规范规定的沥青混合料技术标准的沥青用量范围 $OAC_{min} \sim OAC_{max}$[图 5-7(f)]。选择的沥青用量范围必须涵盖设计空隙率的全部范围,并尽可能涵盖沥青饱和度的要求范围,并使密度及稳定度曲线出现峰值。

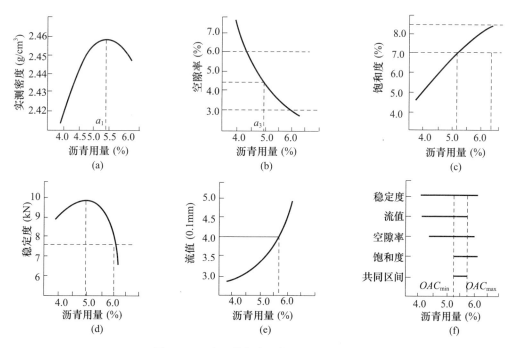

图 5-7 沥青用量与各项指标的关系曲线

(1)根据试验曲线的走势,按下列方法确定沥青混合料的最佳沥青用量$OAC_1$。

求取相应于密度最大值、稳定度最大值、目标空隙率(或中值)、沥青饱和度范围的中值的沥青用量$a_1$、$a_2$、$a_3$、$a_4$。按式(5-7)取平均值作为$OAC_1$。

$$OAC_1 = (a_1 + a_2 + a_3 + a_4)/4 \tag{5-7}$$

如果在所选择的沥青用量范围未能涵盖沥青饱和度的要求范围,按式(5-8)取3者的平均值作为$OAC_1$。

$$OAC_1 = (a_1 + a_2 + a_3)/3 \tag{5-8}$$

对所选择试验的沥青用量范围,密度或稳定度没有出现峰值(最大值经常在曲线的两端)时,可直接以目标空隙率所对应的沥青用量$a_3$作为$OAC_1$,但$OAC_1$必须介于$OAC_{min}$ ~ $OAC_{max}$的范围内。否则应重新进行配合比设计。

(2)以各项指标均符合技术标准的沥青用量范围$OAC_{min}$ ~ $OAC_{max}$的中值作为$OAC_2$,见式(5-9)。

$$OAC_2 = (OAC_{min} + OAC_{max})/2 \tag{5-9}$$

(3)通常情况下取$OAC_1$及$OAC_2$的中值作为计算的最佳沥青用量$OAC$,见式(5-10)。

$$OAC = (OAC_1 + OAC_2)/2 \tag{5-10}$$

检查相应于$OAC$的各项指标是否均符合马歇尔试验技术标准。另外还要根据实践经验和公路等级、气候条件、交通情况,调整确定最佳沥青用量$OAC$。

对于沥青混合料,需在配合比设计的基础上按要求进行各种使用性能的检验,不符合要求的沥青混合料,必须更换材料或重新进行配合比设计。检验的性能主要包括:高温稳定性检验、水稳定性检验、低温抗裂性能检验、渗水系数检验、钢渣活性检验等。

**2. 生产配合比设计**

在目标配合比确定之后,应进行生产配合比设计。因为在进行沥青混合料生产时,虽然所用的材料与目标配合比设计时相同,但是实际情况较实验室还是有一定差别;另外,在生产时,砂、石料经过加热、筛分,与实验室的冷料筛分也可能存在差异。对间歇式拌和机,应从两次筛分后进入各热料仓的材料中取样、筛分,确定各热料仓的材料比例,使所组成的级配与目标配合比设计的级配一致或基本接近,供拌和机控制室使用。同时,应反复调整冷料仓进料比例,使供料均衡,并取目标配合比设计的最佳沥青用量,最佳沥青用量±0.3%等3个沥青用量进行马歇尔试验,确定生产配合比的最佳沥青用量,供试拌试铺使用。

**3. 生产配合比验证**

生产配合比确定后,还需要铺试验路段,并用拌和的沥青混合料进行马歇尔试验,同时钻芯、取样,检验生产配合比。如符合标准要求,则整个配合比设计完成,由此确定生产用的标准配合比;否则还需要进行调整。

标准配合比即为生产的控制依据和质量检验的标准。标准配合比的矿料合成级配中,0.075mm、2.36mm、4.75mm三档筛孔的通过率应接近要求级配的中值。

## 复习思考题

1. 试述石油沥青三大组分及其特征。
2. 石油沥青的主要技术性质是什么？各用什么指标表示？
3. 何谓沥青混合料？路用沥青混合料是如何分类的？
4. 试述马歇尔试验的试验条件、试件尺寸和评价指标。
5. 沥青混凝土配合比设计中，最佳沥青用量是如何确定的？

# 第六章 建筑钢材

金属材料一般分为黑色金属和有色金属两大类。黑色金属如铁、钢和合金钢等,其主要成分是铁元素。有色金属是指以其他金属元素为主要成分的金属,如铝、铜、锌、铅等金属及其合金。土木工程中常用的金属材料主要有钢材、铝合金等。

钢材具有较高的抗拉强度、形成板材型材及线材的能力、可焊性和容易与其他金属焊接等特性,随着高层和大跨度结构迅速发展,钢材在土木工程中的应用也越来越广泛。

## 第一节 钢材的化学成分及其对钢材性能的影响

钢材中主要的化学成分是铁(Fe),此外还含有少量的碳(C)、硅(Si)、锰(Mn)、磷(P)、硫(S)、氧(O)、氮(N)、钛(Ti)、钒(V)、铌(Ni)等元素,这些元素虽含量较少,但对钢材性能的影响很大。这些成分可分为两类:一类能改善优化钢材的性能,称为合金元素,主要有 Si、Mn、Ti、V、Ni 等;另一类能劣化钢材的性能,属钢材的杂质,主要有氧、硫、氮、磷等。

### 一、碳

碳是决定钢材性能的最重要元素。钢与生铁的主要成分都是铁和碳,区别在于含碳量的大小。含碳量小于2%的铁碳合金称为钢,含碳量大于2%的铁碳合金称为生铁。

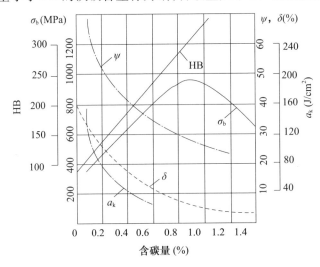

图 6-1 含碳量对碳素钢性能的影响

含碳量对钢材的强度、塑性、韧性等力学性能及工艺性能的影响如图 6-1 所示。从图中可以看出,钢材的强度、硬度随着含碳量的增加而提高。塑性、韧性和冷弯性则随着含碳量的增大而下降。随含碳量的增加,钢材的工艺性能也随之下降。当含碳量增至 0.8% 左右时,强度最大,但当含碳量超过 0.8% 时,强度反而下降,这是钢材变脆所致。钢中含碳量增加,还会使钢的焊接性能变差(含碳量大于 0.3% 的钢可焊性显著下降),冷脆性和时效敏感性增大,并使钢耐大气锈蚀能力下降。一般工程用碳素钢均为低碳钢,即含碳小于 0.25%,工程用低合金钢含碳小于 0.52%。

## 二、硅

硅在钢中是有益元素,炼钢时起脱氧作用。硅是我国钢筋用钢的主加合金元素,它的作用主要是提高钢的强度、疲劳极限、耐腐蚀性及抗氧化性,对塑性和韧性影响不大。但由于硅在钢中的含量很低,因此这一效果并不明显。若作为合金元素加入钢中,使含量提高到 1.0%~1.2% 时,钢材的抗拉强度可提高 15%~20%,但塑性和韧性明显下降,焊接性能变差,并增加钢材的冷脆性。通常碳素钢中含硅量小于 0.3%,低合金钢含硅量小于 1.8%。

## 三、锰

锰是有益元素,是我国低合金钢的主加合金元素,炼钢时能起脱氧去硫作用,可消减硫所引起的热脆性,改善钢材的热加工性能,同时能提高钢材的强度和硬度。锰含量一般在 1.0%~2.0% 范围内,当含锰小于 1.0% 时,对钢的塑性和韧性影响不大;当含锰量达 11%~14% 时称为高锰钢,具有较高的耐磨性。

## 四、磷

磷是钢中的有害元素之一,在常温下磷含量增加,可提高钢材的强度、硬度,但塑性和韧性显著下降。特别是温度越低,对塑性和韧性的影响越大,从而显著增加钢材的冷脆性。

磷也使钢材可焊性显著降低,但磷可提高钢的耐磨性和耐蚀性,故在低合金钢中可配合其他元素如铜作合金元素使用。建筑用钢一般要求含磷小于 0.045%。

## 五、硫

硫是有害元素之一,呈非金属硫化物(FeS)存在于钢中,会加大钢材的热脆性,降低钢材的各种机械性能,使钢的可焊性、冲击韧性、耐疲劳性和抗腐蚀性等均降低。建筑钢材要求硫含量应小于 0.045%。

## 六、氮

氮对钢材性质的影响与碳、磷相似,使钢材强度提高,但塑性特别是韧性显著下降。

氮还会加剧钢的时效敏感性和冷脆性,使可焊性变差。在钢中,氮若与铝或钛元素反应,生成的化合物能使晶粒细化,改善钢的性能。故在有铝、钒等元素的配合下,氮可作为低合金钢的合金元素使用。钢中氮含量一般小于0.008%。

### 七、氧

氧是钢中的有害杂质。含氧量增加,会使钢的力学性能降低,塑性和韧性降低、促进时效作用,还能使热脆性增加,焊接性能较差。通常要求钢中含氧量应小于0.03%。

## 第二节 钢材的技术性质

钢材作为土木工程中主要的受力结构材料,主要是承受拉力、压力、弯曲、冲击等外力作用,因此要求其要具有良好的力学性能和易加工性能。

### 一、抗拉性能

抗拉性能是钢材最重要的性能,在设计和施工中广泛使用。通过拉伸试验,可以测得屈服强度、抗拉强度和断后伸长率,这些是钢材的重要技术性能指标。钢材的抗拉性能和低碳钢的抗拉性能可用受拉时的应力-应变图来阐明(图6-2)。图中曲线明显地可分为弹性阶段($O \rightarrow A$)、屈服阶段($A \rightarrow B$)、强化阶段($B \rightarrow C$)和颈缩阶段($C \rightarrow D$)。

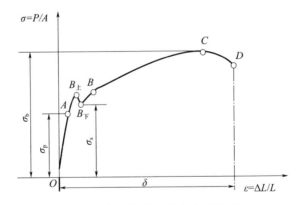

图6-2 低碳钢受拉的应力-应变图

**1. 弹性阶段($OA$ 段)**

在 $OA$ 范围内,随着荷载的增加,应力与应变成正比。如卸去荷载,则恢复原状。

这种性质称为弹性。$OA$ 是一直线,在此范围内的变形,称为弹性变形。$A$ 点所对应的应力称为弹性极限,用 $\sigma_p$ 表示。在这一范围内,应力与应变的比值为一常量,称为弹性模量,用 $E$ 表示,即 $E = \sigma/\varepsilon$。弹性模量反映了钢材的刚度,是钢材在受力条件下计算结构变形的重要指标。土木工程中常用碳素结构钢 Q235 的弹性模量 $E = (2.0 \sim 2.1) \times 10^5 \text{MPa}$,弹性极限 $\sigma_p = 180 \sim 200 \text{MPa}$。

## 2. 屈服阶段(AB 段)

当应力超过弹性极限后,在 AB 曲线范围内,应力与应变非正比例关系变化。应力超过 A 点后,即开始产生塑性变形。应力到达 $B_下$ 之后,变形急剧增加,应力则在不大的范围内波动,直到 B 点,这一阶段称为屈服阶段。在屈服阶段中,外力不增大,而变形继续增加。$B_上$ 是屈服强度上限,$B_下$ 是屈服强度下限,也可称为屈服极限或屈服强度,以 $\sigma_s$ 表示。当应力到达 $B_上$ 上点时,钢材抵外力能力下降,发生"屈服"现象。$\sigma_s$ 是屈服阶段应力波动的最低值,它表示钢材在工作状态允许达到的应力值,即在 $\sigma_s$ 之前,钢材不会发生较大的塑性变形。故在设计中一般以下屈服强度作为强度取值的依据。常用的碳素结构钢 Q235 的 $\sigma_s$ 应不小于 235MPa。

对于在外力作用下屈服现象不明显的硬钢,则规定残余变形为 0.2% 原标距长度时的应力作为该钢材的屈服强度,用 $\sigma_{0.2}$ 表示(图 6-3)。

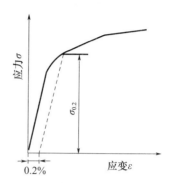

图 6-3 硬钢受拉的应力-应变图

## 3. 强化阶段(BC 段)

在钢材屈服到一定程度以后,由于内部晶格扭曲、晶粒破碎等原因,阻止了塑性变形的进一步发展。钢材抵抗外力的能力重新提高,在应力-应变图(图6-2)上,曲线从 B 点开始上升至最高点 C。这一过程称为强化阶段。这一阶段变形发展速度比较快,随着应力的提高而增加。对应于最高点 C 的应力,称为抗拉强度,用 $\sigma_b$ 表示。抗拉强度不能直接利用,但屈服强度和抗拉强度的比值(即屈强比 $\sigma_s/\sigma_b$)却能反映钢材的安全可靠程度和利用率。屈强比越小,表明材料的安全性和可靠性越高,材料不易发生危险的脆性断裂。如果屈强比太小,则利用率低,造成钢材浪费。常用的碳素结构钢 Q235 的 $\sigma_b$ 应不小于 375MPa,屈强比在 0.58~0.63 之间。

## 4. 颈缩阶段(CD 段)

当钢材受拉时强化达到最高点(C 点)后,试件的变形开始集中于较薄弱区段内,使此段的截面显著缩小,产生"颈缩现象"(图6-4)。由于试件截面积急剧缩小,塑性变迅速增加,拉力也就随着下降,最后发生断裂。

将拉断后的试件于断裂处对接在一起(图6-5),测得其断后标距 $l_1$。标距的伸长($\Delta l$)占原始标距($l_0$)的百分率称为伸长率($\delta$)。即

$$\delta = \frac{l_1 - l_0}{l_0} \times 100\% \tag{6-1}$$

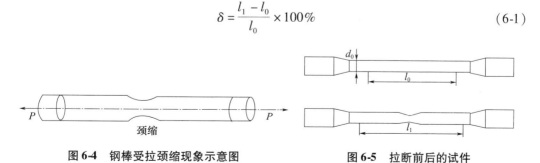

图 6-4 钢棒受拉颈缩现象示意图　　图 6-5 拉断前后的试件

塑性变形在试件标距内的分布是不均匀的,颈缩处的变形最大,离颈缩部位越远其变越小。所以原标距与直径之比越小,则颈缩处伸长值在整个伸长值中的比重越大,计算出来的伸长率就会大些。通常钢材拉伸试件取 $l_0 = 5d$ 或 $l_0 = 10d$,其伸长率分别以 $\delta_5$ 和 $\delta_{10}$ 表示。对于同一钢材 $\delta_5$ 大于 $\delta_{10}$。

伸长率是衡量钢材塑性的重要技术指标,伸长率越大,表明钢材的塑性越好。尽管结构是在弹性范围内使用,但其应力集中处的应力可能超过屈服点。良好的塑性变形能力,可使应力重分布,从而避免结构过早破坏。常用的碳素结构钢的伸长率一般为 20%～30%。

### 二、冷弯性能

冷弯性能是指钢材在常温下承受弯曲变形的能力,是土木工程用钢的重要工艺性质。

钢材的冷弯性能是以试验时的弯曲角度($\alpha$)和弯心直径($d$)为指标表示。钢材冷弯试验是通过直径(或厚度)为 $a$ 的试件,采用标准规定的弯心内径 $d$($d = na$,$n$ 为整数),弯曲到规定的角度(180°或 90°)时,检查弯曲处若无裂纹、断裂及起层等现象,则认为冷弯性能合格。钢材冷弯时的弯曲角度越大,弯心直径越小,则表示其冷弯性能越好。图 6-6 所示为冷弯试验及弯曲角度相同、不同 $d/a$ 时的弯曲情况。

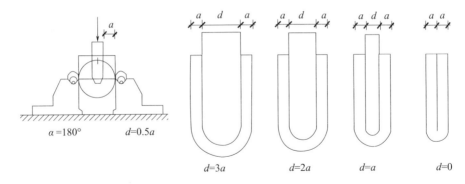

**图 6-6 冷弯试验 $\alpha = 180°$ 时不同 $d/a$ 的弯曲图**

钢材的冷弯性能和伸长率均是塑性变形能力的反映。伸长率反映的是钢材在均匀变形条件下的塑性变形能力。冷弯性能则是钢材在局部变形条件下的塑性变形能力。冷弯性能可揭示钢材内部结构是否均匀、是否存在内应力和夹杂物等缺陷。在土木工程中,还经常采用冷弯试验来检验钢材焊接接头的焊接质量。

### 三、高温性能

钢材是一种耐热而不耐火的材料,当温度小于 200℃,钢材的力学性能基本无变化,但当温度超过 300℃后,其弹性模量、屈服强度及极限强度则显著下降,变形急剧增大,当温度超过 400℃时,强度和弹性模量都急剧下降;温度达到 600℃时,弹性模量、屈服强度和极限强度均接近于零,钢材已失去承载能力。因此,根据钢结构运行环境,在必要时要对其进行防火维护。

# 第三节 钢材的冷加工、时效

## 一、冷加工

冷加工是指将钢材在常温下进行的加工,土木工程所用钢材常见的冷加工方式有:冷拉、冷拔、冷轧,使之产生一定的塑性变形,强度明显提高,塑性和韧性有所降低,这个过程称为钢材的冷加工强化或"三冷处理"。钢筋经冷加工后,屈服强度提高,塑性、韧性和弹性模量则降低。

钢筋经冷拉后性能变化的规律,可从低碳钢试样的拉伸曲线(图6-7)上看到,在图6-7中,$OBCD$为未经冷拉时效试件的变形曲线。将试件拉至超过屈服点的任意一点$K$,然后卸去荷载,则试件产生变形量$OO'$,且曲线沿$KO'$下降,$KO'$大致与$OB$平行。若立即重新拉伸,则可发现屈服点提高到$K$点,以后的发展曲线与$KCD$相似。此现象表明,当钢材受到外力作用时,产生塑性变形,随着变形的增加,金属本身对变形的抗力也随之增加,这可用"晶格的滑移"机理解释:钢

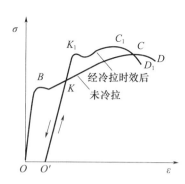

图6-7 钢筋经冷拉及时效后应力-应变图

材在弹性变形阶段,晶体原子排列的位置没有改变,仅在受力方向,原子间距离增大或缩短(拉伸或压缩),直到塑性变形阶段,晶体才沿结合力最差的结晶界面产生滑移。滑移以后的晶体破碎成小晶粒,产生弯扭,不易再滑移变形,提高了钢材抵抗变形的能力。所以就需要更大的外力才能使其继续产生塑性变形,这种现象称作"冷作硬化"或"加工硬化"。

## 二、时效

钢材随时间的延长,强度、硬度提高,而塑性、韧性下降的现象称为时效。钢材在自然条件下的时效是非常缓慢的,若经过冷加工或在使用中经常受到振动、冲击荷载作用时,时效将迅速发展。

如图6-7所示,如果将试样拉到$K$点时,去除荷载后不立即加荷,而经过时效处理,即常温下存放15~20d,或加热到100~200℃,并保持一定时间,再拉伸时则可发现试样的屈服点提高到$K$点,且曲线沿$K_1C_1D_1$发展,这个过程称时效处理,前者称为自然时效,用加热的方法则称为人工时效。冷加工以后的钢材产生时效作用的原因,目前认为是由于溶于铁素体的碳(过饱和)随着时间的增长,慢慢地从铁素体中析出形成渗碳体,分布在晶体的滑移面上阻止滑移,抵抗外力的能力增强,从而产生强化作用。

时效处理方法有两种:在常温下存放15~20d,称为自然时效,适合用于低强度钢筋;加热至100~200℃后保持一定时间(2~3h),称人工时效,适合于高强钢筋。

因时效导致钢材性能改变的程度称为时效敏感性。时效敏感性大的钢材,经时效处理后,其韧性、塑性改变较大。因此,承受振动、冲击荷载作用的重要结构(如吊车梁、桥梁等),应选用时效敏感性小的钢材。建筑用钢筋,常利用冷加工、时效作用来提高其强度,增加钢材的品种规格,节约钢材。

### 三、钢材冷加工和时效处理在工程中的应用

钢筋采用冷加工具有明显的经济效益。钢筋经冷拉后,屈服点可提高20%~25%,冷拔钢丝屈服点可提高40%~90%,由此即可适当减小钢筋混凝土结构设计截面,或减少混凝土中配筋数量,从而达到节约钢材的目的。钢筋冷拉还有利于简化施工工序,如盘条钢筋可省去开盘和调直工序,冷拉直条钢筋时,则可与矫直、除锈等工艺一并完成。

土木工程中对大量使用的钢筋,往往是冷加工和时效处理同时采用。实际施工时,应通过试验确定冷拉控制参数和时效处理方式。冷拉参数的控制,直接关系到冷拉效果和钢材质量。一般钢筋冷拉仅控制冷拉率即可,称为单控,而对用作预应力的钢筋,需采取双控,即既控制冷拉应力,又控制冷拉率。冷拉时当拉至控制应力,可以不达到控制冷拉率;反之,当达到控制的冷拉率而未达到控制应力时,钢筋应降级使用。

需要注意的是,冷加工后的钢材不适用于抗震结构,这是因为经过冷加工的钢材虽然抗拉强度获得提高,但是在加工过程中其弹性和塑性有所下降,易导致结构刚度过高,抗震性降低。因此,在抗震结构中应选用抗震钢筋,抗震钢筋在满足标准所规定普通钢筋所有性能指标外,还应满足:

(1)抗震钢筋的实测抗拉强度与实测屈服强度特征值之比不小于1.25;
(2)钢筋的实测屈服强度与标准规定的屈服强度特征值之比不大于1.30;
(3)钢筋的最大力总伸长不小于9%。

# 第四节　土木工程常用钢种

土木工程中常用的钢材可分为钢结构用钢和钢筋混凝土结构用钢两类,钢结构所用的各种型钢,钢筋混凝土结构所用的各种钢筋、钢丝、锚具等钢材,基本上都是由碳素结构钢和低合金结构钢等钢种经热轧或冷轧、冷拔及热处理等工艺加工而成的。

土木工程常用钢有普通碳素结构钢、优质碳素结构钢与低合金结构钢等。

### 一、碳素结构钢

《碳素结构钢》(GB/T 700—2006)规定,碳素结构钢分为Q195、Q215、Q235、Q275四种牌号。钢的牌号由代表钢材屈服强度的字母Q、屈服强度数值、质量等级符号和脱氧方法符号4个部分按顺序组成。扫码获得《碳素结构钢》(GB/T 700—2006)的详细内容。

符号含义:Q为钢材屈服强度"屈"字汉语拼音首位字母;A、B、C、D为分别为质量等级;F为沸腾钢"沸"字汉语拼音首位字母;Z为镇静钢"镇"字汉语拼音首位字母;TZ为特

殊镇静钢"特镇"两字汉语拼音首位字母。在牌号组成表示方法中"Z"与"TZ"的符号可以省略。

例如:Q235 AF 表示屈服强度为不小于 235MPa,质量等级为 A 级的沸腾碳素结构钢；Q235 A 则表示屈服强度为不小于 235MPa,质量等级为 A 级的镇静或特殊镇静碳素结构钢。

碳素结构钢的化学成分及力学性能应分别符合表 6-1～表 6-3 的规定。

表 6-1　碳素结构钢的牌号和化学成分(GB/T 700—2006)

| 牌号 | 同意数字代号[a] | 等级 | 厚度或直径(mm) | 脱氧方法 | 化学成分(质量分数,%),不大于 | | | | |
|---|---|---|---|---|---|---|---|---|---|
| | | | | | C | Si | Mn | P | S |
| Q195 | U11952 | — | — | F、Z | 0.12 | 0.30 | 0.50 | 0.035 | 0.040 |
| Q215 | U12152 | A | — | F、Z | 0.15 | 0.35 | 1.20 | 0.045 | 0.050 |
| | U12155 | B | | | | | | | 0.045 |
| Q235 | U12352 | A | — | F、Z | 0.22 | 0.35 | 1.40 | 0.045 | 0.050 |
| | U12355 | B | | | 0.20[b] | | | | 0.045 |
| | U12358 | C | | Z | 0.17 | | | 0.040 | 0.040 |
| | U12359 | D | | TZ | | | | 0.035 | 0.035 |
| Q275 | U12752 | A | — | F、Z | | 0.35 | 1.50 | 0.045 | 0.050 |
| | U12755 | B | ≤40 | Z | 0.20 | | | 0.045 | 0.045 |
| | | | >40 | | 0.22 | | | | |
| | U12758 | C | — | Z | 0.20 | | | 0.040 | 0.040 |
| | U12759 | D | | TZ | | | | 0.035 | 0.035 |

a 表中为镇静钢、特殊镇静钢牌号的统一数字,沸腾钢牌号的统一数字代号如下:
　Q195F—U11950;Q215AF—U12150,Q215BF—U12153;Q235AF—U12350,Q235BF—U12353;Q275AF—U12750。
b 经需方同意,Q235B 的碳含量可不大于 0.22%。

表 6-2　碳素结构钢的力学性能(GB/T 700—2006)

| 牌号 | 级 | 屈服强度[a]$R_{eH}$(N/mm²),不小于 | | | | | | 抗拉强度[b] $R_m$ (N/mm²) | 断后伸长率 $A$(%),不小于 | | | | | 冲击试验(V 型缺口) | |
|---|---|---|---|---|---|---|---|---|---|---|---|---|---|---|---|
| | | 厚度或直径(mm) | | | | | | | 厚度或直径(mm) | | | | | 温度(℃) | 冲击吸收功(纵向)(J)不小于 |
| | | 16 | 16~40 | >40~60 | >60~100 | >100~150 | >150~200 | | ≤40 | >40~60 | >60~100 | >100~150 | >150~200 | | |
| Q195 | — | 195 | 185 | — | — | — | — | 315~430 | 33 | — | — | — | — | — | — |
| Q215 | A | 15 | 205 | 195 | 185 | 175 | 165 | 335~450 | 31 | 30 | 29 | 27 | 26 | — | — |
| | B | | | | | | | | | | | | | +20 | 27 |
| Q235 | A | 35 | 225 | 215 | 215 | 195 | 185 | 370~500 | 26 | 25 | 24 | 22 | 21 | — | — |
| | B | | | | | | | | | | | | | +20 | 27[c] |
| | C | | | | | | | | | | | | | 0 | |
| | D | | | | | | | | | | | | | −20 | |

131

续表

| 牌号 | 级 | 屈服强度$^a$ $R_{eH}$(N/mm²),不小于 | | | | | 抗拉强度$^b$ $R_m$ (N/mm²) | 断后伸长率 A(%),不小于 | | | | | 冲击试验(V型缺口) | |
|---|---|---|---|---|---|---|---|---|---|---|---|---|---|---|
| | | 厚度或直径(mm) | | | | | | 厚度或直径(mm) | | | | | 温度(℃) | 冲击吸收功(纵向)(J)不小于 |
| | | ≤16 | >16~40 | >40~60 | >60~100 | >100~150 | >150~200 | | ≤40 | >40~60 | >60~100 | >100~150 | >150~200 | | |
| Q275 | A | 275 | 265 | 255 | 245 | 225 | 215 | 410~540 | 22 | 21 | 20 | 18 | 17 | — | — |
| | B | | | | | | | | | | | | | +20 | 27 |
| | C | | | | | | | | | | | | | 0 | |
| | D | | | | | | | | | | | | | -20 | |

a Q195屈服强度值仅供参考,不作交货条件。
b 厚度大于100mm的钢材,抗拉强度下限允许降低20N/mm²。宽带钢(包括剪切钢板)抗拉强度上限不作交货条件。
c 厚度小于25mm的Q235B级钢材,如供方能保证冲击吸收功值合格,经需方同意,可不作检验。

表6-3 碳素结构钢的冷弯性能(GB/T 700—2006)

| 牌号 | 试样方向 | 冷弯试验180° $B=2a^a$ | |
|---|---|---|---|
| | | 钢材厚度或直径$^b$(mm) | |
| | | ≤60 | >60~120 |
| | | 弯心直径 d | |
| Q195 | 纵 | 0 | — |
| | 横 | 0.5a | |
| Q215 | 纵 | 0.5a | 1.5a |
| | 横 | a | 2a |
| Q235 | 纵 | a | 2a |
| | 横 | 1.5a | 2.5a |
| Q275 | 纵 | 1.5a | 2.5a |
| | 横 | 2a | 2a |

a B为试样宽度,a为试样厚度(或直径)。
b 钢材厚度(或直径)大于100mm时,弯曲试验由双方协商确定。

碳素结构钢随着牌号的增大,其含碳量和含锰量增加,强度和硬度提高,而塑性和韧性降低,冷弯性能逐渐变差。

## 二、优质碳素结构钢

按《优质碳素结构钢》(GB/T 699—2015)的规定,根据其含锰量不同可分为:普通含锰量钢(含锰量小于0.8%,共20个钢号)和较高含锰量钢(含锰量0.7%~1.2%,共11个钢号)两组。扫码获得《优质碳素结构钢》(GB/T 699—2015)的详细内容。

优质碳素结构钢一般以热轧钢供应。其硫、磷等杂质含量比普通碳素钢少,其他缺陷限制也较严格,所以性能好,质量稳定。

优质碳素结构钢的钢号用两位数字表示,它表示钢中平均含碳量的万分数。如 45 号钢,表示钢中平均含碳量为 0.45%。数字后若有"锰"字或"Mn",则表示属较高含锰量钢,否则为普通含锰量钢。如 35Mn 钢,表示平均含碳量为 0.35%,含锰量为 0.7%~1.2%。

若是沸腾钢,还应在钢号后面加写"沸"(F)。

优质碳素结构钢成本较高,建筑上应用不多,仅用于重要结构的钢铸件及高强度螺栓等。如用 30、35、40 及 45 号钢做高强度螺栓,45 号钢还常用作预应力钢筋的锚具。65、70、75、80 号钢可用来生产预应力混凝土用的碳素钢丝、刻痕钢丝和钢绞线。

### 三、低合金高强度结构钢

在碳素结构钢的基础上加入总量小于 5% 的合金元素而形成的钢种。加入合金元素的目的是提高钢材强度和改善性能。常用的合金元素有硅、锰、钛、钒、铬、镍和铜等。根据国家标准《低合金高强度结构钢》(GB/T 1591—2018)规定,低合金高强度结构钢共有 8 个牌号:Q355、Q390、Q420、Q460、Q500、Q550、Q620、Q690。低合金高强度结构钢的牌号由代表屈服强度"屈"字的汉语拼音字母、规定的最小上屈服强度数值、交货状态代号、质量等级符号(分 B、C、D、E 四级)四个部分组成。例如,Q345D,其中:

Q 为钢的屈服强度的"屈"字汉语拼音的首位字;

345 为屈服强度数值(MPa);

D 为质量等级为 D 级。

当需方要求钢板具有厚度方向性能时,则在上述规定的牌号后加上代表厚度方向(Z 向)性能级别的符号,例如:Q345DZ15。

低合金钢不仅具有较高的强度,而且也具有较好的塑性、韧性、可焊性、耐锈蚀等特点。因此,它是综合性能较为理想的土木工程用钢。低合金高强度结构钢主要用于轧制各种型钢(角钢、槽钢、工字钢)、钢板、钢管及钢筋,广泛用于钢结构和钢筋混凝土结构中,尤其是大跨度结构、大型结构、重型结构、高层建筑、桥梁工程、承受动荷载和冲击荷载的结构物中更为适用。扫码获得《低合金高强度结构钢》(GB/T 1591—2018)的详细内容。

# 复习思考题

1. 钢的伸长率如何表示?冷弯性能如何评定?
2. 何谓钢材的屈强比?其大小对使用性能有何影响?
3. 钢的伸长率与试件标距长度有何关系?为什么?
4. 钢的脱氧程度对钢的性能有何影响?
5. 钢材中的有害化学元素主要有哪些?它们对钢材的性能各有何影响?
6. 钢材的冷加工对力学性能有何影响?
7. MnV 表示什么?Q295-B、Q345-E 是哪种结构钢?
8. 土木工程中主要使用哪些钢材?选用的原则有哪些?

# 第七章　其他工程材料

## 第一节　天然石材

天然岩石经不同程度的机械加工后(或不加工)，用于土木工程的材料统称为天然石材。天然石材具有抗压强度高，耐久性、耐磨性好等特点，有些岩石品种经加工后还可获得独特的装饰效果，因此在土木工程中得到广泛的应用。

石材是最古老的土木工程结构和装饰材料，世界上许多古建筑都由天然石材建造而成。如意大利的可里西姆大斗兽场、古埃及的金字塔，我国河北的赵州桥、福建泉州的洛阳桥、明清故宫宫殿基座、人民大会堂、人民英雄纪念碑等都使用了大量的天然石材。天然石材由于脆性大、抗拉强度低、自重大，石结构的抗震性能差，加之岩石的开采加工较困难、价格高等原因，其作为结构材料，已逐步被混凝土等材料所代替。近代，随着石材加工水平的提高，石材独特的装饰效果得到人们的青睐。因此，现代建筑装饰领域中，石材的应用前景十分广阔。另外，天然岩石经过自然风化或人工破碎可得到卵石、碎石、砂等材料，可大量用作混凝土的骨料，是混凝土的主要组成材料之一。有些岩石还是生产人造建筑材料的主要原料，如石灰石是生产硅酸盐水泥、石灰的原料，石英岩是生产陶瓷、玻璃的原料等。扫码了解赵州桥和人民英雄纪念碑的相关知识。

### 一、天然石材的形成和分类

岩石是不同的地质作用所形成的天然固态矿物的集合体。组成岩石的矿物称造岩矿物。不同的造岩矿物在不同的地质条件下，可形成不同性能的岩石。

天然岩石根据其形成的地质条件不同，可以分为岩浆岩、沉积岩、变质岩三大类。

**1. 岩浆岩**

（1）岩浆岩的形成和种类

岩浆岩又称火成岩，是地壳深处的熔融岩浆在地下或喷出地面后冷凝而形成的岩石。根据不同的形成条件，岩浆岩可分为深成岩、喷出岩和火山岩三种。

① 深成岩

深成岩是地壳深处的岩浆在受上部覆盖层压力的作用下经缓慢冷凝而形成的岩石。其结晶完整、晶粒粗大、结构致密，具有抗压强度高、吸水率小、表观密度大、抗冻性好等特点。土木工程常用的深成岩有花岗岩、正长岩、橄榄岩、闪长岩等。

② 喷出岩

喷出岩是岩浆喷出地表时，在压力降低和冷却较快的条件下形成的岩石。由于大部分岩浆来不及完全结晶,因而多呈细小结晶(隐晶质)或玻璃质(解晶质)结构。当喷出的岩浆形成较厚的岩层时,其岩石的结构与性质类似深成岩;当形成较薄的岩层时,由于冷却速度快及气压作用而易形成多孔结构的岩石,其性质近似于火山岩。土木工程常用的喷出岩有辉绿岩、玄武岩、安山岩等。

③ 火山岩

火山岩是火山爆发时,岩浆被喷到空中而急速冷却后形成的岩石。有多孔玻璃质结构的散粒状火山岩,如火山灰、火山渣、浮石等;也有因散粒状火山岩堆积而受到覆盖层压力作用并凝聚成大块的胶结火山岩,如火山凝灰岩等。

(2) 土木工程中常用的岩浆岩

① 花岗岩

花岗岩是岩浆岩中分布较广的一种岩石,主要由长石、石英和少量云母(或角闪石等)组成,具有致密的结晶结构和块状构造。其颜色一般为灰黑、灰白、浅黄、淡红等,大多数情况下同一种花岗岩会呈现多种不同颜色的组合。由于结构致密,其孔隙率和吸水率很小,表观密度大($2600 \sim 2800 kg/m^3$),抗压强度达 $120 \sim 250 MPa$,抗冻性好(F100~F200),耐风化和耐久性好,使用年限约为 75~200 年,对硫酸和硝酸的腐蚀具有较强的抵抗性。表面经打磨抛光加工后光泽美观,是优良的装饰材料。在土木工程中花岗岩常用作基础、闸坝、桥墩、台阶、路面、墙石和勒脚及纪念性建筑物等。但在高温作用下,由于花岗岩内的石英膨胀将引起石材破坏,因此,其耐火性不好。

② 玄武岩

玄武岩颜色较深,常呈玻璃质或隐晶质结构,有时也呈多孔状或斑形构造,硬度高,脆性大,抗风化能力强,表观密度为 $2900 \sim 3500 kg/m^3$,抗压强度为 $100 \sim 500 MPa$。

③ 火山灰

火山灰是颗粒粒径小于 5mm 的粉状火山岩,具有火山灰活性,即在常温和有水的情况下可与石灰[$CaO$ 或 $Ca(OH)_2$]反应生成具有水硬性胶凝能力的水化物。因此,可作水泥的混合材料及混凝土的掺合料。

**2. 沉积岩**

(1) 沉积岩的形成和种类

沉积岩又名水成岩,是由地表的各类岩石经自然界的风化、剥蚀、搬运、沉积、流水冲刷等作用后再沉积,经压实、相互胶结、重结晶等形成的岩石,主要存在于地表及不太深的地下。其特征是呈层状构造,外观多层理,表观密度小,孔隙率和吸水率较大,强度较低,耐久性较差。沉积岩是地壳表面分布最广的一种岩石,面积约占陆地表面积的75%。由于主要存在于地表,易开采、易加工,所以在土木工程中应用较广。根据沉积岩的生成条件,可分为机械沉积岩、化学沉积岩、生物有机沉积岩。

① 机械沉积岩

由自然风化逐渐破碎松散的岩石及砂等,经风、水流及冰川运动等的搬运,并经沉积等机械力的作用而重新压实或胶结而成的岩石,常见的有砂岩和页岩等。

② 化学沉积岩

由溶解于水中的矿物质经聚积、沉积、重结晶和化学反应等过程而形成的岩石,常见的有石灰岩、石膏、白云石等。

③ 生物有机沉积岩

由各种有机体的残骸沉积而成的岩石,如硅藻土等。

(2)土木工程中常用的沉积岩

石灰岩俗称灰石或青石,主要化学成分为 $CaCO_3$,主要矿物成分是方解石,但常含有白云石、菱镁矿、石英、蛋白石、含铁矿物及黏土等。石灰岩的化学成分、矿物组成、致密程度以及物理性质等差别很大,在强度和耐久性等方面不如花岗岩。其表观密度为 2600 ~ 2800kg/m³,抗压强度为 80 ~ 160MPa,吸水率为 2% ~ 10%。如果岩石中黏土含量不超过 3% ~ 4%,其耐水性和抗冻性较好。石灰岩来源广,硬度低,易劈裂,有一定的强度和耐久性,土木工程中可用作基础、墙身、台阶、路面等,碎石可作混凝土骨料。石灰岩也是生产水泥的主要原料。

**3. 变质岩**

(1)变质岩的形成与种类

变质岩是由地壳中原有的岩浆岩或沉积岩,在地层的压力或温度作用下,在固体状态下发生再结晶作用,使其矿物成分、结构构造乃至化学成分发生部分或全部改变而形成的新岩石。

变质岩的矿物成分,除了保留了原来岩石的矿物成分,如石英、长石、云母、角闪石、辉石、方解石和白云石外,还新生成了变质矿物,如绿泥石、精石、蛇纹石等。经过变质过程后,岩浆岩会产生片状构造,强度会下降,沉积岩会变得更加致密。

(2)土木工程中常用的变质岩

① 大理岩

大理岩又称大理石、云石,是由石灰岩或白云岩经过高温高压作用,重新结晶变质而成的。大理石的主要矿物成分是方解石或白云石,化学成分主要为 $CaO$、$MgO$、$CO_2$ 和少量的 $SiO_2$。经变质后,结晶颗粒直接结合呈整体块状构造,抗压强度高(100 ~ 150MPa),质地致密,表观密度为 2500 ~ 2700kg/m³,莫氏硬度为 3 ~ 4,比花岗岩易于雕琢。纯大理石为白色,我国常称汉白玉,分布较少。一般大理石常含有氧化铁、二氧化硅、云母、石墨、蛇纹石等杂质,使石材呈现红、黄、棕、黑、绿等各色斑驳纹理,石质细腻、光泽柔润、绚丽多彩,研磨抛光后装饰性良好,因而是优良的室内装饰材料。扫码了解汉白玉大理石的相关知识。

② 石英岩

石英岩是由硅质砂岩变质形成的,呈晶体结构。结构均匀致密,抗压强度高(250 ~ 400MPa),耐久性好,但其硬度大,加工困难。常用作重要建筑的贴面材料及耐磨耐酸的贴面材料。

## 二、常用装饰石材

**1. 天然大理石板材**

(1)大理石板材产品的分类及等级

大理石板材是用大理石荒料(即由矿山开采出来的具有规则形状的天然大理石块)

经锯切、研磨、抛光等加工而成的石板。

天然大理石板材按《天然大理石建筑板材》(GB/T 19766—2016)规定,根据形状可分为毛光板(MG)、普型板(PX)、圆弧板(HM)、异型板(YX)。其按加工质量和外观质量分为 A、B、C 三级。扫码获得《天然大理石建筑板材》(GB/T 19766—2016)的详细内容。

(2)大理石板材的技术要求

大理石板材的技术要求,按《天然大理石建筑板材》(GB/T 19766—2016)执行。

(3)大理石的应用

大理石板材主要用于室内饰面,如墙面、地面、柱面、台面、栏杆、踏步等。

大理岩加工成的建筑板材,用于装饰等级要求较高的建筑物饰面。经研磨、抛光的大理石板材光洁细腻,白色大理石(汉白玉)洁白如玉,晶莹纯净;纯黑大理石庄重典雅,秀丽大方;彩花大理石色彩绚丽,花纹奇异。大理岩的耐用年限一般为数十年至几百年。

大理岩不宜用作城市建筑的外部饰面材料,因为城市空气中常含有二氧化硫,遇水时生成亚硫酸,继而变成硫酸,与大理岩中的碳酸钙发生反应,生成易溶于水的 $CaSO_4 \cdot 2H_2O$,使表面失去光泽,变得粗糙多孔,从而降低建筑装饰效果。只有含石英为主的砂岩、石英岩、汉白玉等少数致密、纯的品种可用于室外。

**2. 天然花岗岩板材**

花岗岩构造非常致密,质地较硬。矿物全部结晶且晶粒粗大,呈块状构造或粗晶嵌入玻璃质结构中的斑状构造。它们经研磨、抛光后形成的镜面,呈现出斑点状花纹。

(1)花岗岩板材产品的分类及等级

花岗岩板材是用花岗岩荒料加工制成的板材,根据国家标准《天然花岗石建筑板材》(GB/T 18601—2009)规定,花岗石板材按形状分类可分为四种:毛光板(MG)、普型板(PX)、圆弧板(HM)、异型板(YX)。按表面加工程度分类可分为三种:镜面板(JM)、细面板(YG)、粗面板(CM)。扫码获得《天然花岗石建筑板材》(GB/T 18601—2009)的详细内容。

毛光板按厚度偏差、平面度公差、外观质量等级,将板材分为优等品(A)、一等品(B)、合格品(C)三个等级。普型板按规格尺寸偏差、平面度公差、角度公差、外观质量等,将板材分为优等品(A)、一等品(B)、合格品(C)三个等级。圆弧板按规格尺寸偏差、直线度公差、线轮廓度公差、外观质量等,将板材分为优等品(A)、一等品(B)、合格品(C)三个等级。

(2)花岗石板材的技术要求

大理石板材的技术要求,按国家标准《天然花岗石建筑板材》(GB/T 18601—2009)执行

(3)花岗岩的应用

花岗岩板材质感丰富,具有华丽高贵的装饰效果,且质地坚硬、耐久性好,是室内外高级装饰的常用材料,主要用于建筑物的墙、柱、地面、楼梯、台阶、栏杆等表面装饰。另外,花岗石材也可用于重要的大型建筑物基础、堤坝、桥梁、路面、街边石等。

磨光花岗石板材的装饰特点是华丽而庄重;粗面花岗石板材的装饰特点是凝重而粗犷,有镜面感,色彩鲜艳,光泽动人。应根据不同的使用场合选择不同物理性能及表面装饰效果的花岗石。其中剁斧板、机刨板、粗磨板用于外墙面、柱面、台阶、勒脚等部位。磨

光板材主要用于室内、外墙面、柱面、地面等装饰。

近年来花岗石外饰面趋向于以毛面花岗石为主,磨光花岗石仅作一些线条或局部衬托。这种毛面花岗石的制作工艺是先将花岗石磨平,然后用高温火焰烧毛,看不出有色差,色彩均匀,无反光,给人视感舒适、自然美的享受。

(4)天然石材的放射性

天然石材中的放射性是引起人们普遍关注的问题。经检验证明,绝大多数的天然石材中只含有极微量的放射物质,不会对人体造成任何危害。但部分石材产品放射性指标超标,会在长期使用过程中对环境造成污染。放射性水平超标的天然石材产品中的镭、钍等放射元素衰变过程中将生成天然放射性气体氡。氡是一种无色、无味、感官不能觉察的气体,易在通风不良的地方聚集,导致肺、血液、呼吸道发生病变。因此,装饰工程中应选用经放射性测试,且发放了放射性产品合格证的产品。此外,在使用过程中,还应经常打开居室门窗,促进室内空气流通。

# 第二节　木　　材

木材在土木工程中的使用具有十分悠久的历史,是我国古代主要的建筑材料,尤其在唐、宋、辽时期,木结构建筑非常流行,保存至今最有代表性的建筑是位于山西的应县木塔。木材具有轻质高强、易加工、导电导热性低、弹性和韧性较好、花纹多样及装饰性好等优点,因此在现代工程应用中仍是重要的建筑材料。扫码了解应县木塔的相关知识。

## 一、木材的分类

根据树种,木材分为针叶树和阔叶树两大类。

**1. 针叶树**

针叶树多为常绿树,树叶细长如针、树干通直高大、纹理平顺、材质均匀、木质较软、易于加工,故又称软木材。针叶树材质虽软但强度较高、干湿变形较小、耐腐蚀性较强。所以常见的针叶树种如红松、落叶松、云杉、冷杉、柏木等,在建筑工程中被广泛应用,多用作承重构件和门窗、地面板及装饰材料等。

**2. 阔叶树**

阔叶树多为落叶树,树叶宽大、叶脉呈网状、树干弯曲、通直部分较短,材质重而硬,较难加工,所以又称为硬木材。阔叶树表观密度大、强度高、干湿变形大、易于开裂翘曲,适用于尺寸较小的木构件。常见的阔叶树种有:水曲柳、桦木、榉树、柞木、榆树等。这些树种有的加工后木纹和颜色美观,具有很好的装饰性,故常用作建筑装饰材料。

## 二、木材的构造

木材的构造是决定木材性能的主要因素。由于木材的种类和生长环境不同,各种木材在构造上差别很大,通常从宏观和微观两个角度观察。

**1. 木材的宏观构造**

木材的宏观构造是指用肉眼或放大镜所能见到的木材组织特征。从横切面上可以看到树木的髓心、木质部和树皮等三个主要部分，还可以看到年轮、髓线等。

髓心是树干中心松软部分，其木质强度低、易腐朽，是木材的缺陷部位。因此，锯切的板材不宜带有髓心部分。髓心向外的辐射线称为髓线，髓线与周围连接较差，木材干燥时易沿此开裂。

木质部是木材的主体，是指从树皮至髓心的部分。按生长的阶段分为形成层、边材、心材等部分。形成层是指靠近树皮的薄薄的一层树木生长细胞，树木生长是由形成层的不断扩张来实现的，形成层逐年在最外层生长并形成"年轮"。年轮是深浅相间的同心圆环。在同一年轮内，春天生长的木质，色较浅，质松软，称为春材（早材），夏秋两季生长的木质，色较深，质坚硬，称为夏材（晚材）。相同树种，年轮越密且均匀，材质越好；夏材部分越多，木材强度越高。木质部在接近树干中心的部分呈深色，称为心材；靠近外围的部分色较浅，心材比边材的利用价值高些。

树皮是指树干的外围结构层，是树木生长的保护层。一般的树皮均无使用价值，只有极少可加工成高级保温材料。扫码了解木材的宏观构造的相关知识。

**2. 木材的微观构造**

在显微镜下观察到的木材构造称为微观构造。在显微镜下，可以看到木材是由无数呈管状的细胞紧密结合而成的，绝大部分细胞纵向排列形成纤维结构，少部分横向排列形成髓线。每一个细胞分为细胞壁和细胞腔两部分，细胞壁由细纤维组成，其间具有极小的空隙，能吸附和渗透水分。木材的细胞壁越厚，腔越小，木材越密实，表观密度和强度也越大，但胀缩也大。

针叶树和阔叶树的微观构造有较大的差别。针叶树的显微结构简单而规则，主要由管胞和髓线组成，针叶树的髓线较细而不明显，在某些树种中，如松树在管胞间有树脂道，用来储藏树脂。阔叶树主要由导管、木纤维及髓线等组成。其髓线粗大而明显，导管壁薄而腔大。因此，有无导管以及髓线的粗细是鉴别阔叶树或针叶树的有效方法。

### 三、木材的性质

木材的主要物理力学性质是含水率、湿胀干缩、强度等。

**1. 含水率**

木材中所含水的质量占木材干燥质量的百分比，即含水率。

木材所含水分包括存在于细胞壁内的吸附水和细胞腔内的自由水，以及木材中的化学结合水。

吸附水存在于细胞壁内各木纤维中，这部分水对木材的干湿变形和力学强度有明显的影响。当干燥的木材从大气中吸收水分时，通常会先由细胞壁吸收成为吸附水；达到饱和后，水分进入细胞腔和细胞间隙，成为自由水。自由水是存在于细胞腔和细胞间隙中的水分，对木细胞的吸附能力很差。自由水的变化只影响木材的表观密度、导热性、抗腐朽能力和燃烧性等，而对木材的变形和强度影响不大。化学结合水是木纤维中有机高分子形成过程中所吸收的水分,是构成木材的必不可少的组分,正常状态下木材中的结合水应是饱和的,在常温下对木材没有太大影响。

潮湿的木材在干燥蒸发时,首先脱去自由水,然后再脱去吸附水。当木材细胞壁中的吸附水达到饱和,而细胞腔和细胞间隙中尚无自由水时的木材含水率,称为木材的纤维饱和点。木材的纤维饱和点往往是木材性能变化规律的转折点,对一般木材来说多为 25%~35%。

潮湿的木材能在较干燥的空气中失去水分,干燥的木材也能从周围的空气中吸收水分。当木材长时间处于一定温度和湿度的空气中,则会达到相对稳定的含水率,亦即水分的蒸发和吸收趋于平衡,这时的木材含水率称为平衡含水率。平衡含水率随大气的温度和相对湿度的变化而变化。

**2. 木材的湿胀干缩**

木材具有显著的湿胀干缩性。当木材从潮湿状态干燥至纤维饱和点时,自由水蒸发,其尺寸不改变,继续干燥,当细胞壁中的吸附水蒸发时,则木材发生体积收缩。反之,干燥木材吸湿时,将木材发生体积膨胀,直至含水量达到纤维饱和点时为止,此后木材含水量继续增大,但体积不再膨胀。

木材的湿胀干缩对木材的使用有着严重影响,干缩使木结构构件连接处产生缝隙而致结合松弛,湿胀则造成凸起。由于木材的湿胀干缩明显,因此在加工前应尽量将其进行干燥处理,使含水率达到当地年平均温度和湿度所对应的平衡含水率,以减少木制品中在使用过程中的干缩变形。另外,木材存放时间也影响湿胀干缩变形。存放时间长,木质细胞老化,相应的变形就小。

**3. 强度**

(1)木材的强度

木材根据受力状态的不同分为抗拉强度、抗压强度、抗弯强度和抗剪强度。由于木材是一种非均质材料,具有各向异性,使木材的强度有很强的方向性,因此木材的强度有顺纹强度和横纹强度之分。木材的顺纹与横纹强度有很大的差别,它们之间的比例关系见表 7-1。

表 7-1 木材各强度之间的关系

| 抗压强度 | | 抗拉强度 | | 抗弯强度 | 抗剪强度 | |
|---|---|---|---|---|---|---|
| 顺纹 | 横纹 | 顺纹 | 横纹 | | 顺纹 | 横纹切断 |
| 1 | 1/10~1/3 | 2~3 | 1/20~1/3 | 3/2~2 | 1/7~1/3 | 1/2~1 |

木材的顺纹抗压强度为作用力方向与木材纤维方向平行时的抗压强度。其受压破坏是管状细胞受压失稳的结果,而不是纤维的断裂。顺纹抗压强度较高且木材的疵点对其影响较小,因此,这种强度在工程中应用最广,常用于柱、桩、斜撑及桁架等承重构件。木材的横纹抗压强度为作用力方向与木材纤维方向垂直时的抗压强度。这种受压作用,开始时变形与外力成正比,当超过比例极限时,细胞壁失去稳定,细胞腔被压扁,产生大量变形。因此木材的横纹抗压强度比顺纹抗压强度低得多。

木材的顺纹抗拉强度为拉力方向与木材纤维方向一致时的抗拉强度。这种受拉破坏,往往木纤维未被拉断,而纤维间先被撕裂。木材顺纹抗拉强度是木材所有强度中最大的,通常介于 70~170MPa 之间。木材的疵点如木节、斜纹等对木材顺纹抗拉强度影响极为显著,而木材又多少都有一些缺陷,因此木材实际的顺纹抗拉能力反较顺纹抗压为低,使顺纹抗拉强度难以被充分利用。

（2）影响木材强度的主要因素

木材强度除本身构造组织因素外,还与含水率、负荷持续时间、温度因素、木材的缺陷有很大关系。

① 含水率。木材的含水率在纤维饱和点以下时,含水率越低,吸附水减少,细胞壁结合越紧密,木材强度增加,反之强度降低;当含水率超过纤维饱和点时,只是自由水变化（即表观密度增加）,木材强度不变。木材含水率对各强度的影响程度不同,影响较大的是顺纹抗压强度和抗弯强度,对顺纹抗剪强度影响小,而对顺纹抗拉强度几乎没有影响。

② 持久强度。木材在长期荷载作用下不致引起破坏时的最大强度,称为持久强度。木材的持久强度比其极限强度小得多,一般为极限强度的 50%~60%。在木结构设计时,一般以持久强度为设计依据。

③ 环境温度。木材随环境温度升高强度会降低。当温度由 25℃ 升至 50℃ 时,针叶树抗拉强度降低 10%~15%,抗压强度降低 20%~24%。当木材长期处于 60~100℃ 温度下时,会引起水分和所含挥发物的蒸发,而呈暗褐色,使强度下降,变形增大。因此,长期处于高温下的建筑物,不宜采用木结构。

④ 木材的缺陷。木材在生长、采伐、储运、加工、使用过程中会产生一些缺陷,如裂纹、节理、腐朽、虫蛀、斜纹、涡纹等,也称为疵点。这些疵点会造成木材构造的不连续性和不均匀性,从而使木材的强度和力学性质下降,甚至失去利用价值。

# 第三节　防水材料

防水材料是指能防止雨水、地下水及其他水渗入建筑物或构筑物的一类功能性材料。防水材料广泛应用于建筑工程,亦用于公路桥梁工程、水利工程等。

防水材料是建筑工程中不可缺少的功能性材料,它对提高建筑构件的质量,保证建筑物发挥正常的工程效益起到重要的作用。目前防水材料的造价约占工程总造价的 15%,地下室建筑则高达 25%~30%。虽然投入较多但效果并不理想。据统计,近年来新建工程当年渗漏的维修费用约十多亿元,因此防水材料的研究和改进是建设部门亟待解决的问题。

传统的防水材料是以纸胎石油沥青油毡为代表,它的抗老化能力差,纸胎的延伸率低、易腐烂。油毡体表面沥青耐热性差,当气温变化时,油毡与基底、油毡之间的接头容易出现脱离和开裂的现象,形成水路联通和渗漏。新型的防水材料,大量应用高聚物改性沥青材料来提高胎体的力学性能和抗老化性。应用合成材料、复合材料能增强防水材料的低温柔韧性、温度敏感性和耐久性,极大地提高了防水材料的物理化学性能。

## 一、防水卷材

在土木工程防水材料中,防水卷材是重要的品种之一。20 世纪 80 年代以前,沥青防水材料是主流产品,20 世纪 80 年代后逐渐向橡胶、树脂基、改性沥青系列发展,形成了沥青防水卷材、高聚物改性沥青卷材和合成高分子防水卷材三大类型。

**1. 沥青防水卷材**

20 世纪 50～60 年代以来,我国防水材料一直以纸胎石油沥青油毡为代表。由于纸胎耐久性差,现在已基本被淘汰。目前用纤维织物、纤维毡等改造的胎体和以高聚物改性的沥青卷材已成为沥青防水卷材的发展方向。

沥青防水卷材按其胎体可分为有胎卷材和无胎卷材。有胎卷材是一种用玻璃布、石棉布、棉麻织品、厚纸等作胎体,浸渍石油沥青,表面撒布粉状、粒状或片状防粘材料制成的卷材,也称作浸渍卷材。无胎卷材是将橡胶粉、石棉粉等混合到沥青材料中,经混炼、压延而成的防水材料,也称辊压卷材。沥青防水卷材是目前土木建筑中常用的柔性防水材料。

(1)石油沥青玻纤胎防水卷材

石油沥青玻纤胎防水卷材也称作玻纤胎沥青防水卷材或玻纤胎油毡,属于"弹性体沥青防水卷材"之一。它采用玻璃纤维薄毡为胎体,浸涂石油沥青,并在表面涂撒矿物粉料或覆盖聚乙烯膜等隔离材料,制成可卷曲的片状防水材料。

玻纤胎油毡具有较高的抗拉强度,防渗漏性能好,可达到 A 级防水标准。该材料防老化、抗腐蚀性、耐候性强。经与改性沥青复合后,弹性、柔软性、抗震性都得到很大的提高,例如经 SBS(苯乙烯-丁二烯-苯乙烯)改性的产品,能够在 -25～-15℃ 低温下保持良好的柔韧性。扫码了解石油沥青玻纤胎防水卷材的标号、等级、规格与应用的相关知识。

(2)铝箔面沥青防水卷材

铝箔面沥青防水卷材也称为铝箔面油毡。它采用玻纤毡为胎体,浸涂氧化石油沥青,表面用压纹铝箔粘面,撒布细颗粒矿物材料或覆盖聚乙烯膜而制成的防水材料。扫码了解铝箔面沥青防水卷材的标号、等级、规格与应用的相关知识。

**2. 高聚物改性沥青卷材**

石油沥青本身不能满足土木工程对它的性能要求,在低温柔韧性、高温稳定性、抗老化性、黏附能力、耐疲劳性和构件变形的适应性等方面都存在缺陷。因此,常用一些高聚物、矿物填料对石油沥青进行改性,如 SBS 改性沥青、APP 改性沥青、PVC 改性沥青、再生胶改性沥青、橡塑改性沥青和铝箔橡塑改性沥青等。

(1)弹性体改性沥青防水卷材

弹性体改性沥青防水卷材是用沥青或热塑性弹性体(如 SBS)改性沥青浸渍胎基,两面涂以弹性体沥青涂盖层,上表面撒以细砂、矿物粒(片)或覆盖聚乙烯膜,下表面撒布细砂或覆盖聚乙烯膜所制成的一类防水卷材。

SBS 防水卷材是弹性体改性沥青防水卷材中使用较广泛的一种。SBS(苯乙烯-丁二烯-苯乙烯)高聚物属嵌段聚合物,采用特殊的聚合方法使丁二烯两头接上苯乙烯,不需硫化成型就可以获得弹性丰富的共聚物。所有改性沥青中,SBS改性沥青的性能是目前最佳的。改性后的防水卷材,既具有聚苯乙烯抗拉强度高、耐高温性好,又具备聚丁二烯弹性高、耐疲劳性和柔软性好的特性,适用于工业与民用建筑的屋面和地下防水工程,尤其适用于较低气温环境的建筑防水。扫码了解SBS防水卷材的分类相关知识。

(2)塑性体改性沥青防水卷材

塑性体改性沥青防水卷材是用沥青或热塑性弹性体(如无规聚丙烯APP或聚烯烃类聚合物APAO、APO)改性沥青浸渍胎基,两面涂以塑性体沥青涂盖层,上表面撒以细砂、矿物粒(片)或覆盖聚乙烯膜,下表面撒布细砂或覆盖聚乙烯膜所制成的一类防水卷材。

APP防水卷材是塑性体改性沥青防水卷材中使用较广泛的一种。APP卷材耐热性优异,耐水性、耐腐蚀性好,低温柔性较好(但不及SBS卷材)。其中聚酯毡的机械性能、耐水性和耐腐蚀性性能优良。虽然玻纤毡的价格低,但强度较低无延伸性,适用于工业与民用建筑的屋面和地下防水工程,以及道路、桥梁等建筑物的防水,尤其适用于较高气温环境的建筑防水。扫码了解APP防水卷材的分类相关知识。

(3)改性沥青聚乙烯胎防水卷材

扫码了解改性沥青聚乙烯胎防水卷材的相关知识。

(4)自黏聚合物改性沥青防水卷材

扫码了解聚合物改性沥青防水卷材的相关知识。

**3. 合成高分子防水卷材**

合成高分子防水卷材是除沥青基防水卷材外,近年来大力发展的防水卷材。合成高分子防水卷材是以合成橡胶、合成树脂或者两者共混体为基料,加入适量的化学助剂、填充料等,经混炼、压延或挤出等工艺制成的防水卷材或片材。

合成高分子防水卷材耐热性和低温柔韧性好,拉伸强度、抗撕裂强度高、断裂伸长率大,耐老化、耐腐蚀、耐候性好,适应冷施工。

合成高分子防水卷材品种很多,目前最具代表的有合成橡胶类三元乙丙橡胶防水卷材、聚氯乙烯防水卷材和氯化聚乙烯-橡胶共混防水卷材。

(1)三元乙丙橡胶防水卷材

三元乙丙(EPDM)橡胶防水卷材是以三元乙丙橡胶为主体,加入一定量的丁基橡胶、软化剂、补强剂、填充剂、促进剂和硫化剂等制成的防水卷材。三元乙丙橡胶防水卷材具有耐老化、耐热性好(>160℃)、使用寿命长(30~50年)、拉伸强度高、延伸率大、冷施工、对基层开裂变形适应性强、质量轻、可单层施工等特点。美国、日本的新建屋面和维修防水工程的三分之一左右都是应用的三元乙丙橡胶防水卷材。三元乙丙橡胶防水卷材,适用于外露屋面、大跨度、振动大、年限要求长、防水质量要求高的工程。

(2)聚氯乙烯防水卷材

聚氯乙烯(PVC)防水卷材是以聚氯乙烯树脂为主要原料,掺加填充料(如铝矾土)和适量的改性剂、增塑剂(如邻苯二甲酸二辛酯)及其他助剂(如煤焦油)制成。PVC防水卷材耐老化性能好(耐用年限25年以上)、拉伸强度高、断裂伸长率极大、原材料丰富、价格便宜。用热风焊铺粘施工方便,不污染环境。适用于我国南北方广大地区防水要求高、耐用年限长的工业与民用建筑的防水工程。用于屋面防水时,可做成单层外露防水。扫码了解聚氯乙烯防水卷材的分类相关知识。

（3）氯化聚乙烯-橡胶共混防水卷材

该卷材以氯化聚乙烯和合成橡胶共混物为主体制得的防水材料。该防水卷材不但具有氯化聚乙烯特有的高强度、优异的耐臭氧、耐老化性能，还具备橡胶和塑料的高弹性、高延伸性和良好的低温柔性。从物理性能上看，氯化聚乙烯-橡胶共混防水卷材接近三元乙丙橡胶防水卷材的性能，最适应屋面单层外露防水。扫码了解氯化聚乙烯-橡胶共混防水卷材的相关知识。

## 二、防水涂料

保护建筑物构件不被水渗透或湿润，能形成具有抗渗性涂层的涂料，称为防水涂料。防水涂料按照分散剂的不同可分为溶剂涂料、水乳型涂料两种。随着科技的发展，涂料产品不仅要求施工方便、成膜速度快、修补效果好，还须延长使用寿命、适应各种复杂工程的需求。

**1. 沥青类防水涂料**

沥青类防水涂料是以沥青为基料，通过溶解或形成水分散体构成的防水涂料。沥青防水涂料除具有防水卷材的基本性能外，还具有施工简单、容易维修、适用于特殊建筑物的特点。

直接将未改性或改性的沥青溶于有机溶剂而配制的涂料，称为溶剂型沥青涂料。将石油沥青分散在水中，形成稳定的水分散体而构成的涂料，称为水乳型沥青防水涂料。扫码了解沥青类防水涂料的分类相关知识。

（1）SL-2溶剂型SBS橡胶改性沥青防水涂料

此类防水涂料是目前国内外第三代以SBS橡胶为改性材料生产的一种冷施工防水材料。它具有优良的耐腐蚀、高弹性、延展性、黏结性强，对基层开裂适应性良好，高温不流淌、低温不开裂、产品性能稳定，可以在负温（-20℃）下施工、冷施工、省工省力、施工方便等特点。主要适用于各种建筑物屋面、地下室、地沟、涵洞的防水、防潮工程；可以作为防水卷材的冷施工黏结剂，也可作为建筑物防水系统维修补漏及管道防腐等工程材料。

（2）石棉乳化沥青防水涂料

石棉乳化沥青防水涂料以石油沥青为基料，石棉作分散剂，在强制搅拌下制成的厚质防水涂料。此防水涂料无毒、不燃、水性施工、无污染，可在潮湿基层上铺设。其耐水、耐热、耐候、抗裂和稳定性都优于一般的乳化沥青。由于填料采用了无机纤维矿物，它的乳化膜比化学膜更坚固。其不足之处在于施工环境温度一般要在15℃以上，但气温过高易粘脚。石棉乳化沥青防水涂料目前在我国的应用较广泛、效果也较好，配以玻纤布、无纺布等，可适用于钢筋混凝土屋面、地下室、厨池的防水层。

（3）水乳型再生橡胶沥青防水涂料

水乳型再生橡胶沥青防水涂料是以石油沥青为基料和再生橡胶为改性材料复合而成的水性防水涂料。该防水涂料是国内外较通用的一种防水涂料。与同类溶剂型产品比较，它以水取代了汽油，其安全性、环境性更胜一筹。这种涂料因以合成胶乳为原料，因而价格昂贵。

水乳型再生橡胶沥青防水涂料,能够在各种复杂表面形成防水膜,具有一定的柔韧性和耐久性。以水为分散剂,无毒、不燃、无异味,安全可靠。可在常温下冷施工,操作简单、维护方便,能够在潮湿无积水的表面施工。其缺点是一次涂刷成膜较薄,产品质量易受生产条件的影响,气温低于5℃不易施工。产品适用于工业、民用混凝土基层屋面、浴厕、厨房间的防水,沥青珍珠岩保温层屋面防水,地下混凝土建筑防潮,旧油毡屋面翻修和刚性自防水屋面的维修。

**2. 其他品种防水涂料**

我国20世纪70年代主要生产氯丁胶和橡胶改性沥青防水涂料,自20世纪80年代推出焦油聚氨酯防水涂料以来,各种高分子材料的防水涂料层出不穷。液态、粉末态、溶剂型、水乳型、反应型、纳米型、快速型、美术型等新产品不断在工程建设中亮相。

(1) 聚氨酯防水涂料

聚氨酯防水涂料是一种化学反应型涂料,它由异氰酸酯基的聚氨酯预聚体和含有多羟基或胺基的固化剂,以及其他助剂按一定比例混合而成。聚氨酯防水涂料属于高档合成高分子防水涂料,它具有很多突出的优点:容易形成较厚的防水涂膜;能够在复杂的基层表面施工,其端头容易处理;整体性强,涂膜层无接缝;冷施工,操作安全;涂膜具有橡胶弹性,延伸性好,抗拉、抗撕裂强度高;防水年限可达10年以上等。聚氨酯防水涂料适用于各种地下、浴厕、厨房等的防水工程;污水池的防漏;地下管道的防水、防腐工程等。

(2) 硅橡胶防水涂料

硅橡胶防水涂料是以硅橡胶乳液和其他高分子乳液配制成的复合乳液,再添加一定量的外加剂而制得的乳液型防水涂料。硅橡胶防水涂料兼有涂膜防水和渗透防水的双重特性,适合于复杂构件表面的施工,无毒、无味、不燃、安全、冷施工、操作简单,可配制成各种颜色具有一定的装饰效果。硅橡胶防水涂料缺点主要有原材料价格高,成本大;要求基层有较好的平整度;固体含量较低,一次涂刷层较薄;气温低于5℃不宜施工等。硅橡胶防水涂料适用于屋面、地下、输水、贮水构建物等的防水、防潮工程。

(3) 水泥基渗透结晶型防水涂料

水泥基渗透结晶型防水涂料是由硅酸盐水泥、石英砂、特殊的活性物质及一些添加剂组成的无机粉末状防水材料,简称为CCCW。水泥基渗透结晶型防水涂料是一种刚性防水材料,与水作用后材料中含有硅酸盐活性化学离子,通过载体向混凝土内部渗透、扩散,与混凝土孔隙中的钙离子进行化学反应,形成不溶于水的硅酸钙结晶体填塞毛细孔道,使混凝土结构致密、防水。这种防水涂料适用于地下工程、水池、水塔等混凝土结构工程的迎水面和背水面的防水处理,它为混凝土工程提供了可以信赖的防水材料和工艺。

### 三、建筑密封材料

土木工程中为了保证建筑物的水密性和气密性,凡具备防水功能和防止液、气、固侵入的密封材料,称为防水密封材料。它的基本功能是填充构形复杂的间隙,通过密封材料的变形或流动润湿,使缝隙、接头不平的表面紧密接触或粘接,从而达到防水密封的作用。

建筑密封材料可应用于建筑物门窗密封、嵌缝,混凝土、砖墙、桥梁、道路伸缩的嵌缝,给排水管道的对接密封,电器设备制造安装中的绝缘、密封,航空航天、交通运输器具、机械设备连接部位的密封和各种构件裂缝的修补密封等。

建筑密封材料的基材主要有油基、橡胶、树脂、无机类等,其中橡胶、树脂等性能优异的高分子材料是建筑密封材料的主体,故称为高分子建筑密封材料。建筑密封材料有膏状、液状和粉状等。扫码了解防水密封材料的分类相关知识。

**1. 非定型防水密封材料**

非定型密封材料是现场成形的密封材料,多数以橡胶、树脂、合成材料为基料制成,它填充于缝隙中起到密封作用。工程中常用的非定型密封材料有硅橡胶防水密封材料和丙烯酸酯防水密封材料等。

硅橡胶防水密封胶是以聚硅氧烷为主要成分的非定型密封材料,可以在室温下固化或加热固化的液态橡胶。此类材料耐热、耐寒、绝缘、防水、防振、耐化学介质、耐臭氧、耐紫外线、耐老化、耐一些有机溶剂和稀酸,贮存性稳定,密封性能持久,硫化后的密封胶在 $-50\sim250℃$ 范围内长期保持弹性,广泛适用于建筑工程的预制构件嵌缝密封、防水堵漏。

丙烯酸酯防水密封材料是以丙烯酸酯类聚合物为主要成分的非定型密封材料,具有橡胶的弹性和柔软性,有良好的耐水、耐溶剂性等特点。由于其冷流动性差,故不能用于伸缩大的变形缝,嵌缝时要用热施工方法。此类材料适用于钢、铝、木门窗与墙体、玻璃间的接缝密封;以及刚性屋面、内外墙、管道、混凝土构件的接缝密封。

**2. 定型防水密封材料**

建筑定型密封材料的要求有具有良好的水密性、气密性和耐久性;具有良好的弹性、塑性和强度;有耐热、耐低温、耐腐蚀的性能;要求制作尺寸精度高,不致在构件振动、变形等过程中脆裂、脱落。工程中常用的非定型密封材料有聚氨酯遇水膨胀橡胶密封材料和刚性止水带等。

聚氨酯遇水膨胀橡胶材料主要以聚醚多元醇为原料制成的亲水性聚氨酯预聚体。聚氨酯材料中存在大量的极性链节,容易旋转,采用适当的交联固化能产生较好的回弹性能。与水相遇后其链节和水能生成氢链,从而导致材料体积的膨胀。国产 821AF 和 821BF 遇水膨胀密封材料,都属于这类产品。扫码了解国产 821AF 和 821BF 的相关知识。

刚性止水带也称金属止水带,它用钢、铜、铝、合金钢板等制成。钢止水带和铜止水带主要应用于水坝及大型构建物,金属防水材料采用焊接拼接,金属焊接质量至关重要。

# 第四节 绝热材料

建筑物在使用中常有保温、隔热等方面的要求,可采用绝热材料来满足这些建筑功能的要求。在土木工程中,习惯上把用于控制室内热量外流的材料叫作保温材料,把防止热量进入室内的材料叫作隔热材料。保温、隔热材料统称为绝热材料。

在建筑中合理采用绝热材料,能提高建筑的使用效能,更好地满足节能要求,保证人们正常的生产、工作和生活。保温和隔热良好的建筑物,还可以大大降低采暖和空调的能耗,在"双碳时代"的背景下,绝热材料的应用具有越来越重要的意义。

## 一、导热性

**1. 传热方式**

热量的传递方式有三种:传导转换、对流换热和辐射换热。

热量以上述三种方式在建筑物和环境中传递,其传递方式主要是导热,同时也有对流和热辐射存在。主要的散热区域是墙体、顶棚和屋顶、楼板、门窗,建筑物的缝隙和开着的门窗会大大增加热量的散发。

在北方,散热问题是个严重的经济问题。使用绝热材料可以避免建筑物在夏季过多吸收热量,在冬季过分散失热量。

**2. 热阻和导热系数**

当材料的两表面存在温度差,热量就会自动地从温度高的一面向温度低的一面传递。在稳定状态下,通过测量热流量、材料两表面的温度及其有效传热面积,可以按下式计算材料的热阻:

$$R = \frac{A(T_1 - T_2)}{Q} \tag{7-1}$$

式中　$R$——热阻($m^2 \cdot K/W$);

　　　$Q$——平均热流量(W);

　　　$T_1$——试件热面温度平均值(K);

　　　$T_2$——试件冷面温度平均值(K);

　　　$A$——试件的有效传热面积($m^2$)。

如果热阻与温度呈线性关系,且试件能代表整体材料,试件具有足够的厚度,则材料的导热系数可用下式计算:

$$\lambda = \frac{d}{R} = \frac{Q \cdot d}{A(T_1 - T_2)} \tag{7-2}$$

式中　$\lambda$——导热系数[W/(m·K)];

　　　$d$——试件平均厚度(m)。

材料导热系数 $\lambda$ 的物理意义是,厚度为1m的材料,当温度差为1K时,在1s内通过$1m^2$面积的热量。材料的导热系数越小,表示其绝热性能越好。

实际材料常含有孔隙,同时存在热传导、热对流和热辐射,所测量的并非真正的导热系数,而是表观导热系数,或称为当量导热系数和等效导热系数。材料的导热系数是设计建筑物围护结构、进行热工计算的重要参数。选用导热系数小而比热容大的材料,可提高围护结构的绝热性能并保持室内温度的稳定。

在土木工程中,常把导热系数小于0.175W/(m·K)的材料称为绝热材料,选用绝热材料时,一般要求其导热系数不大于0.175W/(m·K),表观密度小于600kg/$m^3$,抗压强度不小于0.30MPa。在实际应用中,由于绝热材料抗压强度一般较低,常将绝热材料与承重材料复合使用。另外,由于大多数绝热材料都具有一定的吸水、吸湿能力,故在实际使用时应注意防潮防水,需要在其表层加防水层或隔气层。

**3. 影响导热系数的因素**

材料的热传导性能是由材料的导热系数的大小决定的,导热系数越小,保温隔热性能越好。影响材料导热系数的主要因素有材料的物质构成、微观结构、孔隙结构、温度、湿度和热流方向等。

(1)物质构成:不同的材料其导热系数是不同的,一般来说,金属材料的导热系数最大,无机非金属材料次之,有机材料的导热系数最小。

(2)微观结构:相同化学组成的材料,结晶结构的导热系数最大,微晶结构次之,玻璃体结构导热系数最小。

(3)孔隙结构:固体物质的导热系数比空气的导热系数大得多,材料的孔隙率越大,导热系数越小。在孔隙率相近的情况下,孔径越大,孔隙相通将使材料导热系数有所提高,这是由于孔内空气流通与对流的结果。对于纤维状材料,当压实达到某一表观密度时,其导热系数最小,称该表观密度为最佳表观密度。当小于最佳表观密度时,材料内空隙过大,由于空气对流作用,会使导热系数有所提高。

(4)湿度:因为固体的导热最好、液体次之、气体导热最差,因此,材料受潮会使导热系数增大,若水结冰,导热系数会进一步增大。为了保证保温效果,绝热材料要特别注意防潮。

(5)温度:材料的导热系数随温度升高而增大。因此绝热材料在低温下的使用效果更佳。

(6)热流方向:对于木材等纤维状材料,热流方向与纤维排列方向垂直时,材料的导热系数要小于平行时的导热系数。

## 二、绝热材料的类型

**1. 多孔型**

对于平板状多孔材料,当热量从高温面向低温面传递时,主要通过以下几种传热方式进行传热:

(1)热量在固相中的传导;
(2)孔隙中高温固体表面对气体的辐射与对流;
(3)孔隙中气体自身的对流与传导;
(4)热气体对低温固体表面的辐射与对流;
(5)热固体表面与冷固体表面之间的辐射。

这几种传热方式同时存在,在多孔材料中进行传热。

在常温下,对流和辐射在总的传热中所占的比例很小,以导热为主。密闭空气的导热系数仅为 $0.025\text{W}/(\text{m}\cdot\text{K})$,远小于固体的导热系数,故热量通过密闭孔隙传递的阻力较大,而且密闭孔隙的存在使热量在固体中的传热进程大大降低,从而大大减缓传热速度。这就是含有大量密闭孔隙的材料能起绝热作用的原因。

**2. 纤维型**

纤维型绝热材料的绝热机理基本上和多孔材料的情况相似,当传热方向和纤维方向垂直时，其绝热性能比平行时要好。

**3. 反射型**

当外来的热辐射能投射到物体上时，通常会将其中一部分能量反射掉，另一部分被吸收。根据能量守恒原理，被吸收的能量与被反射的能量之和为总的辐射能。由此可以看出，凡是反射能力强的材料，吸收热辐射的能力就小，故利用某些材料对热辐射的反射作用，在需要绝热的部位表面贴上这种材料，就可以将绝大部分的外来热辐射反射掉，从而起到绝热的作用。

### 三、常用绝热材料

绝热材料按化学成分可分为有机和无机两大类，按材料的构造可分为多孔组织材料、纤维状和松散粒状三大类，通常可制成板、片、卷材或管壳等多种形式的产品。一般来说，无机绝热材料的表观密度较大，不容易腐蚀，不会燃烧，有的能耐高温。有机绝热材料质轻，保温隔热性能好，但耐热性较差。

**1. 多孔状绝热材料**

常用多孔状绝热材料有膨胀蛭石、膨胀珍珠岩、微孔硅酸钙、发泡硅酸盐、泡沫玻璃、泡沫塑料和加气混凝土等。

**2. 无机纤维状绝热材料**

常用无机纤维状绝热材料主要是指岩棉、矿棉、玻璃棉等人造无机纤维状材料。该类材料在外观上具有相同的纤维状形态和结构，具有密度小、绝热性能好、不燃烧、耐腐蚀、化学稳定性强、吸声性能好、无毒、无污染、防蛀、价廉等优点，广泛应用在住宅建筑和热工设备、管道等的保温、隔热、隔冷和吸声材料中。扫码了解常用无机纤维状绝热材料的相关知识。

**3. 反射型绝热材料**

目前，我国对建筑工程的保温隔热，普遍采用多孔材料和在维护结构中设置空气层的做法，这对改善维护结构的性能有较好的作用。但对于较薄的维护结构，要设置保温层和空气层则较困难，而采用反射型绝热材料往往会有较理想的保温隔热效果。扫码了解常用反射型绝热材料的相关知识。

**4. 其他绝热材料**

软木板、蜂窝板、纤维板等材料也常用作绝热材料，扫码了解相关知识。

## 第五节　吸声材料和隔声材料

吸声材料是一种能在较大程度上吸收由空气传递的声波能量的土木工程材料。在音乐厅、影剧院、大会堂、播音室等室内的墙面、地面、顶棚等部位，采用适当的吸声材料，能改善声波在室内的传播质量，保持良好的音响效果。

## 一、吸声材料的原理

声音来源于物体的振动。它迫使邻近的空气跟着振动而产生声波,并在空气中向四周传播。声波是依靠介质的分子振动而向外传播的能量,介质的分子只能振动而不移动,所以声波是波动。人耳能够听见的声频范围是 128Hz~10kHz。

声音在传播的过程中,一部分随着距离的增大而扩散,另一部分则因空气分子的吸收而减弱。这种减弱现象,在室外空旷处较为明显,而在体积不大的室内,声能的减弱主要是依靠房间四壁的材料表面对声能的吸收。

当声波遇到材料表面时,一部分被反射,另一部分穿透材料,其余部分则传递给材料,引起材料孔隙中空气分子与孔壁的摩擦和粘滞阻力,期间相当一部分声能转化为热能被吸收。这些被吸收掉的能量($E$)(包括部分穿透材料的声能)与传递给材料的全部声能($E_0$)之比称为吸声系数($\alpha$),它是评价材料吸声性能优劣的主要指标,用下式表示为:

$$\alpha = \frac{E}{E_0} \qquad (7\text{-}3)$$

吸声系数与声音的频率及声音的入射方向有关。因此,吸声系数是声音从各个方向入射的吸收平均值,并应指出是对哪一频率的吸收。通常采用六个频率,即 125Hz、250Hz、500Hz、1000Hz、2000Hz、4000Hz。

任何材料都能吸收声音,但吸收程度有很大差别。一般将上述六个频率的平均吸声系数大于 0.20 的材料列为吸声材料,吸声系数越大,吸声效果越好。

## 二、吸声材料的特征

吸声材料按吸声机理的不同可分为两类:一是疏松多孔材料;二是柔性材料、膜状材料、板状材料、穿孔板等。

多孔吸声材料具有大量内外连通的微孔和连续的气泡,当声波入射到材料表面时,声波能很快沿微孔进入材料内部,引起孔隙或气泡内的空气振动。由于摩擦、空气的粘滞阻力和材料内部的热传导作用,使相当一部分声能转化为热能被吸收。柔性材料、膜状材料、板状材料、穿孔板等的吸声原理是材料与声波发生共振,使声能转化为机械能。

## 三、常用吸声材料

建筑上常用的吸声材料及其设置情况,请扫码了解。

许多多孔吸声材料与多孔绝热材料材质相同,但两者对气孔特征的要求则不同。绝热材料要求气孔封闭,以阻止热对流的进行;而吸声材料要求气孔开放,互相连通,且气孔越多,吸声效果越好。这种材质相同而气孔结构不同的多孔材料,主要通过原料的某一组分的差别以及生产工艺中的热工制度和压力不同等来实现。

## 四、隔声材料

建筑将主要起隔绝声音作用的材料统称为隔声材料,隔声材料主要用于外墙、门窗、隔墙、隔断等。

隔声可分为隔绝空气声（通过空气传播的声音）和隔绝固体声音（通过撞击或振动传播的声音）两种。两者的隔声原理截然不同。隔声不但与材料有关,而且与建筑物结构有密切的关系。

**1. 空气声的隔绝**

材料隔绝空气声的能力,可以用材料对声波的透射系数或材料的隔声量来衡量,见式(7-4)和式(7-5)：

$$\tau = \frac{E_t}{E_0} \tag{7-4}$$

$$R = 10\lg \frac{1}{\tau} \tag{7-5}$$

式中　$\tau$——声波透射系数；

　　　$E_t$——透射材料的声能；

　　　$E_0$——入射总声能；

　　　$R$——材料的隔声量(dB)。

材料的 $\tau$ 越小,则 $R$ 越大,材料的隔声性能越好。材料的隔声性能与入射声波的频率有关,常用 125～4000Hz 六个频率的隔声量来表示材料的隔声性能。对于普通教室之间的隔墙和楼板,要求达到不小于 40dB 的隔声量,即透射声应小于入射声能的万分之一。

隔绝空气声,主要服从质量定律,即材料的体积密度越大,质量越大,隔声性能越好,因此选用密实的材料作为隔声材料,如砖、混凝土、钢板等。如果采用轻质材料或薄壁材料,需辅以多孔吸声材料或采用夹层结构,如夹层玻璃就是一种很好的隔声材料。

**2. 固体声的隔绝**

材料隔绝固体声的能力是用材料的撞击声压级来衡量的。测量时,将试件安装在上部声源室和下部受声室之间的洞口,声源室与受声室之间没有刚性连接,用标准打击器打击试件表面,受声室接收到的声压级减去环境常数,即得材料的撞击声压级。普通教室之间的楼板的标准化撞击声压级应小于 75dB。

隔绝固体声音最有效的措施是采用不连续的结构处理,即在墙壁和承重梁之间、房屋的框架和墙板之间加弹性衬垫,如毛毡、软木、橡皮等材料,或在楼板上加弹性地毯。

# 第六节　防火材料

随着我国城市化建设的推进,城市中的防火隐患增多,灾害事故频繁发生,人们对建筑的安全防火提出了更高的要求。防火材料是指在火灾条件下仍能一定时间内保持其使用功能的材料。包括防火板材、防火涂料、防火织物等。

## 一、防火板材

防火板材常以改性无机物作为原料制作,本身具有一定的耐火性,它们大多是不燃性/难燃性材料,可作为建筑物的隔墙、顶棚等构件。防火板材有易加工、质轻、装饰性好、干作业方式施工等优点,广泛应用于各类现代建筑当中。常用的防火板材有水泥刨花板、无机纤维增强水泥板、钢网架水泥夹心复合板、耐火纸面石膏板、硅酸钙板、难燃铝塑建筑装饰板等,请扫码了解。

### 二、防火涂料

防火涂料是涂料的一种,兼具涂料的基本功能和防火保护功能。防火涂料是指涂覆于可燃性基材表面,能降低可燃性基材的火焰传播速率或阻止热量向被保护的构件传递,进而推迟或消除可燃性基材的引燃过程,或者推迟构件失稳或构件机械强度降低的一种功能性涂料。在火灾发生时,防火涂料可以有效地延缓火灾基底材料物理力学性能的变化,从而使人们有充分的时间进行人员疏散和火灾扑救工作,达到保护人们生命财产安全的目的。

### 三、防火分隔设施

建筑物中的电缆、油管、风管、气管等孔洞,在火灾时易成为火和有毒气体的通道,导致火灾的蔓延。防火分隔设施是指在一定时间内能把火势控制在一定空间内,阻止其蔓延扩大的一系列分隔设施。防火分隔设施必须满足我国《建筑设计防火规范》(GB 50016)中不同等级的防火要求。常用的防火分隔设施有防火门、防火卷帘、防火网、阻火包、阻火圈等。扫码了解规范的更多内容。

### 四、阻燃剂

木材、塑料和纺织品等都是可燃或易燃的材料,为阻碍火势的蔓延,常在以上材料内加入一些高聚物进行阻燃处理,使易燃的材料变为难燃或不燃的材料,起到阻燃作用的高聚物称为阻燃剂。一般来讲,阻燃高聚物材料可以通过气相阻燃、凝聚相阻燃以及中断热交换阻燃等几类阻燃机理达到阻燃目的。通过阻止高聚物分解出来的可燃性气体产物的燃烧或对火焰反应产生阻止作用的属于气相阻燃;通过阻止高聚物发生热分解和释放出可燃性气体作用的属于凝聚相阻燃;通过将高聚物燃烧产生的部分热量带走从而实现阻燃的则属于中断热交换阻燃。工业上常用的卤-锑阻燃体系,属于气相阻燃;氢氧化铝,属于凝聚相阻燃;氯化石蜡,属于中断热交换阻燃。

# 复习思考题

1. 某绝热材料受潮后,其绝热性能明显下降。请分析原因。
2. 广州和哈尔滨各有一座在建高档高层建筑需建玻璃幕墙,有吸热玻璃及热反射玻璃两种材料可选用。请选用并简述理由。
3. 请分析用于室外和室内的建筑装饰材料主要功能的差异。

# 参考文献

[1] 赵庆新. 土木工程材料[M]. 北京:中国电力出版社,2010.
[2] 张亚梅. 土木工程材料[M]. 6 版. 南京:东南大学出版社,2021.
[3] 廖国胜,曾三海. 土木工程材料[M]. 2 版. 北京:冶金工业出版社,2022.
[4] MEHTA P K, MONTEIRO P J M. Concrete:Microstructure,Properties,and Materials[M]. 4th edition. New York:McGraw-Hill Education,2014.
[5] CLAISSE P A. Civil Engineering Materials[M]. Oxford:Butterworth-Heinemann,2016.

英文版本

# Contents

**Introduction** ········································································································· 159

**Chapter I  Basic properties of civil engineering materials** ······························· 164

    Section I  Composition, structure and property of civil engineering materials ······ 164
    Section II  Physical properties of materials ································································ 167
    Section III  Mechanical properties of materials ·························································· 175
    Section IV  Durability of materials ················································································ 179
    Questions for review ········································································································ 179

**Chapter II  Air-hardening cementitious material** ·············································· 181

    Section I  Lime ················································································································· 181
    Section II  Gypsum ··········································································································· 185
    Section III  Water Glass ·································································································· 189
    Questions for review ········································································································ 191

**Chapter III  Cement** ································································································ 192

    Section I  Introduction ····································································································· 192
    Section II  Portland cement and ordinary Portland cement ········································ 195
    Section III  Portland cement mixed with SCMs ··························································· 207
    Section IV  Other varieties of cement ·········································································· 215
    Questions for review ········································································································ 217

**Chapter IV  Ordinary concrete and construction mortar** ································· 219

    Section I  Concrete overview ·························································································· 219
    Section II  Component materials of concrete ····························································· 221
    Section III  Workability of fresh concrete ···································································· 237
    Section IV  Structure of hardened concrete ······························································· 243
    Section V  Mechanical properties of hardened concrete ·········································· 245
    Section VI  Quality control and assessment of concrete ··········································· 258
    Section VII  Deformation properties of concrete ························································ 260
    Section VIII  Durability of concrete ············································································· 265
    Section IX  Mix design of concrete ·············································································· 271

Section Ⅹ　Building mortar ································································ 281
Questions for review ·································································· 288

# Chapter Ⅴ　Asphalt and asphalt mixture ········································ 290

Section Ⅰ　Asphalt material ······················································· 290
Section Ⅱ　Asphalt mixture ······················································· 303
Questions for review ·································································· 316

# Chapter Ⅵ　Construction steel ···················································· 317

Section Ⅰ　Chemical composition of steel and its influence on steel properties ······ 317
Section Ⅱ　Technical properties of steel ········································· 319
Section Ⅲ　Cold working and aging of steel ···································· 323
Section Ⅳ　Common category of steel in civil engineering ··················· 326
Questions for review ·································································· 329

# Chapter Ⅶ　Other engineering materials ········································ 330

Section Ⅰ　Natural stone ··························································· 330
Section Ⅱ　Timber ··································································· 335
Section Ⅲ　Waterproof material ··················································· 340
Section Ⅳ　Thermal insulation material ········································· 348
Section Ⅴ　Sound-absorbing and sound-insulating materials ················ 352
Section Ⅵ　Fire resistant material ················································ 355
Questions for review ·································································· 356

**Reference** ················································································ 357

# Introduction

## I. Definition and classification of civil engineering materials

Civil engineering is a catch-all for constructional engineering, road engineering, bridge engineering, geotechnical and underground engineering, port engineering, water conservancy engineering, municipal engineering, etc. A variety of materials used for these projects are collectively referred to as civil engineering materials. In a narrow sense, civil engineering materials generally refer to the various materials applied in the buildings. In a broad sense, they refer to all of the used materials during the construction of engineering buildings. The civil engineering materials can be regarded as the material basis of civil engineering construction. Scan the QR code to find the examples of civil engieernging materials and structures.

There are many kinds of civil engineering materials of different properties, which can be classified in a number of different ways. For example, according to the manufacturing method of materials, they can be divided into natural materials and artificial materials. According to the function of materials, they can be divided into load-bearing materials, thermal insulation and heat insulation materials, sound-absorbing and sound-insulating materials, waterproof materials, heat-resistant fireproof materials, decorative materials, anticorrosive materials, lighting materials, etc. According to the specific use of materials, they can be divided into foundation materials, structural materials, wall materials, roof materials, ground materials, decorative materials, etc. In addition, they can also be divided into inorganic materials, organic materials and composite materials according to the chemical compositions of materials, as shown in the figure 0-1.

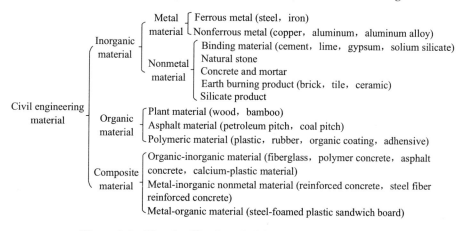

Figure 0-1  The classification of civil engineering materials

## Ⅱ. Development history and trend of civil engineering materials

Civil engineering materials are developed with the continuous progress of human society and the development of social productivity. It can be said that the development history of civil engineering materials is the chronicle of human civilization. In ancient times, humans lived in the natural caves or tree nests. About 10000 ~ 6000 years from now, humans learned how to build their residence. Most of the houses in this period were half-cave-style and constructed out of natural wood, bamboo, reed, grass, mud, etc. At the ruins of Banpo in Xi'an which was built about 6000 years ago, wooden mud walls and ceramic kilns were found. With the improvement of human ability, people began to build houses and commemorative structures by the natural stone. The earliest structure made up of large stones was the Egyptian pyramid, which was built at around 2500 BC. The emergence of artificial burnt earth products has made the ability of human beings to build houses leap a new high level. Clay brick is a typical representation of burnt earth products, which presents a series of good characteristics such as high strength, good water resistance, regular geometry, moderate size and excellent construction. In Qin and Han dynasties, as the most important building materials, the clay bricks were widely applied in engineering. Thus, the appellation "Qin bricks and Han tiles" is widespread enough since.

A new chapter in the history of building materials is opened as the application of cement and steel in civil engineering. In 1824, the cement was prepared by the calcine and grind of the mixed slurry of limestone and clay. And the corresponding invention patent was further obtained by English J. Aspding. Because of the similar color of the congealed cement and the limestone in the isle of Portland, it also be known as Portland cement( i. e. , the silicate cement in our country). In the 19th century, the steel was also applied in the field of civil engineering. For example, the world's first railway was built by English in 1823 and the Eiffel Tower as high as 320m was further built in Paris in 1889. Due to that the steel is very easy to get corrosion and rust in the process of contact with air and water. As a typical brittle material, the compressive strength of concrete material is much larger than its tensile strength. Adding the steel bars in concrete can not only protect steel bars from the erosion of harmful agents in the atmosphere and prevent rusting, but also improve the tensile properties of composites. On this basis, the reinforced concrete composite materials were emerged then. In 1850, the first reinforced concrete boat was made by the Frenchman Langbo. In 1872, the first reinforced concrete house was built in New York. In 1887, the load calculation method of reinforced concrete beam was developed by M. Koenen. In the 20th century, two tremendous progresses about the reinforce concrete have appeared. The first one is the concept of prestressed reinforced concrete proposed by C. R. Steiner in 1908. In 1928, the practical prestressed concrete structure was made by combining the high-tension steel and high-strength concrete by the French E. Fregssinet. The second one is the invention of water reducing agent by Americans in 1934. The adding of a small amount of water reducing agent into the ordinary concrete can greatly improve the workability and durability of material. In addition, since the beginning of the 20th century, the emergence of polymer organic materials, new metal materials, and a variety of composite materials has revolutionized the function and appearance of the buildings. The reliability and service life of buildings were also greatly improved.

High performance, multi-functional, industrial scale and ecological have been the development trends of civil engineering materials in present age. High-performance building materials refer to the building materials of better performance than existing materials. They have the characteristics of high strength, high durability, high impermeability, stable mechanical properties, etc. They can be applied in various super-high, super-long and super-large building structures and harsh conditions. To reduce cost and control quality of materials, the manufacture of building materials should be modernized and industrialized, and the production of them should be standardized, large-scale and commercialized. Moreover, under the background that the resources and environment are the two major problems to be urgently solved in the world. How to reduce environmental pollution, save resources and recycle waste have become a major strategic demand for the development of building materials industry in the 21st century. Green building materials, also known as ecological building materials or healthy building materials, refer to the non-hazardous, non-polluting, non-radioactive and recyclable materials which are produced by the clean production technology. The production of green building material can consume low power, and large quantity of industrial or urban solid waste and little natural resources. For example, by using the industrial waste residue (e. g. , fly ash and slag) as an admixture, the concrete materials are prepared. By using the waste foam plastic, the thermal insulation wall plates are also produced. By adopting above methods, a large amount of industrial waste is recycled, which further leads to the reduction of environmental pollution and conservation of natural resources.

## III. Standardization of civil engineering materials

The technical standards of civil engineering materials are the technical documents for the manufacturers and users to inspect the quality of products. In order to ensure the quality of materials, it is very necessary to set the uniform standards for the product manufacturing specification, which includes product specification, classification, technical requirement, inspection method, acceptance criteria, sign, transportation, storage, etc. In our country, the technical standards can be divided into four levels, i. e. , national standards, industry standards, regional standards and enterprise standards. In addition, the standards can also be divided into compulsory standards and recommendatory standards. The compulsory standards must be complied with in the prescribed scope of application. The recommendatory standards are technically authoritative and instructive, which can be implemented voluntarily. The recommendatory standards have also certain legal properties in the prescribed scope of contracts and administrative documents. All above levels of standards have their own department code names, for example, GB—national mandatory standard; GB/T—national recommendatory standard; JGJ—industry standard of the Ministry of Housing and Urban-rural Development; JG—industry standard of construction; SL—industry standard of water conservancy; DB—regional standard; QB—enterprise standard, etc. The representation of standards is composed of standard name, department code, serial number, approval year, etc. For instance, the national recommendatory standard *Standard for Design of Concrete Structure and Durability* can be represented by the code name "GB/T 50476—2019".

Each country in the world has developed its own national standards. For example, "ANSI" is the abbreviation for American National Standard ("ASTM" is the abbreviation for American Society for Testing and Materials Standard), "JIS" is the abbreviation for Japanese National Standard, and "BS" is the British Standard. In addition, the standards that can be implemented uniformly in the world are referred to as the international standards, which is codenamed "ISO".

The standards are generally set according to the technical level in a period of time. With the development of building materials and scientific technology, the technical standards also need to be gradually revised and improved based on the requirements of technological development.

## IV. Objective, methodology and feature of this course

"Civil engineering materials" is a professional basic course for civil engineering. It is established based on the classes like mathematics, mechanics, physics, chemistry, etc., and provides the basics of material science for the study of a series of follow-up professional courses such as architecture, structure, construction and so on. And this course is also necessary for students to be engaged in engineering practices and scientific research. A large body of knowledge about the materials including the content of raw materials, production, composition, structure, property, application, test, transportation and storage, as well as the currently technical standards is described in the book. The focus of this course is to help students to develop the ability of mastering the basic properties of materials and selecting them reasonably. To achieve this, it is very necessary to understand the characteristics of a variety of materials, which not only includes the basic properties of each material, but also contains the performance comparison of different kinds or types of materials. Only by mastering the characteristics of various materials, the proper materials can be accurately selected. In addition, it is also the requirement to know the basic principle why the materials possess the certain properties as well as the external conditions that affect the change of material properties.

Experimental course is the important part of this curriculum. Through the experiments, a variety of testing methods for the civil engineering materials should be learned and the eligibility of materials can also be judged and checked. Besides, it can also improve the ability of practices, correctly analyzing the experimental data and experimental results, and further cultivate the scientifically rigorous attitude and realistic work style.

The characteristics of this course mainly include the great practicality, strong comprehensiveness, etc. The content is very close to the engineering practice, which is different from the basic theoretical courses( i. e. , the simplification and abstraction of the actual phenomena). Besides, the extensive knowledge and comprehensive ability is further required here for the vast number of complex engineering problems, which is also different from the theoretical courses in which the thorough analysis is more focused.

# Chapter I  Basic properties of civil engineering materials

The reliability, durability and serviceability of engineering structures are heavily affected by the basic properties of civil engineering materials. When targeting different environments and functional requirements, the materials also need have the corresponding properties. For example, the structural materials shall have good mechanical properties and insulation properties, the roofing materials shall have good impermeability waterproof properties, the ground and pavement materials shall have non-slip and wear-resistance properties. According to the principle of material science, the compositions and structural characteristics of materials are the two determining factors for material properties. In the chapter, the relationship among the compositions, structures and properties of materials is analyzed and then their physical, mechanical and durable properties are further emphatically discussed. On one hand, they can be regarded as the important foundation for the selection, application and analysis of materials. On the other hand, they are also the vital knowledge for all the engineers in the engineering design and construction. Overall, mastering the characteristics and properties of materials is the essential foundation for the study, selection and application of various civil engineering materials.

## Section I  Composition, structure and property of civil engineering materials

### I. Composition of materials

The compositions of civil engineering materials are usually regarded as a joint name of chemical composition and mineral composition of the components.

The chemical composition of materials refers to the types and relative amount of the chemical elements and compounds that make up the materials. Mineral composition of materials refers to the relative amount of the mineral components in materials. The minerals are the basic elements of the earth's crust and they mainly refer to the natural elemental substances and compounds that are composed of a variety of chemical components in the special geologic condition. Generally, both the chemical compositions and interior structures of them are relatively stable. The mineral can also be regarded as the combination forms of elemental substances and compounds that make up the materials. In the field of civil engineering materials, the concept of mineral is further extended to the inorganic non-metal materials that have the specified crystalline structures and similar physically mechanical properties to the natural minerals.

The mineral composition of materials that have the same chemical composition does not necessarily identical. For example, both the diamond and graphite are the allotropes of carbon in chemistry, however, their mineral properties are very distinct. The chemical composition of materials that have different mineral composition is possible to be identical. For instance, on one hand, the hemi-hydrate gypsums in civil engineering include a variety of mineral compositions, e. g., $\alpha$-phase, $\beta$-phase and $\gamma$-phase. On the other hand, the chemical composition of them can be chemically symbolized by $CaSO_4 \cdot 0.5H_2O$. The chemical composition of materials that have the same mineral composition must be the same. Besides, the macroscopic performance of materials having different compositions in chemistry generally appears remarkedly different. For example, the conductivity of metallic materials is obviously superior to the non-metallic materials, the insulation performance of organic materials is also superior to the inorganic materials. Moreover, the effective properties of materials may be changed by the minor change of their chemical compositions. For instance, due to the difference of carbon contents, the pure iron, steel and pig iron appear the very different mechanical performance in engineering. The strength of pure iron is relatively low and soft. The strength of steel is relatively flexile. And the strength of pig iron is relatively hard and brittle.

In a word, the composition of materials directly affects the macroscopic properties. During the production and usage of materials, it is very important to determine the special components of materials and their relative proportions according to the requirement of structural performance.

## II. Structures of materials

The structures of materials can be regarded as another important factor that determines the properties of materials and they can usually be divided into three categories, i. e., macroscopic structures, mesoscopic structures and microscopic structures.

ⅰ. Macroscopic structures—the bulky structures that are distinguished and recognized by the eyes or magnifying glass.

1. According to the pore characteristics, the materials can be classified as

(1) Compact structure: the structure type of materials that are only able to absorb little water, e. g., metallic materials, glass, plastic, rubber, etc.

(2) Porous structure: the structure type of materials containing many bulky pores, e. g., aerated concrete, foamed concrete, foamed plastic, artificial lightweight porous materials, etc.

(3) Microporous structure, i. e., the structure type of materials containing many microfine pores, e. g., gypsum products, sintered clay products, etc.

2. According to the structural features, the materials can be classified as

(1) Stacking structure: the structure type of materials that are composed of granular particles and binding components, e. g., concrete, mortar, asphalt mixture, etc.

(2) Fibrous structure: the structure type of materials that are composed of natural or synthetic fibrous components, e. g. , wood, fiberglass, etc.

(3) Layer structure: the structure type of materials that possess the character of laminated layers, e. g. , plywood, plasterboard, etc.

(4) Granular structure: the structure type of materials that is composed of particulate substances, e. g. , expanded perlite, expanded vermiculite, fly ash, etc.

(5) Texture structure: the structure type of materials that possess the characters of textures, e. g. , wood, marble, etc.

ii. Mesoscopic structures (i. e. , submicroscopic structures)—the structures that can be distinguished and recognized by the optical microscope, and their size range is about $10^{-3} \sim 10^{-6}$ mm. For instance, the concrete materials can be deemed to be composed of matrix, aggregates and interfacial transition zone. The wood is composed of wood fiber, duct medullary ray and resin canal.

iii. Microscopic structures—the structural scale of materials is at the molecular, ionic, atomic, and subatomic levels. The macroscopic properties of materials such as elastoplasticity, hardness and strength are closely related to their microscopic structures. The microscopic structures can be divided into crystalline structure and non-crystalline structure.

1. Crystalline structure

Crystalline structure is a description of the ordered arrangement of atoms, ions or molecules in a crystalline material, such as the quartz, bluestone, rock candy, crystal and so on. The crystalline grains show the anisotropic characteristics, while the corresponding crystalline materials composed of these ordered crystal grains are generally isotropic. The concentration of atoms, ions or molecules, the interaction between them and the geometries of crystalline grains play a critical role in determining the properties of materials. The greater the concentration of particles (i. e. , atoms, ions or molecules) in the crystal is, the larger the plastic deformation of materials is. The mechanical strength of materials is larger when the crystalline grains are smaller and their distribution is more homogeneous. The properties of materials can be generally improved by the method of adjusting the size and structural characteristics of crystalline grains. Take the steel for example, the size of crystalline grains in steel shall be refined after cool forming, and the distortion and slip of crystal grains occur after heat treatment, which can further lead to the improvement of the mechanical strength of steel.

2. Non-crystalline structure

Non-crystalline structure is a description of a solid in which the atoms, ions or molecules is disorderly arranged, such as plastics, paraffin, asphalt, rubber and so on. The vitreum is a typically non-crystalline structure. Thus, the amorphous state is also called glass-like state. The vitreum is generally formed once the cooling speed of materials is too fast to make the inner particles disorderly arrange when the liquid melt is cooled down. The vitreous materials show the isotropic characteristics and have no specified melting point. The critical temperature when the materials begin to soften is generally defined as the softening point. Besides, when the liquid melt is cooled down, the energy between inner particles would be stored in the form of internal energy, which can make the vitreum the characteristic of chemical instability.

The crystalline structure and non-crystalline structure can be transformed into each other under a certain condition. For example, the quartz crystal can be transformed into quartz glass when they are melted and then cooled quickly. For the blast furnace, the stale crystalline structure of low chemical activity can be formed when they are naturally cooled in the air. However, once they are cooled quickly by using the high-pressure water, the vitreous structure can be obtained due to the inner particles are not arranged orderly. Furthermore, these pulverized blast furnace slag with vitreous structure can be taken as a high-quality admixture of concrete in engineering.

## Section II    Physical properties of materials

### I. Density, visual density, apparent density and bulk density

i . Density—the mass per unit volume of absolutely dense materials, which can be expressed by

$$\rho = m/V \tag{1-1}$$

where $\rho$, $m$ and $V$ are the density, mass and volume of absolutely dense materials, respectively.

The volume of materials at the absolutely dense state refers to the volume of solid matter in materials, which excludes the inner pore regions. Apart from the steel, glass, asphalt, etc., most of building materials in engineering contain a certain content of pores or cracks. When measuring the density of porous materials, the volume of solid matter can be obtained by the Lee's bottle method after the materials are grinded finely and dried.

ii . Visual density—the mass per unit volume of materials including solid matter and closed pores. It can be expressed by

$$\rho' = m/V' \tag{1-2}$$

where $\rho'$ and $m$ are the visual density and mass of materials at the absolutely dry state, $V'$ is the total volume of materials including the regions of solid matter and closed pores.

Generally, the total volume of solid matter and closed pores is defined as the visual volume. For the dense materials such as steel and glass, the visual density of them is approximately equal to the density. Thus, the visual density is also called the approximate density.

iii. Apparent density—the mass per unit volume of materials at the natural state, which can be expressed by

$$\rho_0 = m/V_0 \tag{1-3}$$

where $\rho_0$, $m$ and $V_0$ are the apparent density, mass and apparent volume of materials at the natural state, respectively.

The apparent volume of materials at the natural state refers to the volume sum of solid matter and all the pores. For the materials of regular geometry, the apparent density can be measured easily after the mass and apparent volume of materials are obtained. However, for the materials of irregular shapes, their volumes need to be obtained by the drainage method after the surface of materials is coated with the wax.

The apparent density of material is closely related to its water content, so the water-bearing state of material must be clearly sure when measuring the apparent density of material. In general, the water-bearing states of granular materials can be classified into four types, i. e., the dry condition, air-dried condition, saturated and surface-dried condition and wet condition, as shown in Figure 1-1. The commonly called apparent density refers to the corresponding value in the air-dried condition. Moreover, the apparent density of materials in the dry condition is called the absolute dry apparent density.

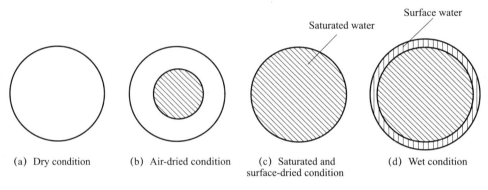

(a) Dry condition  (b) Air-dried condition  (c) Saturated and surface-dried condition  (d) Wet condition

**Figure 1-1  The water-bearing states of granular materials**

ⅳ. Bulk density—the mass per unit volume of granular materials at the natural state, which can be expressed by

$$\rho_0' = m/V_0' \tag{1-4}$$

where $\rho_0'$ and $m$ are the bulk density and mass of granular materials, $V_0'$ is the volume of granular materials at the natural state.

The bulk volume of granular materials here includes not only the volume of solid matter, but also the volume of the inner pores in materials and the empty space among the solid particles. Scan the QR code to find the related contents of experiments about density, visual density, apparent density and bulk density.

## Ⅱ. Porosity and Compactness of materials

Most of civil engineering materials contain pores. The content of pores in materials is generally denoted by the porosity.

The porosity usually refers to the ratio of the volume of inner pores in materials to the apparent volume of materials and it can be expressed by

Chapter Ⅰ  Basic properties of civil engineering materials

$$P = \frac{V_P}{V_0} \times 100\% = \frac{V_0 - V}{V} \times 100\% = \left(1 - \frac{\rho_0}{\rho}\right) \times 100\% \tag{1-5}$$

In contrast to the porosity, the compactness refers to the ratio of the volume of solid matter to the apparent volume of materials, which can be expressed by

$$D = \frac{V}{V_0} \times 100\% = \frac{\rho_0}{\rho} \times 100\% = 1 - P \tag{1-6}$$

The porosity which reflects the compact degree of materials has a direct impact on the mechanical, thermal and durable properties of materials. Besides, the materials with the same porosity may also present different pore characteristics. The properties of materials are closely dependent on the size and characteristics of inner pores. For example, the closed pores have good effect on the thermal insulating properties, and then promoting the frost resistance of materials with limits. The open or connected pores have an adverse effect on the heat preservation and impermeability. The mechanical properties of materials are adversely affected by the larger pores. However, the gel pores with the sizes less than 20nm have little influence on the mechanical strength.

In light of the influence extent of pores on the properties of materials, the inner pores in materials can be classified as harmful pores, harmless pores and beneficial pores. In light of the pore sizes, the pores in materials can be classified as nanometer pores, micron pores and millimeter pores. In light of the pore characteristics, they can be classified as open pores $V_K$ and closed pores $V_B$. And their volume sum is equal to the total volume of inner pores. Moreover, the porosity for open pores is expressed by $P_K = V_K/V_0$ and the porosity for closed pores is expressed by $P_B = V_B/V_0$.

## Ⅲ. Voidage and Fillrate of materials

The content of voids in granular materials is generally represented by the voidage, which is equal to the ratio of the volume of voids $V_V$ to the packing volume of granular materials $V'_0$, as below:

$$P' = \frac{V_V}{V'_0} \times 100\% = \frac{V'_0 - V_0}{V'_0} \times 100\% = \left(1 - \frac{\rho'_0}{\rho_0}\right) \times 100\% \tag{1-7}$$

In contrast to the voidage, the fillrate refers to the ratio of the volume of solid particles to the packing volume of materials, which is expressed by

$$D' = \frac{V_0}{V'_0} \times 100\% = \frac{\rho'_0}{\rho_0} \times 100\% = 1 - P' \tag{1-8}$$

The voidage which reflects the compact degree of solid particles in granular materials can be regarded as a critical indicator for assessing the size gradations of aggregates in concrete and the mixture design of concrete materials. In the process of the mixture design of concrete, the content of cement paste and sand ratio need to be calculated based on a series of parameters (e. g., bulk density of aggregates, voidage, etc.).

## IV. Water-related physical properties of materials

i . Hydrophilicity and hydrophobicity

When the materials contact with water directly, the materials which can be wetted by water is generally defined as hydrophilic materials, the materials which can not be wetted by water is defined as hydrophobic materials. The hydrophilicity of materials can be attributed to that the chemical affinity between the molecule of material and the molecule of water is larger than the cohesion of the molecules of water. Conversely, if the chemical affinity between the molecule of material and the molecule of water is smaller than the cohesion of the molecules of water, the materials is shown to be hydrophobic.

The wetting degree of materials by the water can be expressed by the wetting angle $\theta$, as shown in Figure 1-2. The smaller the value of $\theta$ is, the more easily the material is wetted by water. As displayed in Figure 1-2(a), when $\theta \leqslant 90°$, the surface of material is very easy to absorb water and be wetted by water, which presents the hydrophilicity. As shown in Figure 1-2(b), when $\theta > 90°$, the surface of material is hard to absorb water, which presents the hydrophobicity. In addition, when $\theta = 0°$, the appearance of $\theta = 0°$ indicates that the surface of material can be absolutely wetted by water.

The hydrophilic materials are very easy to be wetted by water and the water can be penetrated into the materials through the capillary action. The hydrophobic materials can prevent the water penetrate into the capillary, and then reduce the water absorbency of materials, which are generally used as the waterproof materials in engineering. Most of civil engineering materials (e.g., cement, concrete, sand, gravel, brick and wood) are hydrophilic, and only a handful of materials (e.g., asphalt, paraffine and some plastics) are hydrophobic.

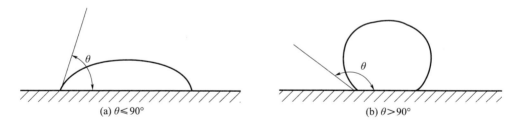

(a) $\theta \leqslant 90°$    (b) $\theta > 90°$

**Figure 1-2  The schematic diagram of hydrophilicity and hydrophobicity**

ii . Water absorbency and moisture absorption of materials

1. Water absorbency

Water absorbency refers to the ability of a body to absorb the water in water, which is generally expressed by the water absorption.

(1) Mass water absorption—the ratio of the mass of absorbed water to the mass of materials in the absolutely dry state, which is expressed by

$$W_m = \frac{m_{sw}}{m} \times 100\% \qquad (1-9)$$

where $W_m$ is the mass water absorption, $m_{sw}$ is the mass of absorbed water in material and $m$ is the mass of material in the absolutely dry state.

(2) Volume water absorption—the ratio of the volume of absorbed water to the apparent volume of materials in the absolutely dry state, which is expressed by

$$W_V = \frac{V_{sw}}{V_0} \times 100\% \qquad (1\text{-}10)$$

where $W_V$ is the volume water absorption, $V_{sw}$ is the volume of absorbed water in material and $V_0$ is the apparent volume of material.

According to Eqs. (1-9) and (1-10), the relationship between mass water absorption and volume water absorption is expressed by

$$W_V = W_m \cdot \rho_{0d} \qquad (1\text{-}11)$$

where $\rho_{0d}$ is the apparent density of materials in the absolutely dry state.

The water is generally penetrated into materials through the open pores. Thus, the larger the porosity of open pores is, the higher the content of absorbed water is. The volume water absorption when the inner pores are saturated can be viewed as the porosity of open pores. The water absorbency of materials is closely dependent on the porosity and pore characteristics. The larger the porosity of tiny connected pores is, the larger the water absorption would be. Due to that the water is impossible to penetrate into materials through the closed pores. Besides, the water is also not able to preserve in the open big pores. Thus, the corresponding water absorption is still very small. Actually, the difference of water absorption for different materials is quite great. For example, the water absorption for granite, concrete, sintered clay brick, and wood are approximately equal to 0.50% ~0.70%, 2% ~3%, 8% ~20% and more than 100%, respectively. Scan the QR code to find the experiments of basic properties of materials about water absorption.

2. Moisture absorption

Moisture absorption refers to the ability of a body to absorb moisture in the wet air, which is expressed by the moisture content. The moisture content is the ratio of the mass of water in material in the wet state to the total mass of material in the dry state, as below:

$$W_h = \frac{m_s}{m} \times 100\% \qquad (1\text{-}12)$$

where $W_h$ is the moisture content, $m_s$ is the mass of water in materials in the wet state, and $m$ is the total mass of materials in the absolutely dry state.

The moisture absorption of materials may vary with the change of air humidity and ambient temperature. With the increase of air humidity and the decrease of ambient temperature, the moisture content of materials would also increase. A specified moisture content (i.e., equilibrium moisture content) is defined to represent the critical state that the content of moisture in materials and the air humidity is in equilibrium. The moisture absorption of materials containing tiny open pores (e.g., wood and some insulation materials) may be especially great. This is because the inner surface area of these materials is very large, which in turn leads to their strong ability to absorb the water.

Both the water absorption and moisture absorption have adverse effect on the performance of materials. As the water is absorbed into materials, the mass of materials may be larger and the thermal insulation, mechanical strength and durability of materials would be reduced in different levels. The moisture absorption of materials can be used to remove the dampness and then maintain the good dry environment.

iii. Water resistance of materials

Water resistance refers to the ability of a body to resist the damage and strength reduction in water for a long time, which is generally expressed by the softening coefficient as below:

$$K_R = \frac{f_b}{f_g} \times 100\% \qquad (1\text{-}13)$$

where $K_R$ is the softening coefficient, $f_b$ and $f_g$ are the compressive strength of materials in the water-saturated and dry states, respectively.

The parameter $K_R$ is used to represent the reduction degree of materials strength after they are immersed in water. The reduction of mechanical strength for materials can be attributed to that: when the materials is immersed in water, a coating of water would be formed on the fine grains, which would in turn lead to the weakening of the binding force between these granular compositions. The smaller the value of $K_R$ is, the larger the reduction degree of the mechanical strength of materials is. The range of $K_R$ is generally in the range of 0 ~ 1 (e.g., $K_R = 0.0$ for clay and $K_R = 1.0$ for metal). In the field of civil engineering, the materials with $K_R > 0.85$ are defined as the waterproof materials. For the important buildings in water or wet conditions for long time, the value of $K_R$ for the materials must be larger than 0.85. For the buildings of minor importance, the value of $K_R$ for the materials should also not be less than 0.75.

iv. Impermeability of materials

Impermeability refers to the ability of a body to resist the penetration of pressure water, which is generally represented by the permeability coefficient. The physical interpretation of permeability coefficient can be described as below: the quantity of transferred water through the materials with 1cm thickness and 1cm$^2$ area in one second under the action of unit head pressure.

$$K_S = \frac{Qd}{AHt} \qquad (1\text{-}14)$$

where $K_S$ is the permeability coefficient, $Q$ is the quantify of permeable water, $d$ and $A$ are separately the thickness and area of materials, $t$ is the time, and $H$ is the head of hydrostatic pressure.

The larger value of permeability coefficient $K_S$ indicates the impermeability of this material is worse. The impermeability of materials can be described by the impermeability grade. The grade is ranked according to the maximum water pressure the specimens can stand in the specified condition and test method, which is symbolized by "P$n$". In the sign "P$n$", $n$ represents the value of the 10 times of the maximum water pressure the specimens can stand. For example, the symbol P4 indicates that maximum water pressure the materials can stand is 0.4MPa.

The impermeability of materials is closely dependent on the porosity and pore characteristics. The water penetrates more easily through the connected pores. Thus, the impermeability of materials with a large number of connected pores would be worse. Comparatively, the water can not penetrate through the closed pores. Thus, the impermeability of materials with a large number of closed pores may be better. In addition, the impermeability of materials with great and connected pores is the worst because the water is the easiest to penetrate through these pores in materials.

V. Frost resistance of materials

Frost resistance refers to the ability of a body to resist the damage and strength reduction in water after many freeze-thaw cycles, which can be expressed by the frost resistance grade. These grades are ranked in right of the number of maximum freeze-thaw cycles the specimens can stand in the specified condition and test method where the reduction of mechanical strength is not more than a specified value and there is no obvious damage and spalling on the surface of materials. The frost resistance grade here is symbolized by "$Fn$", in which $n$ represents the number of maximum freeze-thaw cycles (e. g., F25).

The freeze-thaw damage of materials can be attributed to that the ice is formed from the water when the temperature is lower than 0℃. When the water freezes, its volume would increase about 9 percent. Once the inner pores in materials are filled with water, the great stress may occur due to the volume dilation and the local cracking of hole wall may be further generated when the stress is larger than the tensile strength of materials. Furthermore, the damage degree of materials would also increase with the increase of the number of freeze-thaw cycles. Overall, the frost resistance of materials is strongly dependent on the porosity, pore characteristics and saturated extent of inner pores. If the pores are not saturated by water, the great stress is not generated even if the materials are suffering cold because of the sufficient free space in the pores. For the tiny pores filled with water, the inner water in these pores will freeze when the temperature is very lower than 0℃, which is because that freezing points for the water absorbed onto the pore walls is very low. The capillary pores are likely to be saturated by water, which can also be frozen easily. Thus, these capillary pores would have greatly adverse influence on the performance of materials. The higher the proportion of closed pores is, the better the frost resistance of materials are. Based on above reasons, the frost resistance of materials can be improved by some methods in engineering. For instance, the content of closed pores in materials can be increased by the air entraining agent, which in turn leads to the improvement of frost resistance of materials. In addition, the frost resistance of materials will also be increased with the increase of all the deformability, mechanical strength and softening coefficient of materials.

From the aspect of external environmental factors, the freeze-thaw damage degree of materials is related to the freeze-thaw temperature, icing speed, freeze-thaw frequency and so on. The freeze-thaw damage of materials would be more serious with the decreasing environmental temperature, faster cooling rate and more frequent freeze-thaw cycling.

## V. Thermal properties of materials

The thermal properties of civil engineering materials mainly include the thermal conduction, heat capacity, specific heat, etc.

i . Thermal conduction

Thermal conduction refers to the ability of a body to transfer heat from the side of materials with high temperature to the other side of them with low temperature, which can be represented by the thermal conductivity. The physical interpretation of thermal conductivity can be described as below: the quantity of transferred heat through the materials with 1m thickness and 1m$^2$ area in one second when the temperature is changed by 1K.

$$\lambda = \frac{Q \cdot d}{A \cdot \Delta T \cdot t} \tag{1-15}$$

where $\lambda$ is the thermal conductivity, $Q$ is the total quantify of conducted heat, $d$ and $A$ are the thickness and area of materials, respectively, $t$ is the time, and $\Delta T$ is the temperature difference between the two sides of materials.

The smaller the thermal conductivity of materials is, the better the heat-insulating performance of them is. The thermal conductivity varies enormously depending on the types of materials. For example, the thermal conductivities of foam and marble are separately 0.03W/(m·K) and 3.48W/(m·K). The materials with $\lambda$ <0.23W/(m·K) are commonly known as insulation material. The thermal conductivity of materials is not only dependent on the compositions of materials, but also related to the content, characteristics of inner pores and its water-bearing state. For instance, the thermal conductivities of air and water are 0.025W/(m·K) and 0.6W/(m·K), respectively. For the porous materials, once there exists a large number of closed tiny pores, and the space of these pores is filled with air in the dry state, the thermal conductivity of these materials can be effectively reduced. On the contrary, if these pores are filled with water, the thermal conductivity of materials may be increased, which in turn leads to the reduction of the heat preservation.

ii. Heat capacity and Specific heat

Heat capacity refers to the ability of a body to absorb the heat when heated and release the heat when cooled, which can be represented by the specific heat of materials. The physical interpretation of specific heat can be described as below: the quantity of absorbed (or released) heat in the materials with 1kg when the temperature is changed by 1K.

$$C = c \cdot m \tag{1-16}$$

where $C, c$ and $m$ are separately the heat capacity, specific heat and mass of materials.

The heat capacity and specific heat of materials are the two important parameters for evaluating the thermal properties of materials in the design of building envelopes (e.g., the walls, roofs, etc.). To maintain the stability of indoor temperature in buildings, the materials of small thermal conductivity and large specific heat are generally selected and applied.

# Section III Mechanical properties of materials

The mechanical properties of materials refer to the ability of a body to resist the deformation and damage under the external loads, which is the fundamental factor for the material selection. In the design of engineering structure, the maximum bearing capacity and minimum deformation are generally regarded as the principles for the selection of materials.

## I. Strength, specific strength and theoretical strength of materials

The mechanical strength of materials refers to the ability of a material or a mechanism to withstand an applied stress/strain or mechanical load without structural failure. There are many kinds of stresses due to the form of external loads. As shown in Figure 1-3, the basic forms of mechanical strengths can be divided into compressive strength, tensile strength, flexural strength and shear strength.

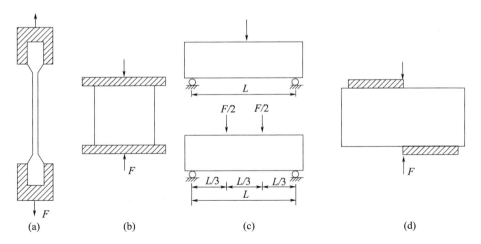

**Figure 1-3** The schematic diagram of the external loads applied on the specimens of materials

All the tensile strength, compressive strength and shear strength can be calculated by Equation (1-17).

$$f = \frac{P}{A} \quad (1\text{-}17)$$

where $f$ is the tensile strength, compressive strength or shear strength, $P$ is the maximum external load when the specimen is damaged, and $A$ is the stressed area.

The bending test of materials is generally carried out based on the specimens of rectangular sections and the flexural strength can be obtained by two methods. The first one is that a concentrated load is applied on the middle position of specimen and the corresponding calculation is expressed as below:

$$f_{tm} = \frac{3PL}{2bh^2} \qquad (1\text{-}18)$$

where $f$ is the flexural strength, $P$ is the maximum load when the specimen is damaged, $L$ is the distance between the two fulcrums, $b$ and $h$ are separately the width and height of the cross section of specimen.

The other method is two equivalent concentrated loads are simultaneously applied on the two positions between the two fulcrums and the corresponding calculation is given as below:

$$f_{tm} = \frac{PL}{bh^2} \qquad (1\text{-}19)$$

Apart from the compositions of materials, there are also many external factors that affect the mechanical strengths of materials. For example, with the increase of material porosity, moisture content and temperature, the mechanical strength of materials may decrease. In addition, the strength of the specimen of materials with larger sizes is lower than the ones with smaller sizes. The strength of prism specimen is usually lower than the strength of cubic specimen. When the oil is coated on the surface of specimen, its strength can be decreased. Moreover, both the lower loading speed and rough surface of materials may also lead to the reduction of the test strengths of materials.

The mechanical strengths of materials can be viewed as the measured values under the specified condition. In order to ensure of the accuracy and comparability of tested results, the unified testing standards of materials were developed in many countries, which should be strictly enforced when the strengths of materials are tested.

Except for the external loads, the materials have also to withstand their own gravities. To compare the strengths of different materials, a new parameter (i. e. , the specific strength) is adopted. The specific strength refers to the ratio of material strength to its apparent density, which is deemed to be an indicator for measuring the relative strength of materials. For instance, the specific strengths of steel, wood and concrete are separately equal to 0. 054, 0. 069 and 0. 017, which indicate that the wood is the high-strength and light-weight material, and the concrete is the low-strength and high-weight material. How to make the concrete materials higher strength and lighter weight is an important and urgent task in modern times.

Above mechanical strengths which are obtained by the experiments are referred to as actual strength. The actual strength of materials is much lower than their theoretical strength. The theoretical strength refers to the maximum theoretical stress the material can bear and it is derived by theoretical analysis based on the structural features of materials. The theoretical strength is the corresponding force required to form two new surfaces in materials when the binding force between inner particles is broken. The damage of materials here is mainly attributed to the occurrence of the cracking or displacement of the inner particles in materials. There are many theoretical formulas for deriving the theoretical strengths of materials, as following:

$$\sigma_L = \sqrt{\frac{E\gamma}{d}} \qquad (1\text{-}20)$$

where $\sigma_L$ is the theoretical tensile strength, $E$ is the elastic modulus, $\gamma$ is the surface energy of solids and $d$ is the interatomic spacing.

In engineering, there are some inevitable defects (e.g., the crystalline defects and micro-fissures) in materials. The existence of crystalline defects may induce the migration of lattices under the relative low stress. The existence of micro-cracks can also lead to the stress concentration at the tips of cracks. As the isolated cracks are expanded and connected, the mechanical strengths of materials would be seriously reduced. For example, the theoretical tensile strength of steel is up to 30000MPa. However, the actual tensile strength of carbon steel is only about 400MPa (the strength of high tensile steel is about equal to 1800MPa).

## II. Elasticity and plasticity of materials

The elasticity of materials refers to the ability of a body to resist a distorting influence or stress and to return to its original size and shape when the stress is removed. This recoverable deformation is called elastic deformation. If the relationship between stress and strain is linear, as described by Eq. (1-21). This body is defined as Hook's elastomer, in which the proportional constant $E$ is the elastic modulus.

$$\sigma = E \cdot \varepsilon \qquad (1\text{-}21)$$

Where $\sigma$ and $\varepsilon$ are the stress and strain of materials, respectively, and $E$ is the elastic modulus.

Elastic modulus is an index to evaluate the deformation resistance of materials. The larger the value of $E$ is, the less easily the deformation of materials is formed, which indicates the high rigidity. The elastic modulus $E$ is an important parameter for the design of structures.

The plasticity refers to a property of solids whereby the solids irreversibly change their dimensions and shape—that is, are plastically deformed—under the action of mechanical loads. Above deformation is called the plastic deformation, which is irreversible.

Actually, there is no pure linear elastic material in engineering. A great deal of materials (e.g., mild steel) show the characteristics of elasticity when the external load is low, and further show the characteristics of plasticity when the value of load exceeds a certain limit. Moreover, the elastic and plastic deformation may be presented simultaneously when the materials (e.g., concrete) are under the action of loads. Once these external loads are removed, the corresponding elastic deformation can be recovered, whereas, the plastics deformation may be still preserved then.

## III. Brittleness and toughness of materials

### i. Brittleness

The brittleness refers to the characteristic of a material that is manifested by sudden or abrupt failure without appreciable prior ductile or plastic deformation. The materials with this property are called brittle materials. In light of that the compressive strength of brittle materials is much larger than the tensile strength (up to several times), they should not be applied in the tensile region and subjected to the impulse load. That is to say, the brittle materials only apply to the pressure bearing structures. Most of inorganic non-metallic materials in civil engineering (e. g. , natural rock, ceramics, glass, ordinary concrete, etc. ) are the brittle materials.

ii. Toughness

The toughness refers to the property of a material capable of absorbing energy by plastic deformation before fracture and to resist shock or impact. Compared with the brittle materials such as stone and concrete, the toughness of steel is very high and generally applied in the structural components subjected to the impact or having the seismic requirements in civil engineering.

## IV. Hardness and abrasion resistance of materials

i. Hardness

The hardness refers to the ability of material surface to resist the scratching by hard substance. The greater the hardness of materials is, the higher both the mechanical strength and abrasion resistance of materials would be.

There are many methods for quantitative determination of material hardness, such as the scratch method, the indentation method and the rebound method. The scratch method is mainly applied to measure the hardness of natural minerals, which can be released with 10 certificates (i. e. , talcum, gypsum, calcite, fluorite, apatite, orthoclase, quartz, topaz, adamantine spar and diamond). The hardness of steel, wood, concrete, etc. can be generally measured by the indentation method. In addition, the rebound method is usually adopted to measure the hardness of the surface of concrete structures, and then estimate the compressive strength of concrete materials.

ii. Abrasion resistance

The abrasion resistance refers to the ability of material surface to resist the abrasion, which is expressed by the wear rate as below:

$$N = \frac{m_1 - m_2}{A} \qquad (1\text{-}22)$$

where $N$ is the wear rate of materials, $m_1$ and $m_2$ are the mass of specimens before and after they are scuffed, respectively. $A$ is the area of specimens.

The abrasion resistance of materials is closely related to the compositions, structure, mechanical strength and hardness of materials. In the water projects, the spillway face, piers and floor on the dam are generally damaged by the flushing and far-reaching impact of high-velocity flow. Thus, the materials in these regions should have good ability to resist the abrasion of external substances. Besides, the materials in the stair steps, ground and pavements should also possess the higher abrasion resistance.

## Section IV  Durability of materials

The durability of materials refers to the ability of civil engineering materials to resist various environmental factors (e. g., weathering action, chemical attack, abrasion and other conditions of service) and retain their characteristics and performance for long after an extended period of time and usage.

### I. Effects of environment on the materials

In the service process of engineering buildings, apart from the intrinsic influencing factors, the materials would also be affected by the serve conditions and various natural factors, which can be classified into the following classes.

i. Physical actions—including the alterations of temperatures, humidity, freeze-thaw and so on. After being subjected to above factors, the phenomenon of expansion, shrinkage or internal stress may occur in materials, which in turn lead to the damage of structures with the prolonging of time.

ii. Chemical actions—including the erosion of acid, alkali, salt or other harmful substances in the atmosphere and water as well as the irradiation of sunlight, ultraviolet light, etc.

iii. Mechanical actions—including the continuous action of loads, which can induce the fatigue and abrasion of materials.

iv. Biological actions—including the invasion of fungi, insects, etc., which may induce the decay and rottenness of materials.

### II. The test of the durability of materials

The most effective way to judge material durability is to make the long-term observation and measurement of materials exposed to the actually natural environment. However, they need to take a long time. Besides, to surmount above deficiencies, the accelerated test methods are generally applied in reality. The essence of these methods is that, the accelerated tests are carried out in the laboratory by simulating the actual environmentsand the durability of materials is further analyzed based on the experimental results (e. g., the tests of drying-wetting cycling, freezing-thawing cycling, carbonation, etc.).

# Questions for review

1. Please describe the effect of the composition and structure change of materials on the properties.
2. Please describe the relationship among the density, apparent density and porosity.
3. Please describe the effect of the pore characteristics in materials on the properties.

4. What are the influencing factors that determine the frost resistance of materials and how to improve the frost resistance of materials.

5. What are the thermal conductivity and specific heat, respectively. And what are the influencing factors that determine the thermal conductivity.

6. Please describe the definition of material strength and why the mechanical strength of materials should be measured by the standard methods.

7. What is the difference between brittle materials and ductile materials and what questions should be noticed in the use of them.

8. Please describe the definition of the durability of materials and what all does that include. How to measure the durability of different kinds of materials.

9. There is a common clay brick with its sizes of 240mm × 115mm × 53mm. The mass of brick after drying and being saturated by water are separately equal to 2420g and 2640g. After it is dryed and grinded, the measured volume of the material with 50g by the Lee's bottle method is equal to 19.2cm$^3$. Please calculate the porosities of open pores and closed pores in the brick, respectively.

# Chapter II  Air-hardening cementitious material

Generally, cementitious material is a substance that can combine dispersed grain material (such as sand and gravel) or blocky material (such as bricks and stones) through a series of physical and chemical interactions. Cementitious material can be divided into inorganic and organic cementitious material according to their chemical composition, the former includes cement, lime, gypsum, magnesite, water glass, etc., and the latter includes asphalt, organic polymer, etc. Among them, inorganic cementitious material is more widely utilized in engineering with abundant dosage.

According to different hardening conditions, inorganic cementitious material can be separated into two categories: air-hardening cementitious material and hydraulic cementitious material. The air-hardening cementitious material can only be hardened in the air, and its strength can maintain and continue to develop only in the air as well, such as lime, gypsum, magnesite, water glass, etc. Hydraulic cementitious material can not only be hardened in the air, but also better in the water, and maintain and continue to develop strength, such as various cements. Therefore, air-hardening cementitious material is only applicable to the ground or dry environment, neither suitable for wet environment, nor for water; nevertheless, hydraulic cementitious material can be applied both on the ground and underground or underwater environment.

## Section I  Lime

Lime is one of the earlier cementitious materials used in construction engineering. It is a general term for quicklime, slaked lime and hydraulic lime with different chemical compositions and physical forms. It still has wide application in civil engineering so far owing to its extensive distribution of raw materials, simple production process and low cost.

### I. Raw materials and production of lime

The main raw materials for preparing lime are limestone whose principal component is calcium carbonate, dolomite, chalk, shell lamp and other natural rocks. Through calcination, calcium carbonate decomposes into calcium oxide, which is quicklime. The chemical reaction formula is as follows:

$$CaCO_3 \xrightarrow{900 \sim 1000^\circ C} CaO + CO_2 \uparrow \quad -178 kJ/mol$$

Due to different compactness, block size and impurity content of limestone, and considering the heat loss, calcination temperature is usually controlled at $1000 \sim 1100^\circ C$ in order to fully calcine $CaCO_3$. Quicklime is in blocky shape, which is also called blocky lime. As production raw materials often contain some magnesium carbonate, calcined quicklime is generally accompanied by magnesium oxide, a minor component. According to China's building materials industry standard *Building Quicklime* (JC/T 479—2013), when magnesium oxide content is less than or equal to 5%, it is called calcareous lime, and when magnesium oxide content is more than 5%, it is called magnesian lime. Magnesian lime reacts slowly, however, its strength is slightly higher after hardening.

Calcined quicklime can be processed in different ways to obtain three other products of lime:

Quicklime powder: it is made of finely ground blocky quicklime.

Slaked lime powder: a powder made by slaking and drying quicklime with appropriate amount of water, and the main component is $Ca(OH)_2$, which is also called hydrated lime.

Lime Paste: A paste of certain consistency made by slaking blocky lime with overdose water (about 3~4 times the volume of quicklime), or mixing slaked lime powder with water, where the main components are $Ca(OH)_2$ and water.

## II. Hydration and hardening of lime

i. Hydration of lime

The hydration of quicklime, also known as slaking, refers to the process of hydration of quicklime with water to produceslaked lime $[Ca(OH)_2]$. The reaction equation is as follows:

$$CaO + H_2O = Ca(OH)_2 \quad +64.9 kJ/mol$$

Slaking reaction of Quicklime is exothermic, where not only the hydration heat is large during slaking, but also the exothermic rate is fast. Whilst, 1.5~2 times volume expansion occurs significantly.

Owing to uneven fire during calcination, lime often contains under-burnt lime and over-burnt lime. Under-burnt lime is generated when the calcination temperature is too low, the calcination time is too shot or the kiln temperature is uneven. Meanwhile the limestone is not completely decomposed, and the resulting lime has more pores, large specific surface area, and an internal core of undecomposed $CaCO_3$. Over-burnt lime is generated under the circumstance of too long calcination time or high calcination temperature. Its internal structure is dense, and CaO grains are coarse. Moreover, the activity of over-burnt lime is greatly low, and the reaction rate is extremely slow. After lime hardening, it begins to react, generating volume expansion and causing bulging or cracking of the hardened lime. In order to prevent bulging and cracking caused by volume expansion of over-burnt lime, the lime paste should be stored in the ash pits for more than two weeks to make it fully hydrated, and this process is also called lime "Chen Fu". During the "Chen Fu" period, the surface of lime paste should be covered in a layer of water, isolated from outside air to avoid carbonation.

ii. Hardening of lime

The hardening of hydratedmortar in air is accomplished by two simultaneous processes.

Crystallization process: During drying process of lime paste, free water is evaporated, causing the crystallization and precipitation of $Ca(OH)_2$ from the saturated solution.

Carbonation process: $Ca(OH)_2$ reacts with $CO_2$ and water in the air to form water-insoluble calcium carbonate crystals, and the formed water is gradually evaporated, as seen in the following reaction:

$$Ca(OH)_2 + CO_2 + nH_2O \longrightarrow CaCO_3 + (n+1)H_2O$$

The carbonation mainly occurs in the surface layer in contact with air, and the resulting $CaCO_3$ film is dense, preventing the infiltration of $CO_2$ from the air and the evaporation of water from the interior to the outside, which causes slow hardening.

## III. Characteristics and technical requirements of lime

### i. Characteristics of lime

1. Good plasticity and water retention

Lime paste formed by the slaking of quicklime is a highly dispersed colloid, generating extremely fine $Ca(OH)_2$ particles (about $1\mu m$ in diameter), which has a large specific surface area and can adsorb a large amount of water; and a thick water film is formed on the surface, reducing the friction between the particles. Lime paste has good plasticity and water retention, and is easy to spread into a uniform thin layer. Taking advantage of this characteristic, the addition of lime mortar to cement mortar can significantly ameliorate its plasticity and water retention.

2. Slow setting and hardening, low strength

The hardening of lime paste can only take place in air. Owing to the low $CO_2$ content in the air, the carbonation process is decelerated, and the hardened surface layer impedes the internal harde-ning, thus the hardening process of lime paste will take long time. The strength of hardened lime is very low and the 28d-strength of a lime mortar with a ratio of 1 : 3 is usually only 0.2~0.5MPa.

3. Poor water resistance

Owing to the slow hardening and low strength of lime paste, the non-carbonated $Ca(OH)_2$ tends to dissolve when lime is exposed to moisture, and the hardened lime will collapse when exposed to water, so lime is not suitable to use in humid environments.

4. Large volume shrinkage during hardening.

A large amount of water is evaporated during the bonding and hardening process of lime paste. On account of capillary water loss shrinkage which causes volume shrinkage, the hardened lime generates cracks, so the lime paste should not be used alone. It is usually mixed with a certain amount of aggregate (sand) or fiber materials (sisal, paper tendons, etc.) in engineering construction.

### ii. Technical requirements for lime

Building quicklime shall comply with the requirements of *Building quicklime* (JC/T479—2013), whose details can be acquired from the QR code.

Building slaked lime shall comply with the requirements of *Building slaked lime* (JC/T 481—2013), whose details can be acquired from the QR code.

Table 2-1  Classification of building slaked lime

| Category | Name | Code |
| --- | --- | --- |
| Calcareous slaked lime | Calcareous slaked lime 90 | HCL 90 |
|  | Calcareous slaked lime 85 | HCL 85 |
|  | Calcareous slaked lime 75 | HCL 75 |
| Magnesian slaked lime | Magnesian slaked lime 85 | HML 85 |
|  | Magnesian slaked lime 80 | HML 80 |

Table 2-2  Chemical composition of building slaked lime

| Code | CaO + MgO | MgO | $SO_3$ |
| --- | --- | --- | --- |
| HCL 90 | ≥90 | ≤5 | ≤2 |
| HCL 85 | ≥85 |  |  |
| HCL 75 | ≥75 |  |  |
| HML 85 | ≥85 | >5 | ≤2 |
| HML 80 | ≥80 |  |  |

Note: The values in the table are based on the dry basis of the specimen after deduction of free and chemically bound water.

Table 2-3  Physical properties of building slaked lime

| Code | Free water/% | Fineness | | Soundness |
| --- | --- | --- | --- | --- |
|  |  | 0.2mm sieve residue/% | 90μm sieve residue/% |  |
| HCL 90 | ≤2 | ≤2 | ≤7 | Qualification |
| HCL 85 |  |  |  |  |
| HCL 75 |  |  |  |  |
| HML 85 |  |  |  |  |
| HML 80 |  |  |  |  |

iii. Application and storage of lime

Lime is one of the most widely used building materials in construction projection and its common applications are as follows:

1. Lime emulsion coating

The slaked lime powder or lime paste mixed with a large amount of water can be used to form a lime emulsion coating, which is a cheap paint, easy to construct and white in color, applying for painting interior walls and ceilings.

2. Construction mortar

Slaked lime paste and slaked lime powder can be used in mortar alone or with cement, the former is called lime mortar and the latter is called mixed mortar. Lime mortar can be applied to plastering for brick walls and concrete base, while mixed mortar is utilized for masonry and plastering.

3. Reinforcement of soft ground containing water

Quicklime block can be used directly to strengthen the soft foundation containing water. Foundation is reinforced by filling the lime block in the pile hole and thus generating expansion stress when it absorbs water and is slaked.

4. Silicate products

Building materials made by mixing finely ground quicklime (or slaked lime powder) and siliceous materials (fly ash, granulated blast furnace slag, furnace slag, etc.) with water, after forming, steam curing or autoclave curing and other processes, are collectively known as silicate products, such as grey sand bricks, fly ash bricks, fly ash blocks, silicate blocks, etc.

5. Lime soil and trinity mixture fill

Grey clay is formed by combining slaked lime powder and clay, and then sand was added, forming trinity mixture fill. The compactness of grey clay or trinity mixture fill is greatly improved by tamping or compaction. Moreover, in a humid environment, a small amount of active silicon oxide and aluminum oxide on the surface of the clay particles react with $Ca(OH)_2$ to produce hydraulic calcium silicate hydrate and calcium aluminate hydrate, favoring the clay's impermeability, compressive strength and water resistance. The trinity mixture fill and grey clay are mainly applied to bed course of building foundations, pavements and floors.

6. Finely ground quicklime powder

Finely ground quicklime is utilized in large quantities to replace lime plaster and slaked lime powder in the preparation of grey clay or mortar, or directly in the manufacture of silicate products.

Care should be taken when storing lime, as lime blocks left for too long will absorb moisture from the air and form slaked lime powder, which then reacts with carbon dioxide in the air and leads to the formation of calcium carbonate and loss of cementitious ability. Therefore, it is best to be stored in a closed and tightly sealed warehouse, protected from moisture and water. In addition, the storage period should not be too long, if long term storage is needed, it can be slaked into lime paste and then covered with sand so as to prevent carbonation. When transporting lime blocks, please try to use a shed truck or cover them with canvas to prevent the water shower from slaking on its own and exothermically causing a fire.

# Section II  Gypsum

Gypsum is one of air-hardening cementitious materials with calcium sulfate as the main component. Gypsum is a traditional cementitious material with many excellent building properties and is widely utilized in the field of building materials due to its abundant raw material sources and low energy consumption in production. There are many varieties of gypsum gelling materials, including construction gypsum, high-strength gypsum, anhydrous gypsum cement, high-temperature calcined gypsum, etc.

## I. Raw materials, production and varieties of gypsum

The main raw material of gypsum production is natural dihydrate gypsum and natural anhydride. Gypsum waste products of the chemical industry (such as phosphogypsum, fluorine gypsum, boron gypsum) are also composed of dihydrate gypsum, which can also be used as raw material for the production of gypsum. When using the chemical gypsum, it should be noted that if the waste slag (liquid) contains acidic components, it must be pre-washed with water or neutralized with lime before use.

Gypsum is divided into low temperature calcined gypsum and high temperature calcined gypsum according to the temperature at which it is calcined during production.

### i. Low temperature calcined gypsum

Low temperature calcined gypsum is a product obtained by calcining natural gypsum at low temperature (110 ~ 170℃), the main component of which is hemihydrate gypsum ($CaSO_4 \cdot 0.5H_2O$). At this temperature, dihydrate gypsum is dehydrated and transformed into hemihydrate gypsum.

$$CaSO_4 \cdot 2H_2O = CaSO_4 \cdot 0.5H_2O + 1.5H_2O$$

Products that belong to low temperature calcined gypsum are building gypsum and high strength gypsum.

#### 1. Building gypsum

Building gypsum, also known as plaster or β-hemihydrate (β-$CaSO_4 \cdot 0.5H_2O$), is produced by heating and calcining natural gypsum in a kiln at atmospheric pressure. β-hemihydrate gypsum crystals are irregularly flaky, secondary particles composed of tiny individual grains, which are less crystalline and more dispersed.

#### 2. High strength gypsum

High strength gypsum is produced by steaming natural gypsum under pressure steaming conditions (0.13MPa, 125℃) and is called α-hemihydrate gypsum (α-$CaSO_4 - 1/2H_2O$). α-hemihydrate gypsum is dense, complete and coarse primary particles, whose crystallization is more complete and its dispersion is lower. Compared to β-hemihydrate gypsum, α-hemihydrate gypsum is slower to hydrate, has a lower heat of hydration, requires less water and has a higher strength of hardened paste.

### ii. High temperature calcined gypsum

High temperature calcined gypsum is a product obtained from natural gypsum after calcination at 800~1000℃ and grinding. At high temperatures, dihydrate gypsum is not only completely dehydrated into anhydrous calcium sulfate ($CaSO_4$), but also partially decomposed into calcium oxide, a small amount of which is the activator for reaction between anhydrous gypsum and water.

High temperature calcined gypsum is slower to set and harden than building gypsum, but it has high water resistance and strength, good abrasion resistance. It can be used to make plaster, masonry and mortar for the manufacture of artificial marble, and can be applied to paving floors, also known as floor plaster.

## II. Setting and hardening of building gypsum

After mixing with proper amount of water, gypsum initially forms a plastic paste, then gradually thickens and loses its plasticity, but does not yet have strength; and this process is known as "setting". Then, the paste gradually becomes a solid with a certain strength, and this process is known as "hardening".

During the setting and hardening process, buildinggypsum undergoes a hydration reaction with water as follows:

$$CaSO_4 \cdot 0.5H_2O + 1.5H_2O = CaSO_4 \cdot 2H_2O$$

When hemihydrate gypsum is added to water, it is first dissolved and then performs the hydration reaction described above, leading to the formation of dihydrate gypsum. Since the solubility of dihydrate gypsum in water is much smaller than that of hemihydrate gypsum in water (only 1/5 of that of hemihydrate gypsum), dihydrate gypsum is constantly precipitated from the supersaturated solution and precipitates out as colloidal particles. The precipitation of dihydrate gypsum disrupts the equilibrium concentration of the original hemihydrate gypsum, which is then further dissolved. This cycle of dissolution of hemihydrate gypsum and precipitation of dihydrate gypsum continues until the hemihydrate gypsum is completely dissolved. This process proceeds relatively quickly and takes about 7 to 12min.

As hydration proceeds, the amount of dihydrate gypsum becomes larger; water gradually decreases and the paste begins to lose its plasticity, which is called "initial setting". The paste then gradually thickens; the friction and bonding between the particles increases and the structural strength begins to develop, which is referred to as "final setting". After final setting of gypsum, its crystal particles are still gradually growing and interlocking, making its strength developed until the remaining water has completely evaporated, then the strength stops developing, which is hardening process of gypsum (Figure 2-1).

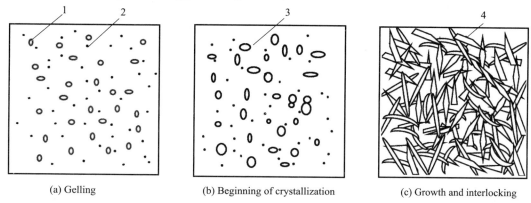

(a) Gelling     (b) Beginning of crystallization     (c) Growth and interlocking

**Figure 2-1 Schematic diagram for the setting and hardening process of building gypsum**
1—Hemihydrate gypsum; 2—Dihydrate Gypsum colloidal particles;
3—Dihydrate gypsum crystals; 4—Grown and interlaced dihydrate gypsum crystals

## III. Characteristics, technical properties and applications of building gypsum

i. Characteristics of building gypsum

1. Fast setting and hardening

The initial setting time of paste after mixing with water is not less than 6min, and the final setting time is not earlier than 30min. It is completely hardened approximately in a week. The operation of molding is difficult owing to the short initial setting time, and in order to delay its setting time, retarder can be added so that the solubility of hemihydrate gypsum is decreased or its dissolution rate can be reduced, slowing down the hydration rate. Retarders such as animal glue, sulphite, alcohol waste have wide application, so do borax, citric acid, etc.

Building gypsum hardens quickly, such as first-class gypsum, whose 1d strength is about 5~8MPa and 7 days will be up to the maximum strength (about 8~12MPa).

2. Slight volume expansion during hardening

Cementitious materials such as lime and cement tend to show shrinkage when they harden, while building gypsum has a slight expansion (expansion rate of 0.05%~0.15%), which enables the surface of gypsum products to be smooth and full, with clear angles and no cracking when dry.

3. Higher void fraction, lower apparent density and strength after hardening

When using building gypsum, the amount of water added to obtain good fluidity is often more than that required for hydration. The theoretical water requirement is 18.6%, nevertheless, the actual amount of water added is about 60%~80%. The evaporation of this excess free water leaves many voids, resulting in small apparent density and low strength.

4. Good insulation and sound absorption

The hardening gypsum has a high void fraction, where the capillary pores are all microscopic, contributing to small thermal conductivity and good thermal insulation ability. Sound conduction or reflexivity is significantly reduced on account of abundant micro-pores of gypsum, especially the surface micro-pores, and thus the sound absorption ability is strong.

5. Good fire resistance

In case of fire, the crystalline water in dihydrate gypsum, the main component of hardening gypsum, evaporates and absorbs heat, forming a steam curtain on the surface of the product, which can effectively prohibit the spread of fire.

6. A certain degree of temperature and humidity regulation

Building gypsum has a high heat capacity and is highly hygroscopic, and hence indoor temperature and humidity can be regulated.

7. Poor water and frost resistance

Building gypsum is hygroscopic and water-absorbent, so the adhesive force between building gypsum crystals will be weakened in a humid environment. In water, it will also dissolve dihydrate gypsum and cause collapse, hence poor water resistance. Moreover, water in building gypsum will crumble when it freezes, contributing to poor frost resistance.

8. Good processability

Gypsum products can be sawn, shaved, nailed and perforated.

ii. Technical requirements for building gypsum

Building gypsum is white in color, with a density of 2.60 to 2.75 g/cm$^3$ and a bulk density of 8.00 to 11.00 g/m$^3$. According to GB 9776—2022, building gypsum is divided into three grades of 4.0, 3.0 and 2.0 on the basis of its setting time and strength (Table 2-1). Details of *Building Gypsum* (GB 9776—2022) can be acquired from the QR code.

**Table 2-1 Physical and mechanical properties of buildinggypsum**

| Grade | Setting time/min | | Strength/MPa | | | |
| --- | --- | --- | --- | --- | --- | --- |
| | | | Wet strength (2h) | | Dry strength | |
| | Initial setting | Final setting | Flexural strength | Compressioe strength | Flexural strength | Compressioe strength |
| 4.0 | ≥3 | ≤30 | ≥3.0 | ≥6.0 | ≥7.0 | ≥15.0 |
| 3.0 | | | ≥2.0 | ≥4.0 | ≥5.0 | ≥12.0 |
| 2.0 | | | ≥1.6 | ≥3.0 | ≥4.0 | ≥8.0 |

iii. Application of building gypsum

Gypsum has a wide range of applications in construction, and can be used to make plasterboard, various architectural art accessories and building decoration, colored plaster products, plaster bricks, hollow plaster blocks, plaster concrete, stucco plaster, artificial marble, etc. Besides, gypsum is an important admixture in cement, cement products and silicate products.

1. Preparation of plastering gypsum

Plastering gypsum is an air-hardened cementitious material made from building gypsum or a mixture of building gypsum and $CaSO_4$ with admixtures, fine aggregates, etc.

2. Building gypsum products

There are many types of building gypsum products, such as paper-faced gypsum board, hollow gypsum strips, fiber gypsum board, gypsum blocks and decorative gypsum board, which are mainly used as partition walls, internal partition walls, ceiling and decoration, etc.

Gypsum with fiber reinforcement and adhesives can also be made into plaster corners, line boards, corner flowers, roman columns, sculptures and other artistic decorative plaster products.

Building gypsum needs to be waterproof and moisture-proof in storage. The storage period is not more than three months and the construction product strength will be degenerated if any expiry or moisture exists.

# Section III  Water Glass

Water glass, commonly known as sodium silicate, is a water-soluble silicate, which is composed of different proportions of alkali metal oxides and silica, and the chemical formula is $R_2O \cdot$

$nSiO_2$, where $n$ is molar ratio between silica and alkali metal oxides, and is the modulus of water glass, generally between 1.5 and 3.5. Common water glass are sodium silicate [$Na_2 \cdot nSiO_2$] and potassium silicate ($K_2 \cdot nSiO_2$), etc. Sodium silicate water glass is commonly applicable for the construction.

## I. Hardening of water glass

Water glass absorbs carbon dioxide in air and forms an amorphous silica gel, which gradually dries and dehydrates into silicon oxide and hardens. The reaction is as follows:

$$Na_2 \cdot nSiO_2 + CO_2 + mH_2O \longrightarrow Na_2CO_3 + nSiO_2 \cdot mH_2O$$

Since the above process proceeds very slowly, an appropriate amount of sodium fluorosilicate ($Na_2SiF_6$) is often added to accelerate the precipitation and hardening of silicate gel.

$$2(Na_2 \cdot nSiO_2) + Na_2SiF_6 + mH_2O \longrightarrow 6NaF + (2n+1)SiO_2 \cdot mH_2O$$

The suitable amount of sodium fluorosilicate is 12% ~ 15% of water glass. Slow hardening and low strength will occur if too little is used and the unreacted water glass is easily soluble in water, accounting for poor water resistance. If the dosage is too much, it will set too fast and thus lead to construction difficulty, making for large permeability and low strength.

## II. Characteristics of water glass

1. Water glass has good binding ability, and the silica gel precipitated during hardening prevents water penetration by blocking capillaries. The larger themodulus of water glass, the more colloidal components, the more difficult to dissolve in water, and the stronger the bonding ability. The higher the concentration of water glass with the same modulus, the greater the density, and the stronger the bonding force. The modulus of water glass commonly used in engineering is 2.6 ~ 2.8, and the density is 1.3 ~ 1.4 $g/cm^3$.

2. Water glass does not burn. Silica gel dries quickly at high temperatures, and the strength does not decrease, or even increase.

3. Water glass has high acid resistance and can resist most inorganic acids (except hydrofluoric acid) and organic acids.

## III. Application of water glass

The above properties of water glass are mainly applied in the following aspects in construction projects:

1. Water glass heat-resistant concrete. It is made up of water glass, sodium fluorosilicate, finely ground admixture and coarse and fine aggregates in a certain proportion, of which heat resistance is 600 ~ 1200℃, strength grade is C10 ~ C20, high temperature strength is 9.0 ~ 20MPa, and the highest use of temperature can be up to 1000 ~ 1200℃.

2. Water glass acid-resistant concrete. A corrosion and wear-resistant material formulated

from water glass, sodium fluorosilicate, diabase powder, granite, quartz sand or quartzite and other raw materials. It has strong acid resistance and can resist the corrosion of various concentrations of three acids, chromic acid, acetic acid (except hydrofluoric acid, hot phosphoric acid, fluorosilicic acid) and organic solvents and other media.

3. Alkali activated materials. Calcium silicate hydrate gel with cementitious ability was formed by alkali activated reaction between water glass and granulated blast furnace slag powder, red mud, fly ash, kaolin and other materials.

## Questions for review

1. What are air-hardening cementitious materials or hydraulic cementitious materials? How to use these two types of materials correctly?

2. Ancient lime is more resistant to water than modern lime, use your knowledge to analyze.

3. What is "Chen Fu" when lime is being slaked? What are the notices? Why can the finely ground quicklime be used directly without "Chen Fu"?

4. Since lime is not water resistant, why can lime soil or trinity mixture fill be used for wet areas such as foundation bedding?

5. What is the difference between $\alpha$-hemihydrate and $\beta$-hemihydrate gypsum?

6. Why is the building gypsum a good material for interior decoration? Why is it not suitable for outdoor use?

7. What is the modulus of water-glass? Briefly describe the process of setting and hardening of water glass.

# Chapter III  Cement

Cement is a powdery cementitious material, and after mixing with water, it becomes paste, and then turns into hardened specimen through a series of physical and chemical changes, cohering bulk materials together. Cement is a hydraulic cementitious material, and it can not only harden in the air, but also better harden in the water and maintain strength growth.

More than 200 kinds of cement are produced and used at present. According to chemical composition, cement can be divided into Portland cement, calcium aluminate cement, calcium sulfoaluminate cement, etc., among which Portland cement is the most fundamental cement and has the widest application.

This chapter takes Portland cement as the main content, and introduces other kinds of cement.

## Section I  Introduction

### I. Development of cement

The development history of cement represents that of cementitious material. As early as 3000 BC, the ancient Egyptians began to utilize calcined gypsum as building cementitious material, while the ancient Greeks used lime made by calcining limestone as building cementitious material. In 146 BC, the use of lime is improved by the ancient Romans, which was not only mixed with sand, but also mixed with fine pozzolana or fine crushed bricks to form a three-component mortar with better performance and partial hydraulicity called "Roman mortar". By the 18th century, owing to Britain's need for nautical lighthouse construction, the hydraulic lime made from clay-bearing limestone was present, which was later calcined at high temperatures and ground into "Roman cement". In 1824, Portland cement was invented and patented by a British engineer called Joseph. Aspdin, marking the birth of modern cement.

The development of cementitious material in China has a long history. As early as 5000 years ago, "white lime" made of limestone with high silica content was used. In the Zhou Dynasty of the 7th BC, lime began to appear; in the Northern and Southern Dynasties of the 5th century, "trinity mixture fill" composed of lime, clay and fine sand appeared, which was similar to the "Roman mortar". Since the Qin and Han Dynasties, cementitious material made of bricks and stones with glutinous rice juice, "lime-tung oil" and "lime-blood" appeared, which was in a leading position in the development of cementitious material worldwide. Due to various reasons, the development of cementitious material in modern China lags far behind the world. Modern cement began to be produced in the early 19th century. Since China's reform and opening up, Cement industry in China has gained great vitality and entered an important new historical stage, while cement varieties have increased from several at the beginning of liberation to nearly 100 types currently. In 2021, China's total annual cement production was 2.363 billion tons, ranking first in the world.

However, cement industry is an important contributor to carbon emissions, and producing a ton of cement emits about 0.8~0.9 tons of $CO_2$. Currently, the carbon emissions in cement industry have accounted for approximately 26% of the total industrial carbon emissions, which is about 7% of total human living carbon emissions. $CO_2$ is one of the culprits of global warming, directly affecting the survival and development of human beings, animals and the whole society. China has solemnly announced that $CO_2$ emissions will peak before 2030 and strive to achieve carbon neutrality before 2060. The current cement production in China has developed steadily and the application of low-carbon green cement-based materials need to be vigorously promoted.

## II. Production of Portland cement

The main raw materials for the production of Portland cement are calcareous raw materials and clay raw materials. Calcareous raw materials mainly provide CaO, e.g., limestone, chalk and limestone tuff, etc. Clay raw materials mainly provide $SiO_2$, $Al_2O_3$ and $Fe_2O_3$ such as clay, clay shale and loess, etc. Sometimes the chemical composition of the two raw materials cannot meet the requirements, so a small amount of correction materials need to be added, such as iron ore powder, a waste product from the production of sulfuric acid from pyrite. In order to improve the calcination conditions and clinker quality, a small amount of mineralizer is commonly added, such as fluorite, gypsum, etc.

When producing cement, the raw materials are first mixed in an appropriate proportion and then ground into raw meals, and the raw meals are put into the kiln (rotary kiln or vertical kiln) for high temperature calcination; then the sintered clinker is mixed with appropriate gypsum and supplementary cementitious materials (SCMs) in the mill to be ground into fine powder to obtain cement, as shown in Figure 3-1.

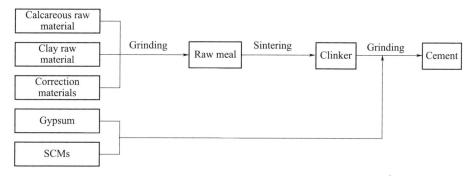

**Figure 3-1 Production process of cement**

The production of Portland cement has three steps: rawmaterial preparation, clinker sintering and cement production, which can be summarized as "two grinding and one sintering". In the cement production process, raw materials can be mixed and ground by adding water(namely wet production) or by dry grinding(namely dry production). In the sintering process, the raw meal is dried, preheated, decomposed, sintered and cooled to generate cement clinkers through a series of physical and chemical changes. In order to make the raw meal fully react, the sintering temperature in the kiln should reach 1450℃. At present, the sintering technique of cement clinker in China mainly includes the production process using new dry rotary kiln based on suspension preheating and precalcining technology, the production process using traditional dry or wet rotary kiln and the production process using vertical kiln. The production process using new dry rotary kiln has become the development direction and mainstream on the strength of its large scale, good quality, low consumption and high efficiency while the production process using traditional rotary kiln and vertical kiln are gradually being eliminated due to backward technology, high energy consumption and low efficiency.

In the Portland cement production, appropriate amounts of gypsum and SCMs required to be added. Adding gypsum is to delay the setting time of cement to meet the use requirement; SCMs are added to improve its variety and performance, expand its range of use, reduce cement costs, and increase cement production.

## III. Composition of Portland cement

Portland cement is generally composed of Portland cement clinker, gypsum and SCMs.

ⅰ. Portland cement clinker

The product produced by sintering the raw material to partial melting with calcium silicate as the main component is called Portland cement clinker. The main compositions in the raw meal are $CaO$, $SiO_2$, $Al_2O_3$ and $Fe_2O_3$. After sintering at a high temperature, the main minerals in the Portland cement clinker are formed: $3CaO \cdot SiO_2$, abbreviated $C_3S$, accounting for about 37% ~ 60%; $2CaO \cdot SiO_2$, abbreviated $C_2S$, accounts for about 15% ~ 37%; $3CaO \cdot Al_2O_3$, abbreviated $C_3A$, accounts for about 7% ~ 15%; $4CaO \cdot Al_2O_3 Fe_2O_3$, abbreviated as $C_4AF$, accounts for about 10% ~ 18%.

In addition to the above main components, Portland cement clinker also contains a small amount of the following compositions:

1. Free calcium oxide. It remains free during the calcination process without completely combination. Excessive free calcium oxide will lead to poor soundness and further cause damage.

2. Free magnesium oxide. Poor cement soundness will occur if its content is high and the crystal particle is large.

3. Alkali-containing minerals and glass, etc. Alkali-aggregate reaction occurs easily in cement with alkali mineral and high content of $Na_2O$ and $K_2O$ in glass when exposed to active aggregates.

ii. Gypsum

Gypsum is an indispensable component material in Portland cement, whose main function is to adjust the setting time of cement. Natural or synthetic dihydrate gypsum and chemical waste slag containing $CaSO_4 \cdot 2H_2O$ can be used.

iii. SCMs

SCMs is widely used in the production of Portland cement, and can be divided into two categories: the active SCMs and the non-active SCMs according to their properties, including active blast furnace slag, pozzolana material (zeolite, pozzolana), fly ash and coal gangue and inactive limestone, quartz sand, steel slag, slow cooling slag, etc.

### IV. Definition and classification of Portland cement

According to GB 175—2007 *Common Portland Cement*, common Portland cement is hydraulic cementitious material made by Portland cement clinker, an appropriate amount of gypsum, and SCMs. Common Portland cement is divided into Portland cement, ordinary Portland cement, Portland slag cement, Portland pozzolana cement, Portland fly ash cement and composite Portland cement according to the types and content of SCMs.

## Section II    Portland cement and ordinary Portland cement

### I. Compositions of Portland cement

According to GB 175—2007 *Common Portland Cement*, Portland cement is composed of Portland cement clinker, 0 ~ 5% limestone or blast furnace slag and an appropriate amount of gypsum. Portland cement is divided into two types: type I, Portland cement without SCMs, which is also called P · I; type II, Portland cement mixed with no more than 5% limestone or blast furnace slag, which is also called P · II.

### II. Hydration, setting and hardening

After cement is mixed with water, the cement paste with plasticity is initially formed, which gradually thickens and loses plasticity as the hydration proceeds, and it is called setting. As the hydration continues, the paste gradually becomes a hardened solid cement paste having strength, which is known as hardening.

i. Hydration of Portland cement

Portland cement clinker consists of four main minerals, whose hydration and hardening properties determine the characteristics of cement. After Portland cement is mixed with water, the minerals in the clinker immediately react with water separately to form hydration products.

The hydration reaction of each mineral is as follows.

1. Hydration of $C_3S$

$C_3S$ is the main mineral of cement clinker, whose hydration, hydration products, setting and hardening have an important impact on the properties of cement. The hydration reaction of $C_3S$ is as follows:

$$2(3CaO \cdot SiO_2) + 6H_2O = 3CaO \cdot 2SiO_2 \cdot 3H_2O + 3Ca(OH)_2$$

It can be abbreviated as follows:

$$2C_3S + 6H = 3C\text{-}S\text{-}H + 3CH$$

where C-S-H is calcium silicate hydrate.

The hydration products of $C_3S$ are C-S-H and CH. C-S-H is a gel with a fibrous morphology. CH is a crystal and soluble in water. The hydration rate of $C_3S$ is very fast with a large hydration heat. The generated C-S-H gel with a high-strength network structure is the main source of cement strength, whose setting time is normal with a high early and late strength.

2. Hydration of $C_2S$

The hydration of $C_2S$ is similar to that of $C_3S$, but the hydration rate is much slower. The hydration reaction is as follows:

$$2(2CaO \cdot SiO_2) + 4H_2O = 3CaO \cdot 2SiO_2 \cdot 3H_2O + Ca(OH)_2$$

It can be abbreviated as follows:

$$2C_2S + 4H = 3C\text{-}S\text{-}H + CH$$

There is no big difference between hydration products of $C_2S$ and those of $C_3S$ in morphology. C-S-H is also known as calcium silicate hydrate gel. The amount of CH with a large crystal size is less than that of $C_3S$. In Portland cement clinker, $C_2S$ hydration rate is the slowest, but the late hydration rate increases at the later age, and the total heat release is small; its early strength is low but the late strength can be close to or even exceed that of $C_3S$, which is the main factor to ensure the growth of late strength.

3. Hydration of $C_3A$

The hydration products of $C_3A$ belongs to a family of calcium aluminate hydrate, and the hydration reaction is as follows:

$$2(3CaO \cdot Al_2O_3) + 6H_2O = 3CaO \cdot Al_2O_3 \cdot 6H_2O$$

It can be abbreviated as follows:

$$2C_3A + 6H = 3C\text{-}A\text{-}H$$

where C-A-H is calcium aluminate hydrate.

Among the clinkers of Portland cement, the hydration rate of $C_3A$ is the fastest with a large hydration heat and a fast heat release rate. Its early strength growth is fast, but the strength is not high with no increase in late strength, which has influences on the early (within 3 days) strength of the cement. The setting rate of C-A-H is fast, making the cement set immediately. In order to control the hydration, setting and hardening rate of $C_3A$, an appropriate amount of gypsum must be added to the cement. The hydration product of $C_3A$ will react with gypsum to form high-sulfur calcium sulfoaluminate hydrate ($3CaO \cdot Al_2O_3 \cdot 3CaSO_4 \cdot 31H_2O$, also known as ettringite). Firstly, the formation rate of ettringite is slower than that of $C_3A$. Moreover, ettringite crystal is precipitated on the surface of the cement, resulting in the formation of barrier called "Semi-Membrane Shell", delaying cement hydration and thus avoiding flash or false set. After the gypsum is completely consumed, part of the ettringite will be converted into monosulfate. However, excessive gypsum will not only reduce the retarding effect but also lead to poor cement soundness.

4. Hydration of $C_4AF$

Hydration and products $C_4AF$ are similar to those of $C_3A$, generating solid solution of calcium aluminate hydrate and calcium ferrite hydrate. The reaction is as follows:

$$4CaO \cdot Al_2O_3 \cdot Fe_2O_3 + 7H_2O = 3CaO \cdot Al_2O_3 \cdot 6H_2O + CaO \cdot Fe_2O_3 \cdot H_2O$$

It can be abbreviated as:

$$C_4AF + 27H = C_3AH_6 + CFH$$

The hydration rate of $C_4AF$ is fast, second only to $C_3A$, and the hydration heat is not high with a normal setting. The strength is low, nevertheless, the flexural strength is relatively high. Cement brittleness can be reduced by increasing the content of $C_4AF$ which is applicable to occasions affected by vibration alternating load such as roads.

The characteristics of hydration, setting and hardening of the above single clinker minerals are shown in Table 3-1 and Figure 3-2.

Table 3-1  Main mineral composition and characteristics of Portland cement

| Performance index | | Clinker mineral | | | |
|---|---|---|---|---|---|
| | | $C_3S$ | $C_2S$ | $C_3A$ | $C_4AF$ |
| $\rho /(g/cm^3)$ | | 3.25 | 3.28 | 3.04 | 3.77 |
| Hydration reaction rate | | Fast | Slow | Fastest | Fast |
| Hydration heat release | | High | Low | Highest | Middle |
| Strength | Early age | High | Low | Low | Low |
| | Later age | | High | | |
| Shrinkage | | Middle | Middle | Big | Small |
| Sulfate corrosion resistance | | Middle | Best | Bad | Good |

The main hydration products generated by the interaction of Portland cement and water are C-S-H gel, calcium ferrite hydrate gel, calcium hydroxide crystal, calcium aluminate hydrate crystal and calcium sulfoaluminate hydrate crystal. In fully hydrated cement, C-S-H accounts for about 70%, calcium hydroxide accounts for about 20%, ettringite and monosulfate for about 7%.

ii. Setting and hardening of Portland cement

The setting and hardening of cement are divided artificially. Actually, setting and hardening are a continuous complex physical and chemical change process. According to the hydration reaction rate and the structural features of cement paste, the setting and hardening process of Portland cement can be divided into three stages: induction period, setting period and hardening period.

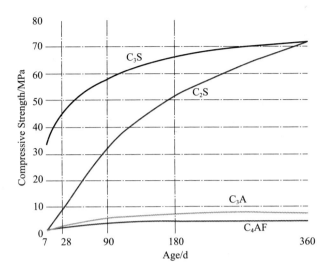

Figure 3-2  Compressive strength of cement clinkers at different ages

The general pattern of strength development of hardened cement paste is as follows: the strength growth at 3~7 days is the fastest; the strength growth within 28 days is faster; strength will continue to develop but grow slowly at more than 28 days.

iii. Main factors affecting the setting and hardening of Portland cement

1. Impact of clinker compositions

The clinker compositions directly affect the hydration, setting and hardening of cement. As shown in Figure 3-3, $C_3S$ initially reacts slowly, but then reacts quickly. On the contrary, the $C_3A$ reaction rate is fast and then slow; the reaction rate of $C_4AF$ is faster than that of $C_3S$ at the beginning and then slows down. The hydration of $C_2S$ is the slowest, but increases steadily later.

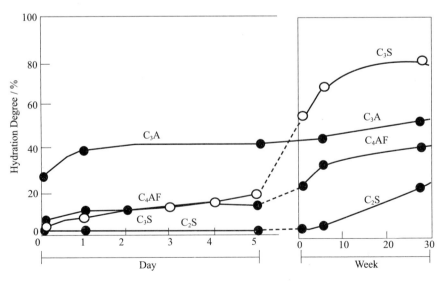

Figure 3-3  Hydration degree of cement clinker minerals at different ages

2. Effect of gypsum content

Gypsum is added to reduce the hydration rate of $C_3A$. With the increase of gypsum, the heat release rate of cement slows down and the heat release peak delays, but excessive gypsum has little effect on setting and hardening.

3. Effect of cement fineness

The fineness of cement particles directly affects the hydration, setting and hardening, strength, drying shrinkage and hydration heat of cement. However, superfine cement is so easy to react with water and carbon dioxide in the air, and thus it should not be stored for a long time. The shrinkage of superfine cement is also larger when it hardens, whose grinding is more energy intensive and costly. It is generally believed that cement has higher activity when particles are less than $40\,\mu m$, and less activity when they are larger than $90\,\mu m$.

4. Effects of curing humidity and temperature

The hydration can be accelerated by increasing temperature. Generally, the early hydration of Portland cement can be accelerated by increasing temperature so that the early strength can develop rapidly, but the late strength may be reduced. In contrast, when hardened at lower temperatures, in spite of low hardening rate, the hydration products are denser, resulting in a higher final strength. The measure that protects the temperature and humidity of the environment to increase cement strength is called curing.

5. Effects of mixing water content

In order to guarantee the plasticity and fluidity of cement paste, the amount of water is usually much more than that needed for full hydration. The excess water forms capillary pores in hardened cement paste. Therefore, the more the mixed water is, the more capillary pores in the hardened cement paste are. When $W/C$ is 0.40, the cement porosity is 29.6% after complete hydration, and when $W/C$ is 0.70, the cement porosity is up to 50.3%. The cement strength decreases linearly with the increase of capillary porosity. Therefore, when clinker composition is similar, the mixing water amount is the main factor that affects the cement strength.

6. Effects of curing age

The hydration and hardening of cement is a continuous and long process. With the increase of the hydration degree of clinker and the continuous formation of gel in the cement, the capillary pores decrease and thus cement strength is improved with the growth of age, which increases rapidly in 3 ~ 14 days but slowly after 28 days.

7. Effects of additives

The hydration and hardening of cement can be promoted by adding accelerator ($CaSO_4$, $CaCl_2$), which improves early strength of cement. On the contrary, the addition of retarder (calcium lignosulfonate, sugar, etc.) will delay cement hydration and hardening, further influencing early strength of cement.

8. Cement moisture and storage

Cement hydrates and forms agglomerates due to dampness, leading to loss of cementitious ability and lower strength. Moreover, cement cannot be stored for too long even under fine storage conditions, since slow hydration and carbonation will occur after water and carbon dioxide in the air is absorbed by cement, which will reduce cement strength about 10% ~ 20% after 3 months, 15% ~ 30% after 6 months and 25% ~ 40% after 1 year.

## III. Technical properties of Portland cement

The main technical properties of Portland cement are regulated by GB 175—2007 *Common Portland Cement*. Details can be acquired from the QR code.

i . Fineness

Fineness is the thickness of cement particles, which influences the hydration activity and setting and hardening rate. Too coarse cement with lower hydration activity is adverse to setting and hardening. Although finer cement leads to faster setting and hardening and higher early strength, faster hydration heat release rate and larger shrinkage of cement will be unfavorable to the properties of hardened cement paste.

Cement fineness can be tested by sieving method and specific surface area method. The sieving method is an experiment that uses a square-hole sieve with a side length of 80μm to screen the cement sample, and the cement fineness is demonstrated by the percentage of sieve residue. The sieve residue of ordinary cement should not exceed 10.0%. Cement fineness screening test can be acquired from the QR code.

The specific surface area method is to determine the specific surface area of cement (the total surface area per unit mass of powder) by air permeability method, with $m^2/kg$ as the unit. Compared with the sieving method, it is a more reasonable method which can better reflect the distribution of coarse and fine particles of cement. According to GB 175—2007, the specific surface area of Portland cement should not be less than $300m^2/kg$.

ii. Setting time

Setting time is the time from adding water to losing plasticity, and it is divided into initial setting time and final setting time. Initial setting time is the time from the addition of water to the cement till beginning of the loss of plasticity of the cement paste, and final setting time is the time from adding water to the cement to beginning to set and gain strength of cement paste. According to GB 175—2007, the initial setting time of Portland cement is not less than 45 min, and the final setting time is not more than 390 min; products with substandard initial setting time are waste, and those with substandard final setting time are unqualified.

Setting time is determined by measuring normal consistency of cement paste at a specified temperature and humidity with a setting time tester. Normal consistency is the amount of mixing water required for cement paste to reach the specified consistency and is marked in percentage of cement mass. *Test Methods for water requirement of normal consistency of cement* and *setting time* can be acquired from the QR codes. The water requirement of normal consistency of cement is generally between 24% ~ 30%, which varies with the mineral composition of cement clinker.

The finer ground cement, the greater water requirement of normal consistency.

Setting time is of great significance for the construction of cement concrete and mortar. In an effort to have enough time to complete operations such as transportation, pouring or masonry of concrete and mortar mixtures, the initial setting time should not be too short. In order to enable the concrete and mortar to set and harden as soon as possible after the completion of pouring or masonry and facilitate the early implementation of the next process, the final setting time should not be too long.

iii. Soundness

Soundness refers to the ability of hardened cement paste to resist volume change without delayed expansion damage after hardening. Cement with poor soundness will lead to concrete cracks caused by expansion and poor construction quality.

Poor soundness is caused by excessive free CaO, free MgO and gypsum in cement clinker. The over-burnt free CaO and MgO was slaked slowly, and begin to hydrate after cement paste has been hardened and shows strength. Expansion of slaking leads to inhomogeneous volume changes and cracks in hardened cement paste. When gypsum is excessive in clinker, gypsum will react with calcium aluminate hydrate to generate ettringite, which will expand more than 1.5 times in volume, resulting in poor soundness of cement and cracks of hardened cement paste.

According to national standard, boiling process is used for testing poor soundness caused by excessive free CaO via accelerating its slaking. Boiling test is divided into two types: the pate test method and the Le Chatelier test method. The Le Chatelier test method should prevail, when there is a dispute between the two methods. *Test Methods for soundness of cement* can be acquired from the QR code.

Boiling test has no effect on free MgO since free MgO slakes more slowly than free CaO, and the soundness caused by free MgO is tested by autoclave method. Poor soundness caused by excessive gypsum can be found after long-term immersion in room temperature water. Therefore, the poor soundness caused by excessive free MgO and gypsum cannot be detected quickly. In order to ensure the soundness of cement, the MgO content in Portland cement should not exceed 5.0%, and the $SO_3$ content should not exceed 3.5% depending on national standard.

The national standard says that the soundness of cement must be qualified; cement with poor soundness is regarded as waste and cannot beapplied to engineering.

iv. Strength and strength grade

Cement strength is a main technical index of cement. Generally, the compressive strength at 28d is adopted to characterize the strength grade of Portland cement.

*Test Method for Strength of Cement Mortar* (ISO Method, GB/T 17671—1994) is used for determining the cement strength, as seen in the QR code. According to the results of 3d and 28d flexural strength and compressive strength shown in GB 175—2007, Portland cement is separated into six strength grades: 42.5, 42.5R, 52.5, 52.5R, 62.5 and 62.5R, where R belongs to early strength cement. Portland cement strength with different types and strength grades at each age should not be lower than the values specified in Table 3-2.

Table 3-2  Strength requirements of Portland cement at each age (GB 175—2007)

| Strength Grade | Compressive Strength/MPa | | Flexural Strength/MPa | |
| --- | --- | --- | --- | --- |
| | 3d | 28d | 3d | 28d |
| 42.5 | ≥17.0 | ≥42.5 | ≥3.5 | ≥6.5 |
| 42.5R | ≥22.0 | | ≥4.0 | |
| 52.5 | ≥23.0 | ≥52.5 | ≥4.0 | ≥7.0 |
| 52.5R | ≥27.0 | | ≥5.0 | |
| 62.5 | ≥28.0 | ≥62.5 | ≥5.0 | ≥8.0 |
| 62.5R | ≥32.0 | | ≥5.5 | |

ⅴ. Alkali content

The alkali content in cement is calculated by the percentage of $NaO + 0.658K_2O$. When the concrete aggregate contains active silicon dioxide, it will react with the alkali in cement to form alkali silicate gel, whose volume expansion will cause concrete cracks and structure damage, also known as "alkali-aggregate reaction", which is an important factor affecting concrete durability. In order to prevent alkali-aggregate reaction, alkali content is restricted in GB 175—2007: alkali content, as an optional requirement, is decided by the buyer and seller through negotiation, but when the low-alkali cement is required, the alkali content in cement should not exceed 0.60%.

ⅵ. Hydration heat

The heat released by cement hydration is called hydration heat. The amount of hydration heat and the heat release rate are determined by the mineral composition of cement clinker and cement fineness, and are also related to the varieties and amounts of SCMs and admixtures.

The hydration heat of cement has both advantages and disadvantages for concrete engineering, which is conducive to low temperature construction, but unfavorable for mass concrete. Owing to the large hydration heat of Portland cement, the internal heat dissipation of mass concrete (e. g. , large infrastructure, dam, pier, and etc. ) is slow with 50 ~ 60℃ internal temperature. However, the external heat dissipation of mass concrete with low temperature is fast. When the difference between the internal and external temperature reaches a certain value, the temperature cracks will occur, resulting in a low strength and poor durability of concrete.

## Ⅳ. Corrosion and prevention

Portland cement has good durability after hardening. However, the strength and durability of hardened cement paste would be reduced when exposure to some certain corrosive environments for a long time, even leading to destruction of the concrete structure, and this phenomenon is known as corrosion. Several typical types of corrosion are described as follows.

ⅰ. Soft water corrosion (dissoluble corrosion)

$Ca(OH)_2$ crystal is one of the main hydration products of cement paste, and other hydration products of cement can only exist stably in a certain concentration of $Ca(OH)_2$ solution, which is soluble in water, especially in soft water containing little calcium bicarbonate. If $Ca(OH)_2$ crystal is dissolved and its concentration is below the minimum requirement for stabilization of other hydration products, the other hydration products will be dissolved or decomposed, leading to destruction of the hardened cement paste, which is known as dissoluble corrosion.

When the hardened cement paste is in soft water such as rainwater, snow water, distilled water, condensate, river water and lake water containing less carbonate for a long time, $Ca(OH)_2$ will dissolve first because of its maximum solubility, and soon reach its saturated solution. When in still water or non-pressure water, the dissolution of $Ca(OH)_2$ has little effect on the properties of hardened cement paste as the dissolution of $Ca(OH)_2$ is limited to the surface of hardened cement paste. But in flowing water or pressure water, $Ca(OH)_2$ will be dissolved and carried away by water, along with a decrease of $Ca(OH)_2$ concentration. On the one hand, the cement porosity is increased, the compactness and strength is decreased, and water can penetrate more easily to the interior; on the other hand, the structure of hardened cement paste is destroyed on account of the continuously reduced alkalinity, which causes the decomposition of hydration products and the formation of products with poor cementitious ability. Loss of 1% $Ca(OH)_2$ will result in a strength decrease of 5% ~ 7%.

ii. Salt corrosion

1. Sulfate corrosion

Sulfates containing sodium, potassium and ammonium often exist in seawater, lake water, salt marsh water, groundwater, certain industrial effluents and water flowing through blast furnace slag or cinder, which undergo a salt metathesis reaction with $Ca(OH)_2$ in the hardened cement paste to form calcium sulfate, which is deposited in the pores on the surface of the hardened cement paste, leading to crystal expansion and cement cracks. If more calcium sulfate is generated and the newly generated calcium sulfate has a high activity, the calcium sulfate easily reacts with calcium aluminate hydrate in the hardened cement paste. The reaction formula is as follows:

$$3CaO \cdot Al_2O_3 \cdot 6H_2O + 3(CaSO_4 \cdot 2H_2O) + 19H_2O = 3CaO \cdot Al_2O_3 \cdot 3CaSO_4 \cdot 31H_2O$$

The structure of hardened cement paste is damaged more seriously due to the formation of ettringite with 31 crystal water, which expands more than 1.5 times of the original volume. Ettringite with a needle-like morphology is also called "cement bacillus".

It is worth noting that anappropriate amount of gypsum is added in order to control setting time during the production of Portland cement. Gypsum reacts with calcium aluminate hydrate to form ettringite, which is formed when cement paste still has plasticity, and thus no destructive effect occurs.

2. Magnesium salt corrosion

Magnesium salts, mainly including magnesium sulfate and magnesium chloride, generally exist in seawater, groundwater and salt marsh water, which react with $Ca(OH)_2$ in hardened cement paste by salt metathesis reaction:

$$MgSO_4 + Ca(OH)_2 + 2H_2O =\!=\!= CaSO_4 \cdot 2H_2O + Mg(OH)_2$$
$$MgCl_2 + Ca(OH)_2 =\!=\!= CaCl_2 + Mg(OH)_2$$

Magnesium hydroxide is loose without cementitious ability; calcium chloride is easily soluble in water, and dihydrated gypsum induces sulfate corrosion. Therefore, magnesium salts have double destructive effect on the hardened cement paste.

3. Acid corrosion

(1) Corrosion of carbonic acid

There is a lot of carbon dioxide dissolved in industrial wastewater and groundwater, which reacts with calcium hydroxide to generate water-insoluble calcium carbonate:

$$Ca(OH)_2 + CO_2 + H_2O =\!=\!= CaCO_3 + 2H_2O$$

The generated calcium carbonate reacts with water containing carbonic acid to generate water-soluble calcium bicarbonate, which is a reversible reaction:

$$CaCO_3 + CO_2 + 2H_2O =\!=\!= Ca(HCO_3)_2$$

More carbon dioxide dissolved in the water makes the aforementioned reaction proceed to the right. $Ca(OH)_2$ in the hardened cement paste is converted into water-soluble calcium bicarbonate and eventually lost. When the concentration of $Ca(OH)_2$ decreases to a certain value, other hydration products will be decomposed successively, further causing the destruction of hardened cement paste.

(2) Ordinary acid corrosion

There are inorganic acids and organic acids in industrial wastewater, groundwater and swamp water. The corrosion degree of various acids to hardened cement paste is different. They react with $Ca(OH)_2$, and the products either are easily soluble in water or crystallize and expand, which will reduce cement strength. Hydrochloric acid, hydrofluoric acid, sulfuric acid and nitric acid are the most serious corrosive among inorganic acids, while acetic acid, formic acid and lactic acid are the most serious corrosive among organic acids. For example, hydrochloric acid and sulfuric acid react with $Ca(OH)_2$ in hardened cement paste, respectively, and the reaction formulas are as follows:

$$2HCl + Ca(OH)_2 =\!=\!= CaCl_2 + 2H_2O$$
$$H_2SO_4 + Ca(OH)_2 =\!=\!= CaSO_4 \cdot 2H_2O$$

The formed calcium chloride is soluble in water. Calcium sulfate crystallizes and expands in pores of hardened cement paste. When more calcium sulfate is generated, it will also react with calcium aluminate hydrate to generate ettringite, which will expand more than 1.5 times, further doing greater harm to the hardened cement paste.

4. Alkali corrosion

Generally, the hardened cement paste will not be corroded when the alkali concentration is not high since the paste itself is in an alkaline condition with a pH value of 12.5 ~ 13.5 due to the existence of calcium hydroxide. If the aluminate content is high in cement, the water-soluble sodium aluminate will be formed with the presence of strong alkaline medium (such as sodium hydroxide), and the reaction formula is as follows:

$$3CaO \cdot Al_2O_3 + 6NaOH == 3NaO \cdot Al_2O_3 + 3Ca(OH)_2$$

When the hardened cement paste is saturated by sodium hydroxide, sodium carbonate is formed by the reaction between sodium hydroxide and carbon dioxide in the air, which crystallizes and expands in the capillary pores of the hardened cement paste, causing cracks and reduced strength, and the corresponding reaction formula is as follows:

$$2NaOH + CO_2 == Na_2CO_3 + H_2O$$

iii. Prevention of corrosion

According to the above analysis, the basic causes of corrosion are: 1) There are components susceptible to corrosion in the hardened cement paste, such as calcium hydroxide and calcium aluminate hydrate; 2) The hardened cement paste is not dense with many capillary channels, and the aggressive medium is easy to enter its interior. Therefore, the following measures can be taken to prevent the corrosion according to the specific situation.

1. Select cement types reasonably according to the characteristics of corrosion environment. Cement with less calcium hydroxide content in hydration products (such as the cement mixed with SCMs) can effectively prevent the corrosion of soft water, magnesium salt and other media because calcium hydroxide and calcium aluminate hydrate are the main contributor to corrosion. Cement with less aluminate content (such as sulfate resistant cement) can significantly improve the ability to resist sulfate corrosion.

2. Improve the compactness of hardened cement paste. The actual mixing water consumption of cement is 30% ~ 60%, which is much larger than the theoretical water demand of its chemical reaction (23%), resulting in the evaporation of excess water and the formation of many pores in hardened cement paste. In actual engineering, the compactness of concrete can be improved by reducing the water-cement ratio, selecting a well-graded aggregate, incorporating SCMs and admixtures, and improving construction technology. In addition, carbonization or fluorosilicic acid treatment can also be carried out on the concrete surface to prevent aggressive media from invading the interior.

3. Use separation layer or protective layer. Highly corrosion-resistant and impermeable protective layer can be added on the concrete surface. For example, acid resistant ceramics, acid resistant stones, glass, plastic are often utilized as acid resistant protective layers in engineering.

## V. Characteristic and application of Portland Cement

Cement has its own characteristics and the corresponding applications, and its characteristics are as follows:

i. Fast setting and hardening, high early strength and later strength.

Portland cement is applicable to cast-in-place concrete, precast concrete, winter construction, prestressed concrete and high strength concrete due to high $C_3S$ content in clinker.

ii. Good frost resistance

Portland cement is suitable for concrete in severe cold regions that suffer repeated freezing and thawing, owing to its fast setting and hardening, high early strength, low bleeding of its mixture, and high compactness.

iii. Good carbonation resistance

After setting and hardening, Portland cement exhibits high concentration of calcium hydroxide in the hydration products, high alkalinity of the cement paste, and high compactness. And calcium carbonate generated from the initial carbonization fills pores on the concrete surface, which makes the concrete surface denser and effectively prevents further carbonation. Therefore, Portland cement is applied to concrete engineering with carbonization requirements.

iv. Poor corrosion resistance

Portland cement is not suitable for the projects under the action of flowing water and pressure water and the projects affected by seawater and other corrosive medium, owing to the high content of tricalcium silicate and tricalcium aluminate in Portland cement clinker and the high content of easily corrosive calcium hydroxide and tricalcium aluminate hydrate in hydration products.

v. High hydration heat

On account of the high content of tricalcium silicate and tricalcium aluminate in Portland cement clinker and high hydration heat, it is not suitable for mass concrete, but can be applied to winter construction.

vi. Poor heat resistance

When the temperature of Portland cement concrete is not high (generally $100 \sim 250℃$), the remaining free water can make the hydration continue, which will further increase the compactness and strength of concrete. When the temperature is higher than $250℃$, calcium hydroxide is decomposed into calcium oxide; If in humid environment, calcium oxide slakes and expands, causing the destruction of concrete. In addition, when cement is heated at about $300℃$, shrinkage occurs and the strength begins to decline. When the temperature reaches $700℃$, the strength will decrease a lot, or even be completely destroyed. Therefore, Portland cement should not be applied to concrete projects with heat resistance requirements.

vii. Good abrasion resistance

Portland cement concrete can be employed in concrete projects such as pavement and airport runway owing to its high strength and good abrasion resistance.

## VI. Ordinary Portland Cement

According to GB 175—2007, ordinary Portland cement, also known as P·O, is composed of Portland cement clinker, 5% ~ 20% SCMs and an appropriate amount of gypsum.

Ordinary Portland cement contains a few SCMs compared with Portland cement, whose main component is still Portland cement clinker, and thus their characteristics are consistent. However, compared with Portland cement of the same strength grade, ordinary Portland cement has a slightly slower early hardening rate, lower 3d strength, worse frost resistance, smaller hydration heat, and better corrosion resistance.

The final setting time of ordinary Portland cement should not be more than 100min, and other technical properties should be the same as those of Portland cement.

According to GB 175—2007, ordinary Portland cement is divided into six strength grades of 42.5, 42.5R, 52.5, 52.5R, 62.5 and 62.5R. The strength of ordinary Portland cement of different types and strength grades at each age should not be lower than the values specified in Table 3-3.

Table 3-3  Strength requirements of ordinary Portland cement at each age (GB 175—2007)

| Strength Grade | Compressive Strength/MPa | | Flexural Strength/MPa | |
|---|---|---|---|---|
| | 3d | 28d | 3d | 28d |
| 42.5 | ≥17.0 | ≥42.5 | ≥3.5 | ≥6.5 |
| 42.5R | ≥22.0 | ≥42.5 | ≥4.0 | ≥6.5 |
| 52.5 | ≥23.0 | ≥52.5 | ≥4.0 | ≥7.0 |
| 52.5R | ≥27.0 | ≥52.5 | ≥5.0 | ≥7.0 |

## VII. Storage of cement

Cement should keep away from moisture and mixing with sundries during transportation and storage, and cement with different types and strength grades should not be mixed as well. Generally, the storage period of cement is three months. Once cement is stored for more than three months, its strength must be retested before use.

# Section III  Portland cement mixed with SCMs

## I. SCMs

The natural or artificial mineral materials, added when grinding cement, are called SCMs, including inactive and active SCMs.

ⅰ. Active SCMs

Active SCMs are substances that can hydrate with lime, gypsum or Portland cement at room temperature to generate hydraulic cementitious materials, among which blast furnace slag, pozzolanic and fly ash is widely used to improve cement properties, adjust the strength grade of cement, reduce hydration heat, lower production cost, increase cement output, broaden cement varieties, and etc.

1. Blast furnace slag

Blast furnace slag is a loose particle obtained by water quenching of molten slag from the smelting of pig iron in blast furnaces, whose main chemical components are $CaO, SiO_2, Al_2O_3$ and a small amount of $MgO$ and $Fe_2O_3$. The structure of quenching slag is an unstable glass phase with great chemical potential, and its main active ingredients are active $SiO_2$ and active $Al_2O_3$.

2. Pozzolana

Pozzolana is natural or artificial mineral materials with pozzolanic properties, whose main active ingredients are active $SiO_2$ and active $Al_2O_3$. There are many varieties, i. e. , the natural ones include pozzolana, pumice, pumice rock, zeolite, diatomaceous earth, etc. ; and the artificial ones include calcined shale, burnt clay, coal cinder, burned coal gangue or spontaneous combustion coal gangue, silica fume, etc.

3. Fly ash

Fly ash is the powder collected from the industrial waste gas discharged by coal-fired power plant through the flue. Fly ash is mostly solid or hollow amorphous pellets with a diameter of 0.001 ~ 0.05mm since fly ash is formed by quenching after burning pulverized coal with a suspended state. The chemical activity of fly ash depends on the content of active $SiO_2$, active $Al_2O_3$ and amorphous phase.

The influences of active SCMs are as follows: 1) Pozzolanic reaction. The reaction is slow due to the low activity of the active SCMs; and this reaction is sensitive to temperature and is suitable for steam curing. 2) Filling effect. Small particle size can be filled in the tiny voids between the unhydrated cement and the hydrated products, improving the compactness of hardened cement paste. 3) Morphological effect. The particle roundness of some SCMs is larger than that of cement, and it is not easy to react with water at early age, which plays a role similar to rolling balls and improves workability.

ii. Inactive SCMs

The natural or artificial mineral materials with no activity or very low activity are called inactive SCMs, which do not react with hydration products of cement. The inactive SCMs is used to adjust the cement strength grade, increase the output, reduce the production cost, reduce the hydration heat, etc. The common inactive SCMs are quartz sand, limestone and slow cooling slag.

## II. Portland slag cement, Portland pozzolana cement and Portland fly ash cement

i. Composition

1. Portland slag cement

Portland slag cement, also known as P·S, is made by grinding Portland cement clinker, blast furnace slag and an appropriate amount of gypsum, among which blast furnace slag content is more than 20% and less than or equal to 70%. P·S can be divided into two types: type A with more than 20% and less than or equal to 50% slag content, also called P·S·A; type B with more than 50% and less than or equal to 70% slag content, also called P·S·B.

2. Portland pozzolana cement

Portland pozzolana cement, also known as P · P, is made by grinding Portland cement clinker, pozzolana and an appropriate amount of gypsum, where pozzolana content is more than 20% and less than or equal to 40%.

3. Portland fly ash cement

Portland fly ash cement, also known as P · F, is made by grinding Portland cement clinker, fly ash and an appropriate amount of gypsum, where fly ash content is more than 20% and less than or equal to 40%.

ii. Technical requirement

The technical requirements specified by the GB 175—2007 are as follows:

1. Fineness: The 80μm spare of square sieve allowance is not more than 10% or 45μm spare of square sieve allowance is not more than 30%.

2. Setting: Initial setting time should not be earlier than 45 min and final setting not later than 600 min.

3. Magnesium oxide: The content of magnesium oxide in cement should not exceed 5.0%, but it is allowed to be widened to 6.0% if the cement passes the autoclave soundness test. In addition, the content of (mass fraction) magnesium oxide in P · S · A, P · P, P · F, P · C should not be more than 6.0%, but it is allowed to be widened to 6.0% if the cement passes the autoclave soundness test. When the content (mass fraction) of magnesium oxide in the cement exceeds 6.0%, the autoclave soundness should be tested and qualified. If there is a lower requirement, the indicator is determined by negotiation between the buyer and the seller.

4. $SO_3$: The content of $SO_3$ in slag cement should not exceed 4.0%, while in pozzolana cement and fly ash cement should not exceed 3.5%.

5. Soundness: It should be tested by boiling.

6. Strength: Cement strength is graded according to the compressive strength and flexural strength of the specified age. The strength of each grade and age of the three kinds of cement should not be lower than the values specified in Table 3-4.

7. Alkali content: The content of alkali in cement is calculated according to $Na_2O$ + 0.658$K_2O$ value. If active aggregates are used and low alkalinity cements are required, the alkali content in the cement should not be more than 0.60% or can be decided by the buyer and seller through negotiation.

Table 3-4  Strength requirements of slag cement, pozzolana cement and fly ash cement at each age (GB 175—2007)

| Strength grade | Compressive strength/MPa | | Flexural strength/MPa | |
|---|---|---|---|---|
| | 3d | 28d | 3d | 28d |
| 32.5 | ≥10.0 | ≥32.5 | ≥2.5 | ≥5.5 |
| 32.5R | ≥15.0 | ≥32.5 | ≥3.5 | ≥5.5 |

continued

| Strength grade | Compressive strength/MPa | | Flexural strength/MPa | |
|---|---|---|---|---|
| | 3d | 28d | 3d | 28d |
| 42.5 | ≥15.0 | ≥42.5 | ≥3.5 | ≥6.5 |
| 42.5R | ≥19.0 | ≥42.5 | ≥4.0 | ≥6.5 |
| 52.5 | ≥21.0 | ≥52.5 | ≥4.0 | ≥7.0 |
| 52.5R | ≥23.0 | ≥52.5 | ≥4.5 | ≥7.0 |

iii. Characteristics and application

1. Similarity of the three types of cement

(1) The early (3d, 7d) strength is low and the late strength is high. Slow hydration of these mixed materials results in low early (3d, 7d) strength. In the late stage, with the increasing products of secondary hydration reaction, the strength can even exceed Portland cement or Ordinary Portland cement of the same strength grade (Figure 3-4). Therefore, three types of cements are not suitable for concrete projects with high early strength requirements, such as winter construction of cast-in-place concrete projects, etc.

(2) Temperature sensibility, suitable for high temperature curing. Three cements hydrate significantly slower at low temperatures and have lower strength. High temperature curing can greatly accelerate the hydration of active SCMs and clinker, and improve early strength with no impact on the development of late strength at room temperature.

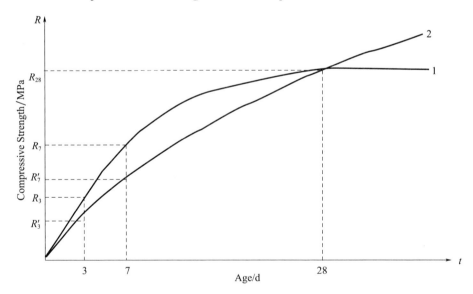

Figure 3-4 Strength development patterns of different cement varieties

1—Portland cement; 2—Portland cement with mixed materials

(3) Excellent corrosion resistance. Three cements have relatively low quantities of clinker and hydration products, calcium hydroxide and calcium aluminate hydrate. What's more, the secondary hydration reaction can further reduce calcium hydroxide content, so they have better

corrosion resistance, and are applicable to environments with sulphate, magnesium salts and soft water erosion, such as hydraulic engineering, harbors, docks, and etc.

(4) Low hydration heat. The low content of clinker in the three cements leads to less hydration heat, especially slow early heat release rate and less heat release, making them suitable for mass concrete engineering.

(5) Poor frost resistance. Slag and fly ash tend to bleeding to form connected pores, and pozzolana generally requires more water, increasing the internal pore content. So, all three types of cement have poor frost resistance.

(6) Poor carbonation resistance. Low amount of calcium hydroxide in the cement stone after hydration and hardening causes poor carbonation resistance. Consequently, none of them are adaptive for industrial plants with high carbon dioxide content.

2. Characteristics of the three cements

(1) Portland slag cement. It has poor impermeability and high dry shrinkage, but preferable heat resistance. Therefore, Portland slag cement has more applications for concrete projects with heat resistance requirements but are not suitable for concrete projects with impermeability requirements.

(2) Portland pozzolana cement. It possesses good water retention and impermeability. However, powdering phenomenon is easy to occur on the surface of hardened cement paste due to its large shrinkage, and its abrasion resistance is also poor. It is applicable for concrete projects with impermeability requirements, instead of the above-ground concrete work in dry environments and concrete work with high abrasion resistance requirements.

(3) Portland fly ash cement. It has poor impermeability, low dry shrinkage and poor abrasion resistance, which is suitable for late loading concrete, not recommended for concrete with impermeability or high abrasion resistance requirements.

## III. Composite Portland cement

According to GB 175—2007, Composite Portland cement, also called P · C, is made by grinding Portland cement clinker, two or more specified SCMs, and an appropriate amount of gypsum. The total amount of SCMs in the cement should be more than 20% and less than or equal to 50%.

Composite Portland cement is available in six strength grades 32.5, 32.5R, 42.5, 42.5R, 52.5 and 52.5R. The age strength requirements for each strength grade are the same as slag cement, pozzolana cement and fly ash cement, and the remaining technical requirements are the same as Portland pozzolana cement.

Composite cement is mixed with two or more specified SCMs, whose effect is not just a simple mixture of various types of SCMs, but to make best use of the advantages and bypass the disadvantages, producing excellent results that a single SCMs cannot play. Hence, the properties of composite cement are between Ordinary Portland cement and three SCMs Portland cement.

The composition, characteristics and selection of the above six common cements are shown in Tables 3-5 ~ 3-7.

**Table 3-5   Composition and properties of six common cement**

| Item | Portland cement | Ordinary Portland cement | Portland slag cement | Portland pozzolana cement | Portland fly ash cement | Composite Portland cement |
|---|---|---|---|---|---|---|
| Composition | Portland cement clinker, small amount (0~5%) of SCMs, appropriate amount of gypsum | Portland cement clinker, small amount (5%~20%) of SCMs, appropriate amount of gypsum | Portland cement clinker, large amounts (>20% and ≤70%) of granulated blast furnace slag, appropriate amount of gypsum | Portland cement clinker, large amounts (>20% and ≤40%) of volcanic, appropriate amount of gypsum | Portland cement clinker, large amounts (>20% and ≤40%) of fly ash, appropriate amount of gypsum | Portland cement clinker, large amounts (>20% and ≤50%) of two or more SCMs, appropriate amount of gypsum |
| Common | Portland cement clinker, appropriate amount of gypsum ||||||
| Differences | None or very little SCMs | Little SCMs | Lots of SCMs (The chemical composition or chemical activity is basically the same) ||| A large number of active or inactive SCMs |
| | | | Blast furnace slag | Volcanic | Fly ash | Two or more active or inactive SCMs |
| Properties | 1. The strength is high in the early and later ages; 2. Poor corrosion resistance; 3. High hydration heat; 4. Good carbonation resistance; 5. Good frost resistance; 6. Good corrosion resistance; 7. Poor heat resistance | 1. The early strength is slightly lower, and the later strength is higher; 2. Corrosion resistance is slightly better; 3. The heat of hydration is slightly less; 4. Good carbonation resistance; 5. Good frost resistance; 6. Good abrasion resistance; 7. Heat resistance is slightly better; 8. Good impermeability | The early strength is slightly lower, and the later strength is higher. ||| High early strength |
| | | | Sensitive to temperature, Suitable for high temperature curing; Good corrosion resistance; The heat of hydration is slightly less; Poor frost resistance; Poor carbonation resistance ||| |
| | | | 1. High bleeding water; 2. Good heat resistance; 3. Large dry shrinkage | 1. Good water retention, Poor impermeability; 2. Large dry shrinkage; 3. Poor abrasion resistance | 1. High bleeding water, Easy to produce water loss cracks, Poor impermeability 2. Low dry shrinkage, good crack resistance; 3. Poor abrasion resistance | Large dry shrinkage |

## Table 3-6  Standard for technical properties of General Portland cement

| Item | | Portland cement | | Ordinary Portland cement P·O | Portland slag cement P·S Portland pozzolana cement P·P Portland fly ash cement P·F | Composite Portland cement P·C |
|---|---|---|---|---|---|---|
| | | P·I | P·II | | | |
| Fineness | | \multicolumn{3}{l|}{Specific surface area >300m²/kg} | | The 80μm spare of mesh sieve allowance is not more than 10% or 45μm spare of mesh sieve allowance is not more than 30% | |
| Time setting | Initial setting | ≥45min | | | | |
| | Final setting | ≤390min | | | ≤600min | |
| Volume stability of cement | Soundness | The boiling test must be qualified, when there is a dispute between the pate test method and the Le Chatelier test method, the Le Chatelier soundness test shall prevail. | | | | |
| | MgO | Content ≤5.0% | | | | |
| | SO₃ | Content ≤3.5% (content ≤4.0% in Portland slag cement) | | | | |

| Strength grade | Age/d | Compressive strength /MPa | Flexural Strength /MPa | Compressive strength /MPa | Flexural Strength /MPa | Compressive strength /MPa | Flexural Strength /MPa | Compressive strength /MPa | Flexural Strength /MPa |
|---|---|---|---|---|---|---|---|---|---|
| 32.5 | 3 | — | — | — | — | 10.0 | 2.5 | 11.0 | 2.5 |
| | 28 | — | — | — | — | 32.5 | 5.5 | 32.5 | 5.5 |
| 32.5R | 3 | — | — | — | — | 15.0 | 3.5 | 16.0 | 3.5 |
| | 28 | — | — | — | — | 32.5 | 5.5 | 32.5 | 5.5 |
| 42.5 | 3 | 17.0 | 3.5 | 17.0 | 3.5 | 15.0 | 3.5 | 16.0 | 3.5 |
| | 28 | 42.5 | 6.5 | 42.5 | 6.5 | 42.5 | 6.5 | 42.5 | 6.5 |
| 42.5R | 3 | 22.0 | 4.0 | 22.0 | 4.0 | 19.0 | 4.0 | 21.0 | 4.0 |
| | 28 | 42.5 | 6.5 | 42.5 | 6.5 | 42.5 | 6.5 | 42.5 | 6.5 |
| 52.5 | 3 | 23.0 | 4.0 | 23.0 | 4.0 | 21.0 | 4.0 | 22.0 | 4.0 |
| | 28 | 52.5 | 7.0 | 52.5 | 7.0 | 52.5 | 7.0 | 52.5 | 7.0 |
| 52.5R | 3 | 27.0 | 5.0 | 27.0 | 5.0 | 23.0 | 4.5 | 26.0 | 5.0 |
| | 28 | 52.5 | 7.0 | 52.5 | 7.0 | 52.5 | 7.0 | 52.5 | 7.0 |
| 62.5 | 3 | 28.0 | 5.0 | — | — | — | — | — | — |
| | 28 | 62.5 | 8.0 | — | — | — | — | — | — |
| 62.5R | 3 | 32.0 | 5.5 | — | — | — | — | — | — |
| | 28 | 62.5 | 8.0 | — | — | — | — | — | — |

| Alkali content | When the user requires low alkalinity cement, the alkali content in the cement should not be more than 0.60% according to NaO + 0.658K₂O value or be decided by the buyer and seller through negotiation. |
|---|---|

**Table 3-7  Selection of different types of cements**

| Concrete engineering characteristics or environmental conditions | | Priority use | Secondary use | Unsuitable use |
|---|---|---|---|---|
| Ordinary concrete | Concrete in ordinary environments | Ordinary Portland cement | Portland slag cement, Portland pozzolana cement, Portland fly ash cement, Composite Portland cement | |
| | Concrete in a dry environment | Ordinary Portland cement | Portland slag cement | Portland pozzolana cement, Portland fly ash cement |
| | Concrete in high humidity environments or permanently underwater | Portland slag cement | Ordinary Portland cement, Portland pozzolana cement, Portland fly ash cement, Composite Portland cement | |
| | Mass concrete | Portland slag cement, Portland pozzolana cement, Portland fly ash cement, Composite Portland cement | Ordinary Portland cement | Portland cement, Rapid-hardening Portland cement |
| Concrete with special requirements | Concrete requiring fast hardening | Rapid-hardening Portland cement, Portland cement | Ordinary Portland cement | Portland slag cement, Portland pozzolana cement, Portland fly ash cement, Composite Portland cement |
| | High strength concrete | Portland cement | Ordinary Portland cement, Portland slag cement | Portland pozzolana cement, Portland fly ash cement |
| | Open-air concrete in severe cold areas, Concrete in cold areas within the range of water level rise and fall | Ordinary Portland cement | Portland slag cement | Portland pozzolana cement, Portland fly ash cement |
| | Concrete in severe cold areas within the range of water level rise and fall | Ordinary Portland cement | | Portland slag cement, Portland pozzolana cement, Portland fly ash cement, Composite Portland cement |
| | Concrete with impermeability requirements | Ordinary Portland cement, Portland pozzolana cement | | Portland slag cement |
| | Concrete with abrasion resistance requirements | Portland cement, Ordinary Portland cement | Portland slag cement | Portland pozzolana cement, Portland fly ash cement |

# Section Ⅳ  Other varieties of cement

In civil engineering, besides the commonly used Portland cement series, other types of cement may be utilized. Here is a brief description for some of these varieties.

## Ⅰ. Road Portland cement

ⅰ. Definition

A hydraulic cementitious material made by grinding road Portland cement clinker, 0 ~ 10% active SCMs and an appropriate amount of gypsum, is called road Portland cement, also known as P · R.

ⅱ. Technical requirements

1. Tricalcium aluminate content: The content of the clinker should not be more than 5.0%.
2. Brownmillerite content: The content of the clinker should not be more than 16.0%.

The content of Tricalcium aluminate ($C_3A$) and Brownmillerite ($C_4AF$) is obtained by the following formula:

$$C_3A = 2.65Al_2O_3 - 0.64Fe_2O_3, \%$$
$$C_4AF = 3.04 Fe_2O_3, \%$$

3. Free calcium oxide content: Its content by Rotary kiln production should be not more than 1.0%, and by vertical kiln not more than 1.8%.

ⅲ. Characteristics and application

Road Portland cement is a special cement with high strength, especially high flexural strength, good abrasion resistance, low dry shrinkage, good impact resistance, relatively better frost and sulfate resistance. It is appropriate for road pavement, airport runway pavement, city square and other projects. On account of low dry shrinkage, good abrasion resistance and impact resistance, concrete pavement cracks, abrasion and maintenance can be reduced and the service life of pavement can be extended.

## Ⅱ. White Portland cement

A hydraulic cementitious material made by grinding clinker with little iron oxide content, a proper amount of gypsum and 0 ~ 10% SCMs, is called white Portland cement, also known as P · W.

The technical properties of P · W should meet regulations of *White Portland Cement* (GB/T 2015—2005), and the whiteness value of cement should not be less than 87. P · W can be utilized to prepare white or colored grout, mortar and concrete.

## Ⅲ. Expansive cement and Self-stressing cement

Expansive cement is a kind of cement that do not show shrinkage during the hardening process with expansion property.

ⅰ. Classification

1. Classification by manufacturing method

(1) CaO-based expansive cement. The hydration of CaO generates volume expansion, when a certain amount of CaO sintered by a proper temperature is added to cement.

(2) MgO-based expansive cement. The hydration of magnesium oxide generates volume expansion, when a certain amount of MgO sintered by a proper temperature is added to cement.

(3) Ettringite-based expansive cement. Ettringite from hardened cement pastes volume expansion.

2. Classification by expansion value

(1) Shrinkage compensation cement. This kind of cement has weak expansion properties, and the compressive stress generated during expansion can roughly offset stress caused by dry shrinkage, which can prevent dry shrinkage cracks in concrete.

(2) Self-stressing cement. Owing to expansion properties of self-stressing cement, steel bar is subjected to higher tensile stress, while concrete is subjected to the corresponding compressive stress when used in reinforced concrete. The tensile stress of the concrete structure caused by external factors can be offset or reduced by pre-existing compressive stress. Prestress achieved by stretching steel bars by hydration to produce expansion is called self-stress.

ⅱ. Application

Expansive cement is applied for shrinkage compensation concrete and impermeable concrete; it can also be used for filling joints and pipe joints of concrete structures or components, structural reinforcement and repair, pouring machine bases and fixing foot screws, etc. And self-stressing reinforced concrete pressure pipes and fittings are usually manufactured by self-stressing cement.

## Ⅳ. Sulfate-resisting Portland cement

A hydraulic cementitious material with good sulfate resistance, made by grinding Portland cement clinker of a specific mineral composition and an appropriate amount of gypsum, is called Sulfate-resisting Portland cement, or simply Sulfate-resisting cement.

Sulfate-resisting cement is separated into two categories according to sulfate resistance: medium and high sulfate-resisting cement.

The sulfate resistance of cement can be improved by reducing the content of $C_3S$ and $C_3A$ and increasing the content of $C_2S$ and $C_4AF$ in clinker.

Sulfate-resisting cement is well employed in seaport, water conservancy, underground tunnel, water diversion, road and bridge foundation and other projects eroded by sulfate, contributing to its high sulfate resistance and low hydration heat.

## V. Rapid-hardening cement

i. Rapid-hardening Portland cement

A hydraulic cementitious material made by grinding Portland cement clinker and an appropriate amount of gypsum and marked by 3d compressive strength is called Rapid-hardening Portland cement, Rapid-hardening cement for short.

The manufacturing method of Rapid-hardening cement is basically the same as that of Portland cement, except for the appropriate addition of fast hardening minerals in the clinker, among which tricalcium silicate accounts for 50% ~ 60%, tricalcium aluminate 8% ~ 14% and total of both 60% ~ 65%. The amount of gypsum is increased (up to 8%) and the fineness of the cement is improved to accelerate hardening.

ii. Calcium aluminate cement

Calcium aluminate cement is made by grinding calcium aluminate cement clinker. It contains 35% ~ 60% monocalcium aluminate (CA), 10% ~ 20% calcium dealuminate ($CA_2$), 20% ~ 40% dicalcium silicaluminate ($C_2AS$) and less than 1% dodecalcium heptaaluminate ($C_{12}A_7$), as well as other trace minerals.

Owing to its high early strength, greater heat of hydration, high resistance to sulfate attack and better resistance to high temperatures, but reduced long-term strength, Aluminate cement is applied to emergency repair work as well as the preparation of refractory materials.

iii. Calcium sulfoaluminate cement

Calcium sulfoaluminate cement is made by grinding cement clinker (calcium sulfoaluminate and dicalcium silicate as main mineral components) mixed with different amounts of limestone, and proper amount of gypsum.

It has been successfully applied in rapid repair and construction, marine engineering and other projects, contributing to its rapid setting and hardening, high early strength, micro-expansion, seepage resistance and seawater erosion resistance.

## Questions for review

1. Briefly describe the production process of Portland cement.

2. What is the mineral composition of Portland cement? What are the effects of minerals on the properties of hardened cement paste?

3. Briefly describe the hydration reaction of various clinkers of Portland cement and the product compositions.

4. What are the factors that affect the setting and hardening of Portland cement? How does it affect?

5. Explain the reasons for the "necessity" of each as follows:

A proper amount of gypsum must be added to produce Portland cement;

Cement grinding must have a certain fineness;

The volume soundness of cement must be qualified.

6. Briefly describe the types, mechanism and prevention of Portland cement corrosion.

7. Briefly describe the technical properties and practical significance of Portland cement.

8. Briefly describe the causes and checking methods of poor soundness of cement.

9. Briefly describe the definition, variety, and role of active and inactive SCMs, as well as the corresponding Portland cement varieties and definitions.

10. Briefly describe the characteristics of various types of Portland cement.

# Chapter IV  Ordinary Concrete and Construction Mortar

## Section I  Concrete overview

### I. Definition of concrete

Concrete is the man-made stone with a certain strength, which is formed through binding granular materials by binders. In the field of engineering construction, concrete is often abbreviated as the Chinese character "砼". As the name implies, "砼" graphically indicates that concrete is an artificial stone.

### II. Classification of concrete

Generally, concrete can be classified in the following ways.

ⅰ. Classification by the binder used

Concrete can be divided into cement concrete, asphalt concrete, polymer concrete, resin concrete, gypsum concrete, water glass concrete, silicate concrete, etc. according to the binder used. At present, the mostly used concrete is cement concrete, which used cement as the binder to bind the sand and rock aggregates distributed according to certain grain sizes together. Cement concrete is today the most versatile and largest amount of man-made civil engineering material used in the world, and is also an important engineering structural material. Concrete in this chapter refers to cement concrete unless otherwise specified.

ⅱ. Classification by apparent density

According to the apparent density, concrete can be divided into heavy-weight concrete, normal weight concrete and lightweight concrete.

1. Heavyweight concrete. The apparent density of heavyweight concrete is larger than $2600kg/m^3$. It is produced with heavy aggregate—barite, iron ore, steel scrap, etc.—as aggregates, and can also be produced with heavy cement—barium cement and strontium cement—as binder at the same time. Heavy-weight concrete has the ability to block the penetration of radiation and thus is mainly used in radiation-proof structures, such as shielding structures for nuclear energy projects, nuclear waste containers, etc.

2. Ordinary concrete. The apparent density of ordinary concrete is 2100 to 2500kg/m$^3$, generally around 2400kg/m$^3$. It is produced with ordinary natural sand and rock as aggregates, usually referred to as concrete. In the field of engineering constructure, ordinary concrete is used in large quantities as a load-bearing material for various buildings and structures.

3. Lightweight concrete. The apparent density of lightweight concrete is less than 1950kg/m$^3$. It is produced with lightweight aggregate, or without aggregate but admixed with air-entraining agent or foaming agent, etc., resulting in a porous structure of concrete, including lightweight aggregate concrete, porous concrete, microporous concrete, etc.

iii. Classification by application

Concrete can be divided into structural concrete, waterproof concrete, heat-resistant concrete, acid-resistant concrete, decorative concrete, mass concrete, expanded concrete, radiation-proof concrete, road concrete, and many others according to their applications.

iv. Classification by production and construction methods

Concrete can be divided into ready-mixed concrete (commercial concrete), pumped concrete, shotcrete, pressure grouted concrete (pre-filled aggregate concrete), extruded concrete, centrifugal concrete, vacuum dewatered concrete, roller compacted concrete, hot-mix concrete, etc. according to the production and construction methods.

## III. Performance characteristics of concrete

Concrete can be widely used in the field of civil engineering mainly because it has the following advantages.

i. The raw materials of concrete are abundant in source and cheap in cost, so they can be obtained locally and cheaply.

ii. Concrete mixture has good plasticity and can be cast into various shapes according to the requirements of the engineering structure, showing a good moldability.

iii. Concrete with different physical and mechanical properties can be formulated by changing the type and proportion of concrete ingredients to meet the different need of various engineering constructions. The strength of concrete in normal engineering project is 20 to 40MPa, while the strength of concrete in special engineering projects can also be as high as 80 to 100MPa, showing a good adaptability.

iv. There is a strong bond between concrete and rebar, and the linear expansion coefficient of concrete and rebar is basically the same. After the two are compounded into the reinforced concrete, they can guarantee to work together, thus greatly expanding the application domain of concrete.

v. Good durability. Concrete does not require maintenance in the normal service environment.

vi. Good fire resistance. The fire resistance of concrete is much better than wood, steel and plastic. Concrete can withstand several hours of high temperature damage and maintain its mechanical properties without significant decline, providing sufficient time for rescue in the event of a fire.

ⅶ. Lower energy consumption in production. The energy consumption of concrete production is much lower than the process of firing soil products and producing metal materials.

The main shortcomings of concrete are as follows.

ⅰ. Heavy self-weight and low strength-to-weight ratio. Each cubic meter of ordinary concrete weighs about 2400kg, resulting in the use of fat beams, fat columns and thick foundations to support the concrete structure in the field of civil engineering construction, which are not conducive for large-span buildings.

ⅱ. Low tensile strength. Generally, the tensile strength of concrete is 1/20 to 1/10 of its compressive strength, so the cracks are easily generated and brittle damage occurs in concrete when under tension.

ⅲ. Large thermal conductivity coefficient. Ordinary concrete has the thermal conductivity coefficient of 1.40W/(m·K), twice as much as the red brick, resulting in an easy thermal conductivity and poor thermal insulation performance of concrete.

ⅳ. Slow hardening and long production cycle. The hardening time of concrete depends on the hardening time of the binder in it. Generally, the hardening time of cement is around several hours after mixing with water.

In summary, concrete material has many advantages, but there are also a number of shortcomings that are difficult to overcome. With the development of modern concrete technology, the shortcomings of concrete have been greatly improved. For example, the use of lightweight aggregate can make the self-weight and thermal conductivity coefficient of concrete significantly reduced; the incorporation of fibers or polymers can greatly reduce the brittleness of concrete; the adoption of fast hardening cement or the addition of early strength agent, water reducer, etc. in concrete can significantly shorten its hardening time. As concrete has these importance advantages mentioned above, while its shortcomings are also easy to properly control, making many other strong and efficient structural materials losing the ability to compete with it. Concrete has been the major civil engineering material, which is widely used in industrial and civil engineering, hydraulic engineering, underground engineering, highways, railroads, bridges and national defense constructions.

## Section Ⅱ  Component materials of concrete

Concrete is composed of binder, water, fine aggregate and coarse aggregate, while a proper quantity of admixtures is also often added. Among them, the binder includes cement and mineral admixtures. Concrete is a multi-phase, multi-component, non-uniform, discontinuous and non-homogeneous material.

## I. The role of component materials

Cement and water in the basic constituent materials of ordinary concrete account for 20% to 30% of the total volume of concrete, while the fine and coarse aggregates accounting for 70% to 80% of the total volume. The mixture of cement and water in concrete is called cement paste, which has a lubricating effect and can bind the aggregates together at the same time, giving the concrete mixture fluidity and plasticity and facilitating the placement and construction of concrete. After the cement paste hardens, it is called cement stone, which has a cementing effect and can bind the loose aggregates together to form a solid whole. Although cement and water account for a small proportion in the concrete, the role they play is crucial. Cement paste is the source of the overall fluidity and plasticity of the concrete mixture, while cement stone is also an important source of the overall strength of the concrete after hardening.

The aggregate in concrete firstly plays the role of acting as skeleton and filling. Secondly, as the aggregate particles are hard and have good volume stability, they can form a solid skeleton by lapping each other and play the role of resisting external forces. At the same time, the aggregate can also limit the shrinkage of cement stone to ensure the volume stability of concrete. Finally, the cost of aggregate, which occupy most of the volume in concrete, is much lower than that of cement, making the overall production cost of concrete can be greatly reduced.

## II. Binder

### i. Cement

Cement is the most important component in concrete, and the reasonable selection of cement includes the following two aspects.

1. Selection of cement type

The type of cement used for concrete production should be reasonably selected according to the characteristics of the project to which the concrete is applied, the environment and construction conditions in which the project is located, and the characteristics of various types of cement.

2. Selection of cement strength grade

The choice of strength grade of cement should be adapted to the design strength grade of concrete. In principle, the preparation of concrete of high strength grade should use cement of high strength grade, and concrete of low strength grade should use cement of low strength grade. Generally, it is appropriate to use the cement with strength grade of about 1.5 times of the strength grade of concrete for ordinary-strength concrete; for high-strength concrete, this multiple can be about 1.

If the low strength grade cement is used to formulate a concrete of high strength grade, a large amount of cement must be used in order to meet the strength requirements, which is not only uneconomical, but also increases the shrinkage and hydration heat in concrete. At the same time, for ensuring the high strength requirements, a low water-to-cement ratio ($W/C$) is required to be adopted in the concrete, which can cause the concrete being too dry during mixing, construction difficulties, and compactness issues, resulting in the loss in guarantee of the construction quality of concrete. Conversely, if the high strength grade cement is used to formulate a concrete of low strength grade, only a small amount of cement can meet the requirements from the strength point of view alone, but in order to meet the workability of concrete mixture and durability requirements of hardened concrete, the amount of cement used needs to be increased, which often produces extra-strong phenomenon of concrete and also is uneconomical.

ii. Mineral admixture

Mineral admixture is an external admixture added directly before or during the mixing process of concrete, like other components of concrete. The vast majority of admixtures used in concrete are industrial solid waste with a certain activity, which comes from the mineral composition in it. Therefore, the admixture is often called mineral admixture. The partially replacement of cement by mineral admixture in concrete can not only recycle the waste, but also reduce the production cost of concrete, while improving the workability of concrete mixture and durability of hardened concrete. As a result, the technical, economic and environmental benefits of adding of mineral admixtures in concrete are very significant.

Commonly used mineral admixtures in concrete include fly ash, ground granulated blast furnace slag, silica fume, etc., in which fly ash and slag are the largest amount and the most widely used mineral admixtures.

1. Fly ash

Fly ash is the flue ash discharged from the combustion of coal in thermal power plants that use pulverized coal furnace to generate electricity, and is a volcanic active admixture, whose main components are oxides of silicon, aluminum and iron, with potential chemical activity. Because of the high fineness of coal powder and the easy formation of glass beads in the high temperature combustion process, most of the fly ash particles are spherical, and their particle size is mostly below 45 $\mu$m, which can be used as concrete admixture directly without grinding.

Fly ash can be divided into Class F fly ash and Class C fly ash according to the source of coal combusted and calcium oxide content. Class F fly ash is collected by calcination of anthracite or bituminous coal, and its CaO content is not more than 10%, also known as low-calcium fly ash; Class C fly ash is collected by calcination of lignite or sub-bituminous coal, and its CaO content is greater than or equal to 10%, also known as high-calcium fly ash. Compared with Class F fly ash, Class C fly ash generally has the characteristics of high activity and good self-hardening. However, Class C fly ash often contains free calcium oxide, so it must be tested for its volume stability when being used in concrete as admixture.

Fly ash is further divided into three grades according to its physical and chemical property requirements: Grade I, Grade II and Grade III. In accordance with the provisions of *Fly Ash*

Used for Cement and Concrete (GB/T 1596—2017), the physical and chemical property requirements of each grade of fly ash are listed in Table 4-1. Scan the QR code for details of *Fly Ash Used for Cement and Concrete* (GB/T 1596—2017).

**Table 4-1 Physical and chemical properties requirements for different grades of fly ash**

| Item | | Technical Requirements | | |
|---|---|---|---|---|
| | | Grade I | Grade II | Grade III |
| Fineness (45μm square-mesh sieve margin)/%, not more than | Class F fly ash | 12.0 | 30.0 | 45.0 |
| | Class C fly ash | | | |
| Water demand ratio/%, not more than | Class F fly ash | 95 | 105 | 115 |
| | Class C fly ash | | | |
| Loss on ignition/%, not more than | Class F fly ash | 5.0 | 8.0 | 10.0 |
| | Class C fly ash | | | |
| Moisture content/%, not more than | Class F fly ash | 1.0 | | |
| | Class C fly ash | | | |
| $SO_3$/%, not more than | Class F fly ash | 3.0 | | |
| | Class C fly ash | | | |
| Free calcium oxide/%, not more than | Class F fly ash | 1.0 | | |
| | Class C fly ash | 4.0 | | |
| Total mass fraction of $SiO_2$, $Al_2O_3$, and $Fe_2O_3$/%, not less than | Class F fly ash | 70.0 | | |
| | Class C fly ash | 50.0 | | |
| Density/g · cm$^{-3}$, not greater than | Class F fly ash | 2.6 | | |
| | Class C fly ash | | | |
| Adequacy Distance increased after boiling of Reynolds clip/mm, not more than | Class C fly ash | 5.0 | | |

### 2. Ground granulated blast furnace slag powder

The molten materials with calcium silicate aluminosilicate as the main component obtained from the blast furnace smelting pig iron in iron smelting plant can be water quenched and granulated to obtain industrial solid waste—granulated blast furnace slag, most of which is glassy. Ground granulated blast furnace slag powder can be produced by drying and grinding of granulated blast furnace slag, and appropriate amount of gypsum and grinding aid can be added during grinding. Ground granulated blast furnace slag powder is referred to as slag powder.

In accordance with the provisions of *Ground Granulated Blast Furnace Used for Cement, Mortar and Concrete* (GB/T 18046—2017), slag powder should meet the technical requirements of Table 4-2. Scan the QR code for details of *Ground Granulated Blast Furnace Used for Cement, Mortar and Concrete* (GB/T 18046—2017).

Chapter IV  Ordinary Concrete and Construction Mortar

Table 4-2  Technical requirements of slag powder

| Item | | Grade | | |
|---|---|---|---|---|
| | | S105 | S95 | S75 |
| Density/g · cm$^{-3}$, not less than | | 2.8 | | |
| Specific surface area/m · kg$^{-1}$, not less than | | 500 | 400 | 300 |
| Activity index/%, not less than | 7d | 95 | 70 | 55 |
| | 28d | 105 | 95 | 75 |
| Flow rate ratio/%, not less than | | 95 | | |
| Moisture/%, not more than | | 1.0 | | |
| SO$_3$/%, not more than | | 4.0 | | |
| MgO/%, not more than | | 13.5 | | |
| Chloride ion/%, not more than | | 0.06 | | |
| Loss on ignition/%, not more than | | 1.0 | | |

The activity index of slag powder in Table 4-2 is the ratio of the strength of the test mortar with 50% cement replaced with slag to the strength of the control mortar with 100% cement. Slag powder is divided into three grades according to its activity index: S105, S95 and S75.

## III. Aggregate

Rock and sand in the concrete play a skeletal role, so they are called aggregate. Aggregates used in ordinary concrete are divided into two types according to the size of the particle size: coarse aggregate with particle size greater than 4.75mm and fine aggregate with particle size between 0.15mm and 4.75mm.

The coarse aggregate usually used in ordinary concrete are crushed stone and pebbles. Crushed stone refers to rock particles with a particle size greater than 4.75mm obtained by mechanical crushing and screening natural rock, pebbles or mine waste rock. Pebbles refers to rock particles with a particle size greater than 4.75mm formed by natural weathering, water transport and sorting, and stacking. The fine aggregate used in ordinary concrete include two types of sand, namely natural sand and manufactured sand. Natural sand refers to rock particles less than 4.75mm in size formed naturally and mined and sieved artificially, including river sand, lake sand, mountain sand and desiccated sea sand, but does not include soft, weathered rock particles. Manufactured sand, whose preparation method is same as crushed stone, refers to the rock, mine trailing or industrial waste particles with size less than 4.75mm made by mechanical crushing and sieving after the soil removal process. Manufactured sand does not include soft and weathered particles, which is also called artificial sand.

The total volume of coarse and fine aggregates generally accounts for 70% to 80% of the volume of concrete, so the quality of the aggregates will directly affect the quality of the concrete properties. To this end, the two national standards *Sand for Construction* (GB/T 14684—2022) and *Pebble and Crushed Stone for Construction* (GB/T 14685—2022) put forward clear quality requirements for sand, pebble and crushed stone. Sand, pebble and crushed stone are divided into three categories: I, II and III. Category I should be used for concrete with strength grade greater than C60; category II should be used for concrete with strength grade C30 ~ C60 and frost resistance, water penetration resistance or other requirements; category III is appropriate for concrete with strength grade less than C30. The following is a brief description of the technical requirements of aggregates used for ordinary concrete. Scan the QR code to get the details of *Sand for Construction* (GB/T 14684—2022) and *Pebble and Crushed Stone for Construction* (GB/T 14685—2022).

i. Apparent density, bulk density, void ratio

The apparent density, bulk density and void ratio of aggregates are the original data necessary for the calculation of mix design of concrete. These properties of aggregates affect the hardness of the concrete internal skeleton and particle gradation, thus influencing the properties of concrete. Specifications require sand used for ordinary concrete features the apparent density greater than $2500kg/m^3$, loose bulk density greater than $1400kg/m^3$, and void ratio less than 44%. The apparent density for pebble and crushed stone used for ordinary concrete should be greater than $2600kg/m^3$, while the void ratios of categories I, II and III rock in a continuous gradation and loose pile state should be less than 43%, 45% and 47%, respectively.

ii. Particle shape and surface characteristics

The particle shape of the aggregate is irregular, there are some particles feature close three-dimensional size, there are also elongated or flat particles; some particles are with angular, while some particles are round and smooth. The elongated or flat aggregate particles in rock are called elongated or flat particles, respectively. These two particle shapes of aggregate are easy to break when stresses, affecting the overall strength of the skeleton. As the same time, these two particle shapes of aggregate are difficult to mix during the mixing, which are not conducive to the mechanical properties and workability of concrete. Therefore, the total content of elongated or flat particles in the coarse aggregate used for concrete should not exceed certain limits, as shown in Table 4-3.

Surface characteristics of aggregates refer to the roughness or smoothness of the particle surface, and the surface roughness of different aggregates varies. Natural pebbles and river sand have a smooth surface due to long-term water washing and abrasion, which is beneficial to the workability of concrete mixture; however, they are poorly bonded to the cement paste and the performance of the formed skeleton is poor, resulting in a negative effect on strength. Crushed stone and mountain sand obtained by mechanical crushing have rough surface, while they can lap each other to form a strong skeleton and strongly bond with the cement paste, facilitating to improve the strength; but they need to be wrapped with more cement pate on the surface, and a thicker payer of cement paste is needed to lubricate between the particles to achieve the required concrete workability. In the actual engineering project, the aggregate should be selected according to the strength and workability indexes required by the project.

## Chapter IV  Ordinary Concrete and Construction Mortar

**Table 4-3  Elongated or flat particles content in coarse aggregate**

| Item | Indicator | | |
|---|---|---|---|
| | Category I | Category II | Category III |
| Elongated or flat particles content (by mass, %) | ≤5 | ≤10 | ≤15 |

iii. Clay content, clay lumps and friable particles content, rock powder content

Clay content refers to the content of particles with size less than 75 μm in the coarse aggregate and natural sand. Rock powder content refers to the content of particles with size less than 75 μm in manufactured sand. Clay lumps and friable particles content in fine aggregate refers to the content of particles with size greater than 1.18mm but less than 0.6mm after washing with water and hand kneading; in coarse aggregate, it refers to the content of particles with size greater than 4.75mm but less than 2.36mm after washing with water and hand kneading.

The clay particles and rock powder particles in the aggregate are very fine, which will adhere to the surface of the aggregate, affecting the binding ability between cement stone and aggregate, reducing the strength and durability of concrete, and increasing the dry shrinkage of concrete. The clay lumps and friable particles in the aggregate has a very low strength and can collapse after immersion in water and shrink after drying, which will form a weak part in the concrete and have greater impact on the quality of concrete. Therefore, the national standards have certain requirements for the clay content, clay lumps and friable particle content and rock power content in various categories of aggregate, while the aggregate that fails to meet the requirements needs to be flushed and other processing. Table 4-4 lists the requirements of clay content and clay lumps and friable particles content of aggregate. For manufactured sand, the methylene blue (MB) value is mainly used to determine whether the particles with size less than 75 μm are clay or rock powder with the same chemical composition as the parent rock, with the value less than 1.4 indicating these particles are mainly rock powder and larger than 1.4 indicating the content of clay gradually increases. Table 4-5 and Table 4-6 list the limit values of rock powder content in manufactured sand when the MB value is less and larger than 1.4, respectively.

**Table 4-4  Clay content and clay lumps and friable particles content**

| Item | | Explanation | Indicator | | |
|---|---|---|---|---|---|
| | | | Category I | Category II | Category III |
| Sand | Clay content (by mass, %) | Natural sand | ≤1.0 | ≤3.0 | ≤5.0 |
| | Clay lumps and friable particles content (by mass, %) | Natural sand, manufactured sand | 0 | ≤1.0 | ≤2.0 |
| Rock | Clay content (by mass, %) | | ≤0.5 | ≤1.0 | ≤1.5 |
| | Clay lumps and friable particles content (by mass, %) | | 0 | ≤0.2 | ≤0.5 |

Table 4-5  Content limit of rock powder in manufactured sand when MB value is less than 1.4

| Item | Indicator | | |
|---|---|---|---|
| | Category I | Category II | Category III |
| MB value | ≤0.5 | ≤1.0 | ≤1.4 |
| Rock powder content (by mass, %) | ≤10.0 | | |

Table 4-6  Content limit of rock powder in manufactured sand when MB value is larger than 1.4

| Item | Indicator | | |
|---|---|---|---|
| | Category I | Category II | Category III |
| Rock powder content (by mass, %) | ≤1.0 | ≤3.0 | ≤5.0 |

iv. Harmful substance content

Coarse and fine aggregates used for ordinary concrete should not be mixed with grass roots, leaves, branches, plastics, furnace slag, cinder and other debris. Mica and light substances (material lighter than 2000kg/m$^3$) in fine aggregate have poor adhesion with cement paste, affecting the strength and durability of concrete. Sulfides, sulfates, chlorides and organic matter in aggregates can impede the hydration of cement, exert rushing effect on rebar in concrete, and react with the cement hydration products to generate harmful expansion products. The content of these harmful substances should comply with the provisions of Table 4-7.

Table 4-7  Content of harmful substances in aggregates

| | Item | Indicator | | |
|---|---|---|---|---|
| | | Category I | Category II | Category III |
| Sand | Mica content (by mass, %) | ≤1.0 | ≤2.0 | ≤2.0 |
| | Light substance content (by mass, %) | ≤1.0 | ≤1.0 | ≤1.0 |
| | Organic matter content (colorimetric method, %) | Qualified | Qualified | Qualified |
| | Sulfide and sulfate content (by mass of $SO_3$, %) | ≤0.5 | ≤0.5 | ≤0.5 |
| | Chloride content (by mass of chloride ion, %) | ≤0.01 | ≤0.02 | ≤0.06 |
| Rock | Organic matter content (colorimetric method, %) | Qualified | Qualified | Qualified |
| | Sulfide and sulfate content (by mass of $SO_3$, %) | ≤0.5 | ≤1.0 | ≤1.0 |

v. Soundness

Soundness of aggregate refers to the ability of aggregate to resist damage caused by the climate, environmental change or other physical factors, which is tested using sodium solution method with the mass loss after 5 cycles should be in line with the provisions of Table 4-8.

Table 4-8  Aggregate soundness indicator

| Item | | Indicator | | |
|---|---|---|---|---|
| | | Category I | Category II | Category III |
| Rock | Mass loss/% | ≤5 | ≤8 | ≤12 |
| Sand | | ≤8 | ≤8 | ≤10 |

ⅵ. Alkaline activity

Aggregates containing reactive silica will react with alkali ($Na_2O$ or $K_2O$) in cement to produce expansive products. If the expansive stress is greater than the strength of the concrete, it will lead to cracking of the concrete. Therefore, when the aggregate is used for important engineering concrete or when there is doubt about the aggregate, it shall be tested for alkali activity by chemical method or length method according to the standard. Standard stipulates that the specimen prepared by the aggregate should be free of crack, crisp cracking, colloidal spillage and other phenomena after the alkali-aggregate reaction, as well as the expansion rate of the specimen should be less than 0.10% at the specified test age.

ⅶ. Strength of aggregate

Coarse aggregate plays the role of the overall skeleton in concrete, and the strength of the coarse aggregate itself directly affects the overall strength of the concrete; therefore, the standard has certain requirements for the strength of coarse aggregate. There are two methods of measuring the strength of coarse aggregate, namely the compressive strength of the parent rock and the crushing index value.

The so-called compressive strength of the parent rock is the compressive strength of the water-saturated cubic specimen with side length of 50mm (or cylinder specimen with both diameter and height of 50mm) made using the parent rock after submerging in water for 48h. Usually, the ratio of compressive strength of rock to the strength grade of concrete is required to be not less than 1.5.

Crushing index value is a direct measurement of the ability of stacked pebble or crushed stone to withstand pressure without breaking, which more directly reflects the state of the aggregate in the concrete under stress, resulting it being an important mechanical property to measure the degree of hardness of the aggregate. During the test, the elongated or flat particles are firstly removed from the air-dried rock with particle size of 9.50 ~ 19.0mm; then 3000g specimen is put into the test mold according to the standard method, and the pressure of 200kN is imposed on specimen according to the specified loading rate, which is kept for 5s and then unloaded; at last, specimen is poured out from the mold and sieved to remove the crushed fines with a 2.36mm standard sieve and weight out the mass of specimen left on the sieve. Calculate the crushing index value according to Equation (4-1):

$$Q_a = \frac{G_1 - G_2}{G_1} \times 100\% \qquad (4-1)$$

where  $Q_a$——crushing index value(%);

$G_1$——total mass of the specimen(g);

$G_2$——the mass of the specimen remaining on the sieve with a mesh size of 2.36mm after crush test(g).

The smaller the crushing index value, the harder the rock it, and the stronger the compression capacity. The crushing index of each category of rock must meet the values specified in Table 4-9.

Table 4-9 Crushing index of coarse aggregate

| Item | Indicator | | |
| --- | --- | --- | --- |
| | Category Ⅰ | Category Ⅱ | Category Ⅲ |
| Crushing index of crushed stone/% | ≤10 | ≤20 | ≤30 |
| Crushing index of pebble/% | ≤12 | ≤14 | ≤16 |

ⅷ. Water content of aggregate

Different water contents of aggregates can lead to large variations in the amount of water used in concrete and the amount of aggregates used in the preparation of concrete, which in turn affects the concrete properties. The saturated surface dry state of the aggregate neither draws water from the concrete, nor releases water into the concrete mixture, so the water demand in concrete can be accurately controlled. The water content of aggregate in the saturated surface dry state is called the saturated surface dry water absorption rate. The smaller the saturated surface dry water absorption rate, the denser the aggregate particles and the better the quality of aggregate. The saturated surface dry water absorption rate of normal solid aggregate is about 1%, and the water content of aggregate in air dry state is less than 1%, which is not much different with the saturated surface dry water absorption rate, so the engineering project usually uses the water content of aggregate in air dry state as a benchmark for mix design of concrete. However, in the construction of the engineering project, the water content of the aggregate must be measured frequently and the actual amount of concrete composition materials must be adjusted in time, ensuring the stability of the quality of concrete.

The volume and bulk density of sand are closely related to its water content. As the water content of sand in air dry state increases, a layer of adsorbed water film is formed on the surface of sand particles, which pushed the sand particles apart and causes an increase in the volume of sand, calling the wet swelling of sand. The wet swelling of sand is related to its particle size, and the wet swelling of fine sand is larger than that of coarse sand. When the water content of sand increases to 5% ~ 8%, its volume can be increased by 20% ~ 30%. If the water content of sand continues to increase, the water film on the surface of sand thickens, and the self-weight of water exceeds the adsorption of water by the surface of sand grains, leading to the rupture and disappear of water film and the decreasing of the volume of sand. When the water content of sand increases to about 20%, the volume of wet sand decreases to be similar to that of dry sand.

ⅸ. Gradation of aggregate

The gradation of aggregate refers to the distribution of different particle sizes in the aggregate. The particle size distribution of aggregate in the same size range will produce a large void ratio, as shown in Figure 4-1(a); when the particle size distribution of aggregate is in two size ranges, the void ratio will be reduced, as shown in Figure 4-1(b); if the particle size distribution of aggregate is in more size ranges, the void ratio will be further reduced, as shown in Figure 4-1(c). It can be seen that only a suitable aggregate particle size distribution can achieve the requirements of good gradation. Good gradation should make the aggregate void ratio and total surface area small, thus reducing the amount of cement paste required and improving the compactness, strength and other properties of concrete; in the consideration of reducing the void ratio while also taking the requirements of concrete flowability into account, fine aggregate should have a certain degree of surplus to enable it to wrap the coarse aggregate and provide the flowability.

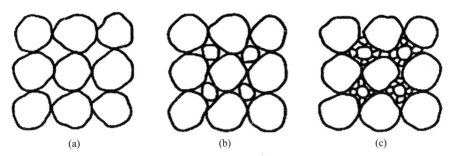

(a)             (b)             (c)

**Figure 4-1 Particle gradation of aggregate**

Aggregate gradation includes macroscopic and mesoscopic gradations, while the macroscopic gradation refers to sand ratio and the mesoscopic gradation includes the respective particle gradation of sand and rock.

1. Sand ratio

Sand ratio $S_p$ refers to the mass of sand(S) in the concrete as a percentage of the total mass of sand and rock(G), i. e. :

$$S_p = \frac{S}{S + G} \times 100\% \qquad (4\text{-}2)$$

2. Particle gradation and fineness of sand

The particle gradation and fineness of sand are determined by sieve analysis. The sieve analysis method of sand uses a set of standard sieves with mesh sizes of 4.75mm, 2.36mm, 1.18mm, 0.6mm, 0.3mm and 0.15mm to sieve 500g of dry sand from coarse to fine, and then weigh the mass of sand on each sieve and calculate the percentage of sieve residue on each sieve $a_1, a_2, a_3, a_4, a_5, a_6$ and cumulative sieve residue $A_1, A_2, A_3, A_4, A_5, A_6$ (the mass sum of the sand residue on each sieve and all sieves coarser than that sieve). The relationship between the cumulative sieve residue and the sieve residue is shown in Table 4-10, with any set of cumulative sieve residue ($A_1$ to $A_6$) characterizes a gradation.

**Table 4-10 Relationship between cumulative sieve residue and sieve residue**

| Sieve mesh size/ mm | Sieve residue/% | Cumulative sieve residue/% |
|---|---|---|
| 4.75 | $a_1$ | $A_1 = a_1$ |
| 2.36 | $a_2$ | $A_2 = a_1 + a_2$ |
| 1.18 | $a_3$ | $A_3 = a_1 + a_2 + a_3$ |
| 0.60 | $a_4$ | $A_4 = a_1 + a_2 + a_3 + a_4$ |
| 0.30 | $a_5$ | $A_5 = a_1 + a_2 + a_3 + a_4 + a_5$ |
| 0.15 | $a_6$ | $A_6 = a_1 + a_2 + a_3 + a_4 + a_5 + a_6$ |

The fineness of sand id expressed in terms of fineness modulus, which is calculated by the following formula:

$$M_x = \frac{(A_2 + A_3 + A_4 + A_5 + A_6) - 5A_1}{100 - A_1} \qquad (4\text{-}3)$$

The larger the fineness modulus, the coarser the sand. The range of fineness modulus of sand used for ordinary concrete is generally 1.6~3.7, of which $M_x$ in 3.1~3.7 is coarse sand, $M_x$ in 2.3~3.0 is medium sand, and $M_x$ in 1.6~2.2 is fine sand. It is appropriate to give priority to medium sand when formulating concrete.

It should be noted that sand with the same fineness modulus can have a different gradation. The fineness modulus of sand does not reflect its gradation, therefore, the gradation curve of sand must be considered when formulating concrete. According to the standard, the sand is divided into three gradation zone according to the cumulative sieve residue of 0.60 sieve, see Table 4-11. With the cumulative sieve residue as the vertical coordinate and the sieve mesh size as the horizontal coordinate, the sieving curves (see Figure 4-2, using natural sand as an example) of the upper and lower limits of three gradation zones (Zone 1, Zone 2 and Zone 3) of sand can be drawn according to the values specified in Table 4-11.

When conducting the test, the cumulative sieve residue of each sieve during the sand sample sieve analysis is marked on the gradation zone figure, which can be connected to observe which gradation area this sieve curve could falling in. The method of determining whether the sand gradation is qualified is as follows.

(1) The gradation zone depends on the cumulative sieve residue of 0.60mm sieve, while the rest of the cumulative sieve residue on each sieve should completely fall in that gradation zone in principle.

(2) No overruns on the 4.75mm and 0.60mm sieves are allowed.

(3) The rest of the sieves is allowed to have a small amount of excess, but the sum of the accumulated excess of the cumulative sieve residue should be less than 5%.

Table 4-11　Range of particle gradation zone of sand

| Square mesh size/mm | Cumulative sieve residue/% | | | | | |
| --- | --- | --- | --- | --- | --- | --- |
| | Natural sand | | | Manufactured sand | | |
| | Zone 1 | Zone 2 | Zone 3 | Zone 1 | Zone 2 | Zone 3 |
| 9.50 | 0 | 0 | 0 | 0 | 0 | 0 |
| 4.75 | 10~0 | 10~0 | 10~0 | 10~0 | 10~0 | 10~0 |
| 2.36 | 35~5 | 25~0 | 15~0 | 35~5 | 25~0 | 15~0 |
| 1.18 | 65~35 | 50~10 | 25~0 | 65~35 | 50~10 | 25~0 |
| 0.60 | 85~71 | 70~41 | 40~16 | 85~71 | 70~41 | 40~16 |
| 0.30 | 95~80 | 92~70 | 85~55 | 95~80 | 92~70 | 85~55 |
| 0.15 | 100~90 | 100~90 | 100~90 | 97~85 | 94~80 | 94~75 |

# Chapter IV  Ordinary Concrete and Construction Mortar

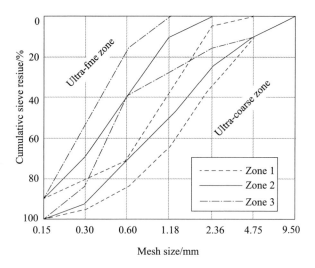

**Figure 4-2  Gradation zone curve of sand**

When preparing concrete, priority should be given to Zone 2 sand; when Zone 1 sand is used, the sand ratio should be increased and sufficient cement content should be maintained to meet the workability of concrete; when Zone 3 sand is used, the sand ratio should be appropriately reduced to ensure the strength of concrete.

3. Particle gradation and maximum particle size of rock

The particle gradation of rock is divided into two types, namely continuous gradation and gap gradation. The gradation of rock is determined by the sieving test using a set of standard sieves, which consists of 12 sieves with mesh sizes of 2.36mm, 4.75mm, 9.50mm, 16.0mm, 19.0mm, 26.5mm, 31.5mm, 37.5mm, 53.0mm, 63.0mm, 75.0mm and 90.0mm. The sieves can be selected for sieving as needed, and then the sieve residual percentages and cumulative sieve residual percentages are calculated for each sieve (the calculation method is the same as for sand). The gradation range requirements for crushed stone and pebble are the same and should conform to the provisions of Table 4-12.

**Table 4-12  Particle gradation range of crushed stone or pebble**

| Nominal particle size (mm) | | Cumulative sieve residue (by mass, %) | | | | | | | | | | |
|---|---|---|---|---|---|---|---|---|---|---|---|---|
| | | Mesh size (square-mesh sieve, mm) | | | | | | | | | | |
| | | 2.36 | 4.75 | 9.50 | 16.0 | 19.0 | 26.5 | 31.5 | 37.5 | 53.0 | 63.0 | 75.0 | 90.0 |
| Continuous gradation | 5~16 | 95~100 | 85~100 | 30~60 | 0~10 | 0 | | | | | | | |
| | 5~20 | 95~100 | 90~100 | 40~80 | — | 0~10 | 0 | | | | | | |
| | 5~25 | 95~100 | 90~100 | — | 30~70 | — | 0~5 | 0 | | | | | |
| | 5~31.5 | 95~100 | 90~100 | 70~90 | — | 15~45 | — | 0~5 | 0 | | | | |
| | 5~40 | — | 95~100 | 70~90 | — | 30~65 | — | — | 0~5 | 0 | | | |

233

continued

| Nominal particle size(mm) | | Cumulative sieve residue(by mass,%) | | | | | | | | | | | |
|---|---|---|---|---|---|---|---|---|---|---|---|---|---|
| | | Mesh size(square-mesh sieve,mm) | | | | | | | | | | | |
| | | 2.36 | 4.75 | 9.50 | 16.0 | 19.0 | 26.5 | 31.5 | 37.5 | 53.0 | 63.0 | 75.0 | 90.0 |
| Gap gradation | 5~10 | 95~100 | 80~100 | 0~15 | 0 | | | | | | | | |
| | 10~16 | | 95~100 | 80~100 | 0~15 | | | | | | | | |
| | 10~20 | | 95~100 | 85~100 | | 0~15 | 0 | | | | | | |
| | 16~25 | | | 95~100 | 55~70 | 25~40 | 0~10 | | | | | | |
| | 16~31.5 | | 95~100 | | 85~100 | | | 0~10 | 0 | | | | |
| | 20~40 | | | 95~100 | | 80~100 | | | 0~10 | 0 | | | |
| | 40~80 | | | | | 95~100 | | | 70~100 | | 30~60 | 0~10 | 0 |

The upper limit of the nominal particle size in the coarse aggregate is the maximum particle size of the aggregate. When the aggregate particle size increases, its specific surface area decreases. Compared with the small size aggregate, the amount of cement paste required to wrap the same quality of large size aggregate is less, so it can save cement. In the case of condition permitting, it should try to use the coarse aggregate with larger maximum particle size. However, the maximum particle size of coarse aggregate is subject to some conditions, such as the minimum size of the cross-section of the member, the spacing of the reinforcement and the thickness of the plate. Usually, the maximum particle size of the aggregate should not exceed 1/4 of the minimum size of the cross-section of the structure, not more than 3/4 of the net distance between the reinforcement; when pouring a solid concrete slab, the maximum particle size of the aggregate should not exceed 1/2 of the slab thickness and not more than 50mm. For the pumping concrete, it also needs to consider the pipe blockage during the pumping process. Usually, the maximum particle size of the aggregate cannot be greater than 1/3 of the inner diameter of the pipe, in order to avoid the pumping blockage.

Scan the QR code to get the content of the testing of aggregates used for concrete.

## IV. Water

Water is one of the important components of concrete. The quality of water not only affects the setting and hardening of concrete, but also affects the strength and durability of concrete. The water used for preparing and curing of concrete shall not contain harmful substances that affect the normal setting and hardening of cement, where all the water that can be drunk and clean natural water can be used to mix and cure concrete.

## V. Chemical admixture

i. Definition and classification of chemical admixture

Chemical admixture is a type of material that can significantly improve one or more properties of concrete before or during mixing, and its dosage is generally not more than 5% of the mass of the binder. Chemical admixture in concrete does not replace cement, but it can significantly improve the workability, strength, durability or adjust the setting time and save cement. The use of chemical admixture promotes the rapid development of concrete, making the production and application of high-strength, high-performance concrete has become a reality and solving many of the technical problems in the actual engineering projects. At present, the chemical admixture has become the fifth important component material (called the fifth component) in addition to the binder, water, sand and rock, and it is increasingly widely used.

Chemical admixtures can be divided into four categories according to its main role.

1. Chemical admixtures that improve the workability of concrete mixture, such as water reducer, air-entraining agent and pumping aid.

2. Chemical admixtures that regulate the setting time and hardening properties of concrete, such as set retarding admixture, quick-setting admixture and accelerating admixture.

3. Chemical admixtures that improve the durability of concrete, such as air-entraining agent, water-repellent admixture and corrosion-inhibition admixture.

4. Chemical admixtures that provide special properties of concrete, such as expansive admixture, anti-freezing admixture and shrinkage reducing agent.

ii. Commonly used chemical admixtures

1. Water reducer

Water reducer is a chemical admixture that can reduce the amount of mixing water under the condition of keeping the concrete flowability basically the same, or can increase the flowability of concrete with the same mix design and materials. Water reducer is a surfactant, whose molecule is composed of two parts: hydrophilic group and hydrophobic group. After the water reducer is added into the cement paste, the hydrophilic group in its molecule points to water, and the hydrophobic group is directed to adsorb on the surface of cement particles, so that the surface of cement particles features the same charge, forming electrostatic repulsion to disperse cement particles; water reducer adsorption film on the surface of cement particles can form a layer of stable solvent water film with water molecules, which has a good lubricating effect and can improve the flowability of concrete; see Figure 4-3. There are many varieties of water reducers, which can be divided into common and high-range types according to the water reducing effect; they can also be divided into lignin sulfonates, polyols, polycyclic aromatic sulfonates, naphthalene, polycarboxylates, etc. according to the chemical composition.

The following technical and economic results can be achieved with the addition of water reducer in concrete:

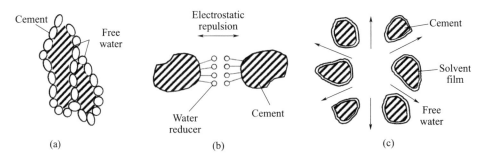

Figure 4-3  Diagram of water reducing mechanism of water reducer

Significantly improve the flowability of concrete at the condition of using the same amount of mixing water, which is beneficial for casting;

Reduce the water required at the condition of keeping flowability of concrete and cement content constant, which can lower $W/C$ and improve the strength and durability of concrete;

Save cement content and reduce cost while maintaining the same flowability and strength of concrete;

Make concrete more homogeneous and improve the pore structure of concrete.

2. Air-entraining agent

Chemical admixture that can introduce a large number of uniformly distributed, stable and closed tiny bubbles and thus improve the workability and durability (mainly frost resistance) of concrete is called air-entraining agent. The diameter of the bubbles introduced by air-entraining agent is in the range of 0.02 ~ 1mm, mostly less than 0.2mm. Commonly used air-entraining agents are rosin, saponins, alkyl sulfonates, etc., whose dosage in the concrete is in general 0.005% to 0.01% of the mass of binder.

3. Set retarding admixture

Set retarding admixture is a chemical admixture that can extend the setting time of concrete. Commonly used set retarding admixture are sugar polyhydroxy compounds (glucose, sucrose, molasses, calcium gluconate, etc.), hydroxy carboxylic acids (citric acid, tartaric acid, gluconic acid, salicylic acid, etc.) and polyols (sorbitol, polyvinyl alcohol, maltitol, xylitol, etc.), etc. The dosage of set retarding admixture is generally 0.01% to 0.1% of the binder.

4. Pumping aid

Pumping aid refers to the chemical admixture that can improve the pumping performance of concrete mixture. Pumping aid generally consists of water-reducing component, set-retarding component and air-entraining component, in which water-reducing component can improve the flowability of concrete mixture, air-entraining component has lubricating effect, and set-retarding component can inhibit the slump loss of the mixture.

5. Accelerating admixture

Accelerating admixture refers to the chemical admixture that can accelerate the development of early strength of concrete. Commonly used accelerating admixtures are salts (chloride, sulfate, sulfate compound, nitrate, nitrite, etc.) and water-soluble organic compounds (triethanolamine, formate, acetate, propionate, etc.), in which the dosage of triethanolamine is less than 0.05% and the dosage of other types of accelerating admixtures is generally less than 2%. It is worth noting that the use of inorganic salt type accelerating admixture will affect the later strength development of concrete, in addition, the use of chloride accelerating admixture is prohibited in reinforced concrete.

6. Anti-freezing admixture

Anti-freezing admixture refers to the chemical admixture that can make concrete harden at negative temperatures and produce sufficient strength to avoid frost damage. Commonly used anti-freezing admixture usually consists of water-reducing component, anti-freezing component and air-entraining component. The role of water-reducing component is to reduce the amount of water used in concrete, thereby reducing the ice expansion stress in concrete; at the same time, the lowered $W/C$ can refine the pore structure and reduce the inherent defects in concrete. The role of the anti-freezing component is to lower the freezing point and ensure that the liquid phase in concrete does not freeze or freezes less under the specified negative temperature conditions. The air-entraining component, on the other hand, improves the frost resistance of concrete at later age.

## Section Ⅲ  Workability of fresh concrete

### Ⅰ. Concept and definition of concrete workability

Concrete in the stage from existing from mixer to finishing the casting and vibrating is called fresh concrete. Workability refers to the ability of fresh concrete to maintain its components uniform and the ability to be easily transported, cast, vibrated and formed without segregation, delamination and water bleeding, which reflects the ease of fresh concrete construction. Workability of concrete includes the meaning of three aspects, namely flowability, cohesion and water retention.

Flowability is the ability of the concrete mixture to flow and fill the mold uniformly and densely under self-weight or external forces.

Adhesion, also known as anti-diffusion, refers to the performance of the concrete mixture to maintain the uniformity of components without delamination and segregation during the transport, pouring and vibrating processes.

Water retention refers to the performance of concrete mixture that has a certain ability to maintain it internal moisture and do not produce severe water bleeding during the construction process.

Flowability, cohesion and water retention of concrete mixture are interrelated and contradictory. When the flowability is large, cohesion and water retention are often poor, and vice versa. Therefore, the so-called good workability refers to all three aspects are good in the specific condition.

## Ⅱ. Evaluation method of concrete workability

The evaluation of the concrete workability is usually done by using certain experimental methods to determine the flowability of concrete mixture and using intuitive empirical visual assessment to determine the cohesion and water retention. According to the *Standard for Quality Control of Concrete* ( GB 50164—2011 ) , the flowability of concrete mixture is indicated by slump or Vebe consistency. Slump is applicable to concrete mixtures with greater flowability, while Vebe consistency is applicable to dry concrete mixtures. Scan the QR code to get the details of the *Standard for Quality Control of Concrete* ( GB 50164—2011 ).

i . Slump

The slump cone (standard truncated cone, no bottom, diameter of the upper opening and the lower opening and height of 100mm, 200mm and 300mm, respectively) is firstly placed and fixed on a horizontal and non-absorbent rigid base plate. The concrete mixture is then put into the cone in three layers, and a slug-type metal pounding rod is used to pound each layer 25 times to the upper surface of the last layer from the edge to the center along the spiral direction. Finally, smooth the upper opening and vertically lift the slump cone, the mixture in the cone under the action of self-weight will slump downward. The measurement of the difference in height between the highest point of the specimen after slumping and the slump cone is the slump value ( mm ) (Figure 4-4). The greater the slump, the larger the flowability of the concrete mixture. Because of the simplicity of this method, it is now commonly used around the world. After measuring the slump, the pounding rod is also used to strike the slumped concrete mixture to observe its sinking, slumping and water bleeding, for determining the cohesion and water retention of concrete mixture by visual inspection.

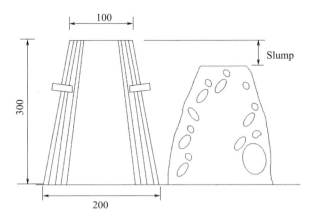

Figure 4-4 Concrete mixture slump measurement

The slump test is only applicable to non-dry concrete with the maximum particle size of aggregate not greater than 40mm and slump value greater than 10mm. According to the size of the slump value, the concrete mixture is divided into 5 levels, as shown in Table 4-13.

# Chapter IV  Ordinary Concrete and Construction Mortar

**Table 4-13  Classification of fresh concrete according to slump and Vebe consistency**

| Level | Name | Slump/mm | Level | Name | Vebe consistency/s |
|---|---|---|---|---|---|
| S1 | Low plasticity concrete | 10~40 | V0 | Ultra-dry concrete | ≥31 |
| S2 | Plastic concrete | 50~90 | V1 | Extra-dry concrete | 21~30 |
| S3 | Flowable concrete | 100~150 | V2 | Dry concrete | 11~20 |
| S4 | Pumpable concrete | 160~210 | V3 | Semi-dry concrete | 6~10 |
| S5 | Fluid concrete | ≥210 | V4 | Low-dry concrete | 3~5 |

ii. Vebe consistency

Concrete with slump value less than 10mm is called dry concrete. Slump value is difficult to reflect the flowability of dry concrete, while Vebe consistency (VB consistency value) is used to reflect its dryness. The instrument used in this method is shown in Figure 4-5, which is called VeBe consistometer. Install the cylindrical container on the vibrating table, load the concrete mixture in the container according to the slump test method, lift the slump cone and place the transparent plate on the concrete mixture, start the vibrating table, and measure the time from the start of vibration to the concrete mixture fully contacting with the transparent plate. This time is called Vebe consistency value, which is used to quantitatively evaluate the consistency of dry concrete. The larger the Vebe consistency value, the drier the concrete mixture and the poorer the flowability. The concrete mixture can be divided into 5 levels according to the size of its Vebe consistency value, as shown in Table 4-13.

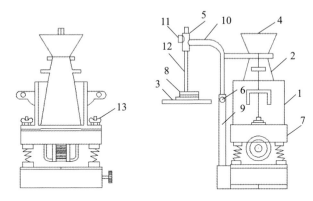

**Figure 4-5  Vebe consistometer**

1—container; 2—slump cone; 3—transparent plate; 4—hopper; 5—sleeve;
6—positioning screw; 7—vibrating table; 8—load; 9—pillar; 10—spiral frame;
11—measuring rod screw; 12—measuring rod; 13—setting screw

iii. Slump flow

For fluid concrete with slump greater than 210mm, slump flow is also needed to be tested in addition to slump.

The slump flow test is based on the traditional slump test, which simultaneously determines the horizontal flow of the mixture and the time used to expand to a certain diameter (generally 50cm) to reflect the deformation capacity and deformation speed of the mixture.

## III. Factors affecting the concrete workability

i. Unit water consumption

In the preparation of concrete, when the type and quantity of coarse and fine aggregates used are constant, the flowability of concrete remains unchanged even if the cement content fluctuates slightly (cement content increases or decreases by $50 \sim 100 \text{kg/m}^3$) at the condition of keeping the amount of water used per cubic meter of concrete unchanged. This law is called the fixed water demand law. The amount of water used in concrete increases, the flowability of concrete becomes better, but the adverse effect of large water consumption is that the cohesion and water retention of the concrete mixture become worse and water bleeding and segregation easily occur.

ii. Water-to-cement ratio

The size of $W/C$ reflects the degree of consistence of the cement paste. When a certain amount of cement paste is used, lowering $W/C$ will result in the thickened cement paste, increased adhesion of cement paste, and improved cohesion and water retention of concrete, but decreased flowability; increasing $W/C$ is the opposite.

iii. Cement paste content

Concrete mixture must overcome its internal resistance to flow under the action of self-weight or external vibration force. The resistance within the mixture mainly comes from two aspects, one is the frictional resistance between the aggregates, the other one is the cement paste adhesion. The size of the frictional resistance between the aggregates mainly depends on the thickness of the cement paste layer on the surface of the aggregate particles, that is, the amount of cement paste; the size of the cement pate adhesion mainly depends on the dryness of the cement paste, that is, the consistency of the cement paste.

In the case of maintaining the same $W/C$ of the concrete mixture, the more cement paste, the thicker the paste layer wrapped on the surface of the aggregate particles, the better the lubrication effect that lower the friction resistance between the aggregates and make the concrete mixture easily flow, and the better the flowability; vice versa, the worse the flowability.

iv. Cement type

In the case of maintaining the same cement content and water consumption, the flowability of concrete mixture with slag Portland cement or pozzolan Portland cement is less than that with Portland cement or ordinary Portland cement. This is because the density of the former is less, then the absolute volume is larger in the case of using the same amount of cement and thus the concrete mixture appears to be drier and thicker in the case of using the same amount of water. In addition, the concrete mixture made of slag Portland cement has poor water retention. Due to the morphological effect of fly ash, concrete mixture with fly ash Portland cement shows better flowability, as well as the better water retention and cohesion.

v. Aggregate property

Aggregate properties refer to the type, gradation, particle size and surface properties of aggregates used in concrete. In the case of maintaining the amount of aggregate used in concrete constant, the flowability of concrete mixture prepared with pebble and river sand is better than that with crushed stone and mountain sand, because the former aggregate has a smooth surface and small friction resistance. In the case of maintaining the cement paste constant, well-graded aggregates have fewer voids, resulting in less cement paste being required to fill the voids and relatively thicker layer of cement paste wrapping around the surface of the aggregate particles, which improves the workability of the concrete mixture.

ⅵ. Sand ratio

Sand ratio indicates the combination relation between sand and rock in concrete. The change of sand ratio will make the total surface area and void ratio of the aggregate change greatly, so it has a significant impact on the workability of the concrete mixture. Increasing the sand ratio can increase the flowability, as well as increase the cohesion and water retention of the concrete mixture. However, when the sand ratio is too large, the total surface area and void ratio of the aggregate both increase, which reduce the thickness of the cement paste layer on the surface of the aggregate particles, resulting in the mixture becoming dry and thick and the flowability becoming poor while the cohesion also becoming poor. Conversely, if the sand ratio is too small, there is too much rock and too little sand in the mixture, resulting in the amount of mortar being not enough to wrap the surface of the rock, which reduces the flowability as well as cohesion of the concrete mixture.

As can be seen from the above, in the preparation of concrete, the sand ratio cannot be too large nor too small, and it should choose the reasonable sand ratio. Reasonable sand ratio refers to the sand ratio that can make the concrete mixture obtain the maximum flowability and maintain the good cohesion and water retention in the condition of keeping the amounts of water and cement constant, as shown in Figure 4-6. On the other hand, when the reasonable sand ratio is used, the cement content is minimized while obtaining the required flowability and good cohesion and water retention conditions for the mixture, as shown in Figure 4-7. In general, the reasonable sand ratio should make the sand fill the rock voids and have a certain amount of surplus.

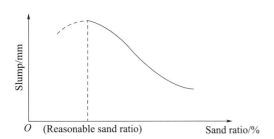

Figure 4-6 Relationship between slump and sand ratio (in the condition of keeping the amounts of water and cement constant)

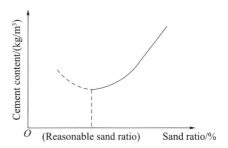

Figure 4-7 Relationship between cement content and sand ratio (in the condition of keeping the slump constant)

ⅶ. Chemical admixture

The flowability of concrete mixture containing water reducer or air-entraining agent can be significantly improved. Air-entraining agent can also effectively improve the cohesion and water retention of concrete mixture.

ⅷ. Storage time and ambient temperature

Concrete mixtures gradually become dry and thick with time, and the slump gradually decreases, a phenomenon called slump loss over time. This is due to the gradual absorption of water in the concrete mixture by the aggregate, evaporation of part of the water, hydration of cement and the gradual formation of the cohesive structure, as well as the adsorption of the water reducer in the concrete mixture by the cement particles.

The workability of concrete mixture is also influenced by the ambient temperature. As the ambient temperature rises, the slump of concrete is lost more quickly, because the water evaporates and the chemical reaction of the cement proceeds more quickly at this time.

## Ⅳ. Performance of fresh concrete after casting

The interval between the completion of concrete casting and setting is about several hours, during which the concrete mixture is in plastic and semi-flowing state. The performance of concrete in this state is called the performance of fresh concrete after casting.

ⅰ. Segregation

Due to the different densities of the components of fresh concrete, the relative movement occurs under the action of gravity, including the aggregate sinks and the paste floats, resulting in the concrete mixture being not uniform and losing of continuity, which is known as segregation. The occurrence of segregation in fresh concrete is mainly associated with the surface area of solid particles and the quality of the mixing water, where the solid particles include aggregates and cementitious materials. The segregation resistance index of the fresh concrete is defined as the ratio of the surface area of the solid particles in the concrete to the mass of the mixing water. The larger the surface area of the solid particles in the concrete and the smaller the mass of the mixing water, the less likely it is that the concrete will segregate.

ⅱ. Bleeding

The phenomenon that solid particle sin concrete sink and water rises and precipitates on the surface of the fresh concrete after it has been cast util the initial setting is called bleeding. The water bled out of the concrete either evaporates outward or is sucked back because of cement hydration, both of which are accompanied by a reduction in concrete volume. Bleeding has two effects on the performance of concrete: first, the top or near the top of the concrete features the loose hydration products structure due to the high content of water, which is very harmful to wear resistance, etc. ; second, part of the rising water accumulates below the aggregate to form water pockets, weakening the interfacial transition zone between cement paste and aggregates and affecting the strength and durability of hardened concrete.

ⅲ. Plastic settlement

The overall settlement of concrete mixture due to water bleeding is called plastic settlement. If the plastic settlement of fresh concrete is impeded (e.g., by reinforcement), then the plastic settlement cracks can be formed from the surface down to the top of the reinforcement.

ⅳ. Setting time

The fundamental reason for concrete to set is the hydration reaction of cement, but the setting time of concrete mixture is not necessarily identical to the setting time of the cement used. The setting time of cement is measured by the cement paste under the specified temperature and consistency conditions, and the conditions of the concrete mixture may not be the same as the conditions for determining the setting time of cement. The $W/C$ of concrete, ambient temperature and the performance of chemical admixtures have a great impact on its setting.

The setting time of concrete mixture is usually determined by the penetration resistance method, and the instrument used is the penetration resistance test. First, the mortar is sieved from the concrete mixture with a 4.75mm square-mesh sieve, which is filled into the specified container according to certain methods, and then the resistance for penetrating 25mm into the mortar is measured every 0.5h. The relationship curve between penetration resistance and time is drawn, and two straight lines parallel to the time coordinate are drawn with the penetration resistances of 3.5MPa and 28.0MPa. The time at the intersection of the straight lines and the curve is the initial and final setting times of the concrete mixture, respectively. The initial setting time indicates that the concrete mixture can no longer be cast and pounded normally, that is, the time limit of construction, while the final setting time indicates that the concrete strength begins to develop at a considerable speed.

Scan the QR code to get the related testing content of the concrete mixture.

## Section Ⅳ  Structure of hardened concrete

### Ⅰ. Structural characteristics

Concrete that has already produce strength after setting and hardening is called hardened concrete. In terms of structural characteristics, hardened concrete is a dispersion system formed by granular coarse and fine aggregates uniformly dispersed in cement stone, as shown in Figure 4-8. Specifically, cement paste wraps sand and fills the void of sand to from cement mortar; cement mortar wraps rock and fills the void of rocks to form concrete.

In terms of composition, concrete is a composite structure consisting of three phases, including the aggregate phase, the cement stone phase and the transition zone phase. The sand and rock in the aggregate phase are the skeleton of the concrete, which are dispersed in the cement stone, also known as the dispersed phase. The cement stone phase is the matrix of the concrete, also known as the continuous phase. Transition zone phase refers to the interfacial transition zone between the aggregate and the cement stone, also known as the interface phase.

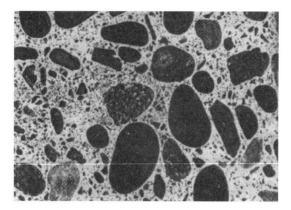

Figure 4-8  Structure of hardened concrete

## II. Interfacial transition zone

The interfacial transition zone (ITZ) is a thin layer of cement stone around the aggregate, with a thickness of about 20~50μm. The number of aggregate particles in the concrete it large, and the volume of the ITZ can reach 1/3 to 1/2 of the hardened cement paste, whose impact on the performance of concrete cannot be ignored.

For the formation of the ITZ in the concrete, it is generally believed that because of the downward settlement of the aggregate particles by gravity and the upward migration of water due to the small density, the relative movement between them makes the water enrich below the aggregate particles, resulting in the gradual formation of a high $W/C$ cement paste film around the aggregate. After the hardening of the concrete, the ITZ is formed in this area.

Compared with the cement stone matrix phase, the $W/C$ is higher in the ITZ. At the same time, the number of C-S-H gel is low, the compactness is poor, and the porosity is large, especially the large pores are more, in the ITZ. Besides, calcium hydroxide, ettringite and other crystals are larger in size and more abundant in the ITZ, and most of them grow in orientation (perpendicular to the surface of the aggregate with directional growth), as shown in Figure 4-9. Due to the differences in deformation modulus and shrinkage properties of aggregate and cement stone, or due to water evaporation and other reasons, there are a large number of pre-existing microcracks in the ITZ, which is

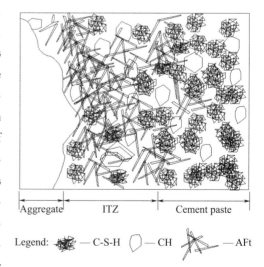

Figure 4-9  Schematic diagram of the ITZ

the weak area in concrete.

### III. Pore structure

Cement stone consists of hydration products of cement(including crystals and gels), incompletely hydrated cement particles, pores of different sizes, and pore water or gas in them. Among them, various pores have an important influence on the mechanical properties, deformation performance and durability performance of concrete. It is generally believed that the pores in concrete can be divided into four categories, namely, gel pores with characteristic sizes of 0.5 to 10nm, capillary pores with average radii of 5nm to 5μm, air-entraining pores with sizes of 20 to 200μm, and large pores formed due to insufficient compactness with size of 1000μm.

China's academician Wu Zhongwei proposed to divide the pores in concrete into four levels, which are harmless pores with size less than 20nm, less harmful pores with size of 20 ~ 50nm, harmful pores with size of 50 ~ 200nm and more harmful pores with size larger than 200nm. Among them, the pores above 50nm are more harmful to the strength and durability of concrete, while the pores below 20nm have an extremely small impact on the performance of concrete.

It is generally believed that the harmful effects of pores on concrete are mainly reflected in the following three aspects: firstly, pores are the channels for erosive substances to invade the interior of concrete; secondly, capillary pores can reduce the strength of concrete; finally, water loss from gel pores leads to shrinkage of concrete. At the same time, pores also have some positive effects on concrete, for example, the presence of pores provides space for the later strength development of concrete and the small and closed spherical pores can improve the frost resistance of concrete.

## Section V  Mechanical properties of hardened concrete

### I. Damage process of concrete under compression

As mentioned earlier, there is already a certain amount of pre-existing microcracks in the ITZ in the hardened concrete before it being subjected to external forces. When the concrete is loaded, these interface microcracks will gradually expand, extent and link up to form visible cracks, resulting in the loss of concrete continuity and the complete destruction of concrete.

When uniaxial static compression test is performed on the concrete cube specimens, the damage process of crack state of the concrete show four different stages, as shown in Figure 4-10 and Figure 4-11. The details of each stage are as follows.

Stage I: Before the load reaches the "proportional limit" (about 30% of the ultimate load), the load and deformation are approximately linear(section $OA$ in Figure 4-10), and there is no significant change in the interface cracks.

Stage II: After the load exceeds the "proportional limit", the rate of deformation is greater than the rate of increase of load, and the relationship between load and deformation is no longer linear(section *AB* in Figure 4-10). At this stage, the number, length and width of interface cracks increase continuously, and the interface continues to share the load by frictional resistance, while no obvious cracks appear in the mortar.

Stage Ⅲ: After the load exceeds the "critical load" (about 70% ~ 90% of ultimate load), the deformation speed is further accelerated and the curve is obviously bent toward the deformation axis(section *BC* in Figure 4-10). In this stage, interface cracks continue to develop, cracks start to appear in the mortar, and some interface cracks are connected into continuous cracks.

Stage Ⅳ: After the external load exceeds the ultimate load, the load decreases while the deformation increases rapidly, and the curve bends down and terminates(section *CD* in Figure 4-10). At this time, continuous cracks develop rapidly, and the load-bearing capacity of concrete decreases to complete destruction.

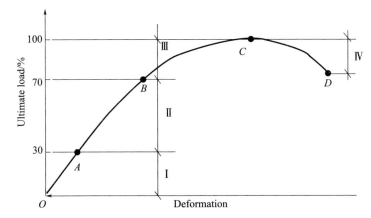

Figure 4-10   Load-deformation curve of concrete under compression

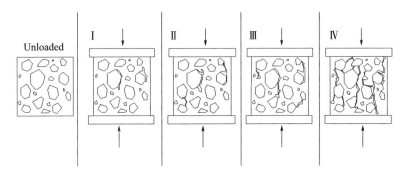

Figure 4-11   Schematic diagram of cracks in different stress stages

Ⅰ—no obvious change of interface cracks; Ⅱ—obvious growth of interface cracks;
Ⅲ—appearance of mortar cracks and continuous cracks; Ⅳ—rapid development
of continuous cracks; Ⅴ—slow growth of cracks; Ⅵ—rapid growth of cracks

## Ⅱ. Strength of concrete

Strength is the most important technical property of hardened concrete, and is also the main indicator for controlling and assessing the quality of concrete in engineering construction. The strength of concrete has compressive, tensile, flexural and shear strengths, of which the compressive strength is the largest, so concrete is mainly used to withstand compression in structural engineering.

i . Cubic compressive strength

China uses cubic compressive strength as the strength characteristic value of concrete. According to the national standard *Standard for Test Methods of Concrete Physical and Mechanical Properties* (GB/T 50081—2019), it is stipulated that a cube standard specimen with a side length of 150mm is made and cured for 28d under standard curing conditions [temperature $(20 \pm 2)$℃, relative humidity of 95% or more], which is then tested by the standard test method, resulting in the obtained compressive strength being called the cube compressive strength of concrete, which is expressed as $f_{cc}$. The compressive strength of the concrete cube specimen is calculated according to the following equation.

$$f_{cc} = \frac{F}{A} \tag{4-4}$$

Where    $F$——specimen damage load(N);

          $A$——specimen compression-bearing area($mm^2$).

The determination of the compressive strength value of cubic specimen shall conform to the following three provisions.

Take the arithmetic mean of the measured values of three specimens as the strength value of the group of specimens, which should be accurate to 0.1MPa;

When the difference between one of the maximum or minimum of the three measured values and the medium value exceeds 15% of the medium value, the maximum and minimum values are eliminated and the medium value is taken as the compressive strength value of the group of specimens;

When the difference between the maximum value and the medium value and the difference between the minimum value and the medium value both exceed 15% of the medium value, the test results of the group of specimens are invalid.

The compressive strength of concrete is determined using standard specimens with a side length under standard conditions for comparability between different batches of tests. In actual construction, specimens with non-standard size are allowed to be used, but their compressive strength should be converted into the compressive strength of standard specimens, with the conversion factors showing in Table 4-14.

Table 4-14    Concrete cube specimen side length and strength conversion factor

| Specimen side length/mm | Compressive strength conversion factor |
|:---:|:---:|
| 100 | 0.95 |
| 150 | 1.00 |
| 200 | 1.05 |

The larger the size of the concrete specimen, the lower the measured compressive strength value. This is due to the hoop effect and the different chances of defects in the specimens during testing. First of all, when the concrete cube specimen is placed on the compression test machine under compression, the concrete deforms longitudinally along the direction of loading application, while the concrete specimen and the upper and lower steel plates also have transverse free deformation due to Poisson's ratio. However, the modulus of elasticity of steel plate is about 10 times greater than the concrete, while Poisson's ratio of the former is only about two times of that of the concrete, resulting in the transverse deformation of the steel plate is less than that of the concrete, which causes the formation of friction resistance between the upper and lower steel plates and the concrete specimen. This friction resistance imposes a restraint effect on the transverse deformation of concrete, thereby increasing the strength of concrete, which is called the hoop effect, as shown in Figure 4-12. This restraint effect becomes smaller with the faring away from the end of the specimen, so the specimen is a pair of top prismatic cones after compression damage, as shown in Figure 4-13. When the size of the concrete cube specimen is larger, the relative role of the hoop effect is smaller, and the measured compressive strength if thus low; conversely, the measured compressive strength is high. In addition, the presence of microcracks and pores in the concrete specimen and other defects reduce the actual compression area of the concrete specimen as well as cause the stress concentration, resulting in the reduced strength. The presence of defects in large-size concrete specimens is more likely to result in lower measured strengths than in small-size concrete specimens. Scan the QR code to get the details of *Standard for Test Methods of Concrete Physical and Mechanical Properties* (GB/T 50081—2019).

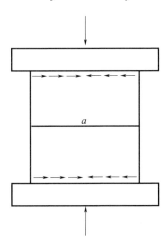

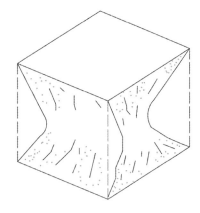

Figure 4-12 The restraint effect of the steel plate in compression test machine on the test specimen

Figure 4-13 Residual prismatic cone of test specimen damaged by compression plate restraint

ii. Axial compressive strength

Concrete axial compressive strength is also known as prismatic compressive strength. In the actual structure, most of the reinforced concrete compression members are prisms or cylinders. In order to make the measured concrete strength can be close to the actual force of the concrete structure, the design of reinforced concrete structures in the calculation of axial compression members (such as columns, trusses, webs, etc.), are required to use the axial compressive strength of concrete as the basis.

According to the *Standard for Test Methods of Concrete Physical and Mechanical Properties* (GB/T 50081—2019), the axial compressive strength of concrete $f_{cp}$ should use prism with size of 150mm × 150mm × 300mm as the standard specimen, if necessary, non-standard size prismatic specimens can be used, but its height to width ratio should be in the range of 2 to 3. Standard prismatic specimens are made under the same conditions as standard cubic specimens, but the measured compressive strength value of the former is smaller. Tests show that when the standard cube compressive strength $f_{cc}$ is in the range of 10 to 50MPa, $f_{cp} = (0.7 \sim 0.8)f_{cc}$, generally taken as 0.76.

iii. Axial tensile strength

The axial tensile strength $f_t$ of concrete is very low, only 1/20 to 1/10 of its compressive strength (usually taken as 1/15), and this ratio is reduced with the increase in the strength grade of concrete. Therefore, the concrete features brittle fracture in tension and no obvious residual deformation when damaged. For this reason, in the design of reinforced concrete structures, concrete is not considered to withstand tensile forces, but rather reinforce concrete with rebar, which is used to bear the tensile forces in the structure.

iv. Splitting tensile strength

The axial tensile strength of concrete is difficult to determine, and the test is subject to many external interferences, such as the line of action of the load is difficult to maintain coincidence with the axis of the specimen, which is prone to eccentricity. At present, both China and foreign countries use the splitting method to reflect the tensile properties of concrete and determine the splitting tensile strength of concrete. China's standard provides that the concrete splitting tensile strength using the cube specimen with side length of 150mm as the standard specimen. The principle of this method is: a pair of uniformly distributed compression force are loaded in the middle of the upper and lower surface of the cube specimen, resulting in a uniform tensile stress in the vertical plane of the specimen (Figure 4-14), which can be calculated according to the theory of elasticity. The concrete splitting tensile strength is calculated by the following equation:

$$f_{ts} = \frac{2F}{\pi A} \qquad (4-5)$$

Where  $f_{ts}$——splitting tensile strength of concrete (MPa);
  $F$ ——breaking load (N);
  $A$ ——splitting area of the specimen (mm$^2$).

v. Flexural strength

The flexural strength of concrete is also known as flexural tensile strength. According to the standard, the standard specimen used for concrete flexural strength test is 150mm × 150mm ×

600mm or 150mm×150mm×550mm prismatic specimen. The test is carried out in the "three-point loading mode", i. e., two equal loads are applied vertically at two points of trisection of the span of the specimen at the same time, as shown in Figure 4-15. According to the theory of material mechanics, the flexural strength of concrete can be obtained from the maximum tensile stress of the specimen at fracture.

$$f_f = \frac{Fl}{bh^2} \tag{4-6}$$

Where $f_f$——flexural strength of concrete(MPa);
$F$ ——breaking load(N);
$l$——span between supports(mm);
$b$ ——width of the specimen cross-section(mm);
$h$ ——height of the specimen cross-section(mm).

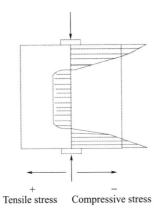

Figure 4-14　Stress distribution perpendicular to the load surface during splitting test

Figure 4-15　Flexural strength test device

Scan the QR code to get the content of the test for hardened concrete mechanical properties.

### III. Strength grade of concrete

According to the *Standard for Evaluation of Concrete Compressive Strength*(GB/T 50107—2010), the strength grade of concrete should be determined by its cube compressive strength standard value. The concrete cube compressive strength standard value is a value in the overall distribution of concrete compressive strength measured by standard test methods at 28d for a cube specimen with the side length of 150mm made and cured in accordance with standard methods, and the probability of the strength being lower that this value is 5%, which is expressed as $f_{cc,k}$. In other words, the probability that the overall distribution of concrete strength being greater than the cube compressive strength standard value is 95%, that is, the strength guarantee rate that the overall distribution of concrete strength being greater than the design strength grade is 95%. Concrete strength grade using the symbol "C" and the cube compressive strength standard value (in N/mm²). At present, the strength of concrete used in China's construction engineering is divided into 14 grades, namely C15, C20, C25, C30, C35, C40, C45, C50, C55, C60, C65, C70, C75 and C80. Scan the QR code to get the details of *Standard for Evaluation of Concrete Compressive Strength* (GB/T 50107—2010).

i. Distribution of compressive strength

Under the normal production and construction conditions, the factors affecting the strength of concrete are randomly varying, so the strength of concrete should also be a random variable. Under certain construction conditions, $n$ sets of specimens ($n \geq 25$) are chosen based on the random sampling for the same type of concrete, whose compressive strength is measured at the curing time of 28d. The concrete strength is plotted as the horizontal coordinate and the probability of occurrence of concrete strength as the vertical coordinate, the probability distribution curve of concrete strength can be produced. The strength distribution curve of concrete generally conforms to the normal distribution, as shown in Figure 4-16. The normal distribution curve of concrete strength has the following characteristics.

1. The curve is bell-shaped and symmetrical on both sides, and the axis of symmetry is at the average strength $\overline{f}_{cc}$, where the highest peak of the curve appears, indicating that the concrete strength close to its average strength value appears most frequently. As the distance from the axis of symmetry is farther, the probability of occurrence of the measured strength value lower or higher than the average value is less.

2. The area enclosed between the curve and the horizontal coordinate is the sum of the probabilities, which is equal to 100%. The probabilities on both sides of the symmetric axis are equal, each being 50%.

3. There is an inflection point on each side of the curve, and the vertical distance from the inflection point to the axis of symmetry if the standard deviation of the normal distribution of strength.

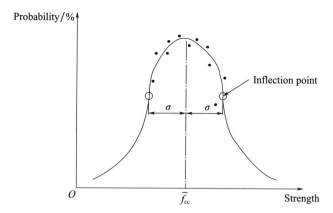

Figure 4-16 Normal distribution curve of concrete strength

ii. Strength guarantee rate

The concrete strength guarantee rate $P(\%)$ is the probability that the overall concrete strength is greater than or equal to the design strength grade value ($f_{cc,k}$), which is represented by the shaded area in the normal distribution curve graph of concrete strength, as shown in Figure 4-17. The strength guarantee rate $P(\%)$ can be obtained by integrating the equation of the normal distribution curve, i. e. :

$$P(t) = \int_{t}^{+\infty} \varphi(t)\,dt = \frac{1}{\sqrt{2}} \int_{t}^{+\infty} e^{\frac{t^2}{2}}\,dt \tag{4-7}$$

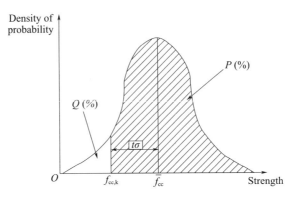

Figure 4-17  Concrete strength guarantee rate

The process of calculating the concrete strength guarantee rate, the probability density coefficient is firstly calculated using the design strength grade value $f_{cc,k}$, the average value of strength $\bar{f}_{cc}$ and the standard deviation $\sigma$:

$$t = \frac{\bar{f}_{cc} - f_{cc,k}}{\sigma} \tag{4-8}$$

In turn, the strength guarantee rate $P(\%)$ can be derived by the standard normal distribution curve equation or be obtained by choosing in Table 4-15. From Table 4-15, it can be seen that the strength guarantee rate of concrete reaches 95% when the probability density coefficient is 1.645.

Table 4-15  Strength guarantee rete values for different $t$ values

| $t$ | 0.00 | 0.50 | 0.80 | 0.84 | 1.00 | 1.04 | 1.20 | 1.28 | 1.40 | 1.50 | 1.60 |
|---|---|---|---|---|---|---|---|---|---|---|---|
| $P(\%)$ | 50.0 | 69.2 | 78.8 | 80.0 | 84.1 | 85.1 | 88.5 | 90.0 | 91.9 | 93.5 | 94.5 |
| $t$ | 1.645 | 1.70 | 1.75 | 1.81 | 1.88 | 1.96 | 2.00 | 2.05 | 2.33 | 2.50 | 3.00 |
| $P(\%)$ | 95.0 | 95.5 | 96.0 | 96.5 | 97.0 | 97.5 | 97.7 | 98.0 | 99.0 | 99.4 | 99.87 |

### IV. Factors affecting the strength of concrete

i . Water-to-cement ratio

Theoretically, the water required for hydration of cement is generally only about 23% of the mass of cement, but it is often necessary to add more water in the preparation of concrete mixture in order to obtain the workability required by the construction. This excess water in the concrete can form bubbles or watercourses, which evaporate as the concrete hardens and finally leave a large number of pores, and the existence of pores reduces the actual stress area of the concrete; at the same time, when the concrete is stresses, it is easy to produce stress concentration around the pores. In the case of fully compacted, the larger the $W/C$, the more excess water, leaving more pores, the lower the strength of concrete; conversely, the higher the strength of concrete, as shown in Figure 4-18. In the case of not fully compacted, if the $W/C$ is too small, the workability of concrete mixture is too poor, resulting in a serious decline in the strength of concrete.

Tests have proved that, in the case of using the same material, the relationship between the strength $f_{cc}$ of concrete and its $W/C$ is in an approximate hyperbolic shape (as shown in the solid line in Figure 4-18), while the relationship between concrete strength and cement-to-water ratio is a linear relationship, as shown in Figure 4-19. Through considering the cement strength and applying the mathematical and statistical methods, the empirical formula between concrete strength and cement strength as well as cement-to-water ratio can be established, that is, the empirical formula for concrete strength (also known as the Bolomey's formula).

$$f_{cc} = \alpha_a f_{ce} \left( \frac{C}{W} - \alpha_b \right) \tag{4-9}$$

where  $f_{cc}$——compressive strength of concrete at the curing age of 28d (MPa);

$C$——the amount of cement in $1m^3$ concrete (kg);

$W$——the amount of water used in $1m^3$ concrete (kg);

$C/W$——cement-to-water ratio of concrete;

$f_{ce}$——measured compressive strength of cement at the curing age of 28 (MPa);

$\alpha_a$、$\alpha_b$——regression coefficients related to the type of aggregates.

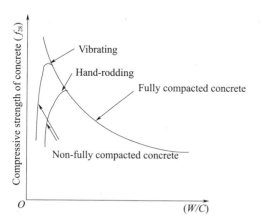

Figure 4-18  Relationship between compressive strength of concrete and water-to-cement ratio

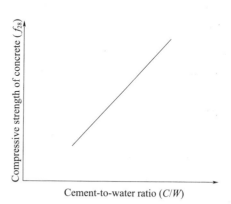

Figure 4-19  Relationship between compressive strength of concrete and cement-to-water ratio

$\alpha_a$ and $\alpha_b$ should be determined by the relationship between the cement-to-water ratio and concrete strength obtained by using the cement and aggregates in the engineering projects. When the above test statistics are not available, then the values in Table 4-16 can be taken.

Table 4-16  Values of regression coefficient $\alpha_a$ and $\alpha_b$

| Type of coarse aggregate | $\alpha_a$ | $\alpha_b$ |
| --- | --- | --- |
| Crushed stone | 0.53 | 0.20 |
| Pebble | 0.49 | 0.13 |

ii. Aggregate

The ratio of the mass of aggregate to the mass of cement in concrete is called the aggregate-to-cement ratio($A/C$). The $A/C$ has a greater impact on concrete with compressive strength above 35MPa. In the condition of using the same $W/C$ and slump, the increase in aggregate content could increase the surface area of aggregate, which also increases the water absorption and reduces the effective water-to-cement ratio. As a result, the strength of concrete increases with the increase of $A/C$.

The surface characteristics of the aggregates also have an impact on the surface of concrete. Crushed stone has a rough surface and is angular, which has good binding with cement stone. Aggregates particles have the embedded effect, so the strength of concrete mixture with crushed stone is higher than that with pebble at the conditions of using the same materials and slump. When the $W/C$ is less than 0.40, the strength of concrete with crushed stone can be about one-third higher than that with pebble. However, as the $W/C$ increases, the difference between the two types of concrete gradually decreases. When the $W/C$ reaches 0.65, the strength of the two types of concrete has no significant difference. This is because when the $W/C$ is small, the main conflict of concrete strength is the strength of the ITZ, while the strength of the cement stone becomes the main conflict when the $W/C$ is large.

iii. Mineral admixture

The active mineral admixture reacts with calcium hydroxide in concrete to produce additional C-S-H gel, which can improve the later-age strength of concrete. In addition, the mineral admixture can interfere with the crystallization process of hydration products, which makes the crystallization size of hydration products smaller, reduces the degree of enrichment and orientation, and reduces the porosity of the ITZ, resulting in the reduction in the number of defects in the ITZ. In addition, the fine mineral admixture can fill the voids between cement particles and make the concrete matrix more compact.

iv. Chemical admixture

The two chemical admixtures that are most closely related to the strength of concrete are water reducer and accelerating admixture. Water reducer can reduce the amount of water used in concrete, which in turn reduces the $W/C$ and increases the strength of concrete. Accelerating admixture can improve the early-age strength of concrete by promoting the process of cement hydration reaction, it is worth noting that the strength of concrete with accelerating admixture may potentially decrease at later ages.

v. Construction method

Concrete mixture prepared with mechanical mixing is more uniform than that with manual mixing. Practice has proved that in the conditions of the same mix design and compactness, the strength of concrete prepared with mechanical mixing is generally about 10% higher than that with manual mixing.

Concrete mixture cast using mechanical vibration is denser than that using manual compaction. The vibration temporarily destroys the cohesive structure of cement paste and reduces the viscosity of cement paste, while the frictional resistance between aggregates is greatly reduced, which is conducive to the improvement of concrete compactness and strength.

ⅵ. Curing condition

The process of artificially changing the temperature and humidity conditions of the environment around the concrete to make its microstructure and performance reach the desired result during the hardening is called the curing of concrete.

Temperature is an important condition to determine the speed of hydration of cement. High curing temperature can result in a fast early-age hydration of cement and thus a high early-age strength of concrete. It is worth noting that the temperature of the early stage of concrete hardening has an impact on its later-age strength, and the higher the initial curing temperature of concrete, the lower the growth rate of its later-age strength. This is because that the cement hydration rate accelerates under the high initial temperature (above 40℃), resulting in the gathering of the high concentration of hydration products around the cement particles that being hydrated and the slowing down of the rate of further hydration of cement. In this case, the hydration products of cement cannot diffuse in time, forming an unevenly distributed porous structure, which becomes a weak zone in the cement paste, and ultimately has a negative impact on the long-term strength of concrete. On the contrary, at a low curing temperature (e. g. ,5 ~ 20℃), although the cement hydrates slowly and the generation rate of hydration product is low, there is sufficient diffusion time to form a uniform structure, which results in a higher final strength. When the temperature drops below 0℃, the cement hydration reaction stops and the concrete strength stops developing; at the same time, the water in concrete freezes to produce volume expansion (about 9%), imposing compression stress (up to 100MPa) to the pore wall, which results in the damage of the structure of hardened concrete. When constructing concrete in winter, special attention should be paid to heat preservation and maintenance to avoid early-age frost damage to concrete.

Humidity is a necessary condition to determine whether the cement can hydrate properly. After casting the concrete in the environment with suitable humidity, cement can hydrate smoothly and the strength of concrete can be fully developed. If the humidity of the environment is low, the cement cannot hydrate normally, or even stop the hydration, seriously reducing the strength of concrete. The relationship between the concrete strength and the moist curing period is shown in Figure 4-20. As can be seen from the figure, the earlier the concrete is subjected to drying, the greater the loss of strength. Lack of water during the hardening of concrete will also lead to its loose structure and the formation of dry shrinkage cracks, which increases the water penetration and affect the durability of concrete. For this reason, the construction specification *Code for Quality Acceptance of Concrete Structure Construction* (GB 50204—2015) stipulates that after the concrete is cast, it should be covered and watered within 12h. For the concrete in the condition of natural curing in summer construction, more special attention should be paid to watering maintenance. Scan the QR code to get the details of *Code for Quality Acceptance of Concrete Structure Construction* (GB 50204—2015).

ⅶ. Curing time

Under normal curing conditions, the strength of concrete increases with the curing age. The strength of concrete develops fast in the initial 7 to 14d, and then gradually develops slow, which tends to stabilize after 28d. However, as long as there is a certain temperature and humidity conditions, the strength of concrete can develop for decades. The relationship between concrete strength and curing age can be seen from the curve in Figure 4-20.

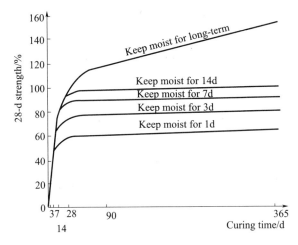

Figure 4-20  Relationship between concrete strength and moist curing time

It has been proved that the strength development of concrete prepared by ordinary cement of medium strength grade is roughly proportional to the common logarithm of its curing age under the standard curing condition, and its empirical estimation formula is as follows.

$$\frac{f_n}{f_{28}} = \frac{\lg n}{\lg 28} \tag{4-10}$$

where  $f_n$——compressive strength of concrete at age $n$d(MPa);

$f_{28}$——compressive strength of concrete at 28d(MPa);

$n$ ——curing time(d), $n \geqslant 3$d.

## Ⅴ. Measures to improve the strength of concrete

In actual engineering, in order to meet the requirements of concrete construction or engineering structure, it is often necessary to improve the strength of concrete. According to the factors affecting the strength of concrete, the following measures can usually be taken to improve the strength of concrete.

ⅰ. Use high strength grade cement or early-strength type of cement

In the case of keeping the mix design of concrete unchanged, the use of high strength grade cement can improve the strength of concrete at the curing time of 28d; the use of early-strength type of cement can improve the early-age strength of concrete, which is conducive to speeding up the progress of the project.

ⅱ. Adopt low $W/C$

Reduce the $W/C$ is the most effective way to improve the strength of concrete. In the concrete mixture with low $W/C$, the free water is less and thus the pores left in concrete after hardening is less, resulting in the increased compactness and significantly increased strength of concrete. If the $W/C$ is reduced too much, it will affect the flowability of the mixture, causing construction difficulties. For this reason, the water reducer is usually added in the mixture at the same time, which can make the concrete still has a good workability in the case of using a low $W/C$.

ⅲ. Add chemical admixture and mineral admixture

The addition of chemical admixture in concrete is one of the important means to make it obtain early-age strength and high strength. Concrete mixed with accelerating admixture can significantlyimprove its early-age strength; when admixing water reducer in the concrete, especially high-range water reducer, the amount of mixing water can be significantly reduced, improving the strength of concrete. For high-strength concrete and high-performance concrete, in addition to high-range water reducer, mineral admixtures, such as fly ash, slag or silica fume, are also needed to mixed to adapt to the needs of high strength and high performance.

ⅳ. Use mechanical vibration

When concrete with low $W/C$ is adopted in the construction, mechanical mixing and mechanical vibrating must be used at the same time, otherwise the concrete is difficult to achieve compacted status and high strength.

ⅴ. Use humid and heat curing

In steam curing, concrete is put in the atmospheric pressure steam with the temperature greater than 90℃ in order to accelerate the hydration of cement, in which the strength of concrete can reach 70% to 80% of the strength under normal conditions at the curing time of 28d. The purpose of steam curing is to obtain sufficient high early-age strength, speed up the demolding process, improve the turnover rate of the formwork and construction site, and thus effectively improve the production efficiency and reduce the cost.

In autoclaved curing, concrete is put in the autoclave with the temperature of 173℃ and the pressure of 8 atmosphere pressure. In this high temperature and high air pressure, the hydration and hardening of cement accelerate, while the calcium hydroxide generated during the cement hydration reacts with silica to form C-S-H with good crystallization, which can effectively improve the strength of concrete. This method is more effective for concrete mixed with mineral admixtures or concrete mixed with cement containing active addition materials.

# Section VI  Quality control and assessment of concrete

During the production and construction process, the quality of concrete is affected by different factor.

(1) The influence of raw materials, such as the changes in types and strength grades of cement, fluctuations in cement strength, changes in impurity content, gradation, particle size and particle shape of sand and rock, changes in water content of aggregate, etc.

(2) The impact of construction operations, such as the error in the measurement of the constituent materials, fluctuations in the $W/C$, variations in mixing time, changes in casting conditions, changes in temperature and humidity during curing, etc.

(3) The influence of the testing conditions, such as the differences in sampling methods, differences in the specimen forming and curing conditions, loading speed during testing, testing operator's own error, etc.

The fluctuation of concrete quality is objective, it is necessary to carry out quality management in order to make the quality of concrete fluctuates within a certain range and achieve the stable quality. Because the compressive strength of concrete has a close correlation with other properties and can reflect the overall quality of concrete well, so the engineering projects often use the compressive strength of concrete as an important quality control index, which is also used as a basis for assessing the quality level of concrete production.

## I. Assessment indicator

The indicators to assess the quality level of concrete construction mainly include the mean, standard deviation coefficient of variance of concrete strength under normal production control conditions.

i. Mean of strength

The mean of the concrete strength represents the average of the overall strength of concrete and is calculated as

$$\bar{f}_{cc} = \frac{1}{n}\sum_{i=1}^{n} f_{cc,i} \qquad (4\text{-}11)$$

where  $f_{cc}$——arithmetic mean of compressive strength of $n$ group of concrete (MPa);

$f_{cc,i}$——compressive strength of $i$th group of concrete specimen (MPa);

$n$——the number of groups of specimens, $n \geqslant 25$.

ii. Standard deviation of strength

The mean of the strength of concrete does not reflect the fluctuation of concrete strength. The standard deviation of strength is used to characterize the discreteness of concrete strength, and a smaller the standard deviation of strength indicates a smaller discreteness of strength and a more stable quality control of concrete, while a larger standard deviation means a larger discreteness of strength and a poor quality control. The formula for calculating the standard deviation of strength is as follows.

$$\sigma = \sqrt{\frac{\sum_{i=1}^{n}(f_{cc,i} - \bar{f}_{cc})^2}{n-1}} \qquad (4\text{-}12)$$

$$\text{or} \quad \sigma = \sqrt{\frac{\sum_{i=1}^{n}f_{cc,i}^2 - n\bar{f}_{cc}^2}{n-1}} \qquad (4\text{-}13)$$

iii. Coefficient of variation

The coefficient of variation is used as an index to assess the uniformity of concrete quality. The smaller the coefficient of variation, the more stable the quality of concrete; the larger the coefficient of variation, the worse the stability of concrete quality. The coefficient of variation is calculated as follow.

$$C_v = \sigma / \bar{f}_{cc} \qquad (4\text{-}14)$$

## II. The assessment of strength

Concrete strength assessment is divided into statistical method and non-statistical method. When the test results meet the provisions of the statistical method or non-statistical method, the strength of the batch of concrete is assessed qualified.

i. Statistical method

When the production conditions of continuously produced concrete can be consistent over a long period of time and the strength variation of the same species and same strength grade of concrete remains stable, a continuous group of three test specimens should form a test batch, the strength of which should meet the following requirements.

$$\bar{f}_{cc} \geqslant f_{cc,k} + 0.7\sigma \qquad (4\text{-}15)$$

$$f_{cc,min} \geqslant f_{cc,k} - 0.7\sigma \qquad (4\text{-}16)$$

When the strength grade of concrete is not higher than C20, the minimum value of its strength should also meet the following requirement.

$$f_{cc,min} \geqslant 0.85 f_{cc,k} \qquad (4\text{-}17)$$

When the strength grade of concrete is higher than C20, the minimum value of its strength should meet the following requirement.

$$f_{cc,min} \geqslant 0.90 f_{cc,k} \qquad (4\text{-}18)$$

Where  $\bar{f}_{cc}$——the average value of the cube compressive strength of the same batch of concrete(MPa);

$f_{cc,min}$——the minimum value of the cube compressive strength of the same batch of concrete(MPa);

$f_{cc,k}$——the cube compressive strength standard value of concrete specimens(MPa);

$\sigma$——standard deviation of the cube compressive strength of the test batch of concrete (MPa).

ii. Non-statistical method

Non-statistical method can be used to assess the strength of concrete when the number of specimens is limited, which does not have the conditions for assessing the strength of concrete according to the statistical method. When the strength of concrete is assessed through the non-statistical method, its minimum value should meet the following requirement.

$$f_{cc,min} \geqslant 0.95 f_{cc,k} \quad (4\text{-}19)$$

When the strength grade of concrete is not higher than C60, the average value of its strength should also meet the following requirement.

$$\bar{f}_{cc} \geqslant 1.15 f_{cc,k} \quad (4\text{-}20)$$

When the strength grade of concrete is higher than C60, the average value of its strength should meet the following requirement.

$$\bar{f}_{cc} \geqslant 1.10 f_{cc,k} \quad (4\text{-}21)$$

# Section VII  Deformation properties of concrete

Due to the physical, chemical and mechanical factors, concrete often deforms in various types in the process of hardening and using. These deformations are one of the main causes that result in the formation of cracks in the concrete, which affects the strength and durability of concrete. The deformation of concrete usually includes the following types.

## I. Chemical shrinkage

During the hardening process of concrete, the volume of solid phase of cement hydration products is smaller than the total volume of reactants before hydration, resulting in volume shrinkage of concrete, which is called chemical shrinkage. The chemical shrinkage of concrete is not recoverable, and the shrinkage increases with the extension of the age of concrete hardening, which generally stabilizes within 40d. The value of chemical shrinkage of concrete is very small (less than 1%), which has no destructive effect on the concrete structure, but microcracks may be generated within the concrete.

## II. Plastic shrinkage

The shrinkage of concrete in the plastic stage due to the rate of water loss on the surface being greater than the rate of internal water migration is called plastic shrinkage. The plastic shrinkage of concrete generally occurs in conditions of poor curing, harsh environment, or large area construction and at times concentrated in 3 ~ 12h after casting. Small cracks could occur on the concrete surface if the plastic shrinkage of concrete is too large.

## III. Drying shrinkage

The phenomenon of volume shrinkage in concrete due to water dissipation in a dry environment is called drying shrinkage of concrete.

i. Drying shrinkage mechanism

There are many pores of different sizes and shapes inside the concrete, and there is usually water in the pores, which will be lost gradually when the ambient humidity drops, leading to drying shrinkage of the concrete.

1. Free water, which presents in the larger pores or gel and crystal surface, evaporates easily and has no effect on the shrinkage of concrete.

2. Capillary pore water, which presents in the capillary pores, can evaporate when the relative humidity of the environment is 40% to 50%, resulting in negative capillary pore pressure and the initiation of drying shrinkage of concrete. The negative capillary pore pressure in concrete is calculated by the following equation.

$$\Delta P = -\frac{2\gamma}{r} = \frac{RT}{M}\ln\frac{P_v}{P_0} \tag{4-22}$$

Where  $M$——molar mass of the aqueous solution (kg/mol);

$\gamma$——surface tension of the aqueous solution (N/m);

$R$——gas constant [8.314J/(K·mol)];

$T$——absolute temperature of the system (K);

$r$——the radius of curvature of the liquid meniscus of the pore water (m);

$P_v$——external vapor pressure;

$P_0$——saturated vapor pressure.

As can be seen from the above formula, the greater the external humidity, the greater the vapor pressure. When the external vapor pressure is close to the saturated vapor pressure, the negative capillary pore pressure is close to zero; when the external environment is drier, the greater the negative capillary pore pressure.

3. Gel absorption water, which is adsorbed on the surface of the cement gel particles under the action of molecular gravity. When the relative humidity drops to 30%, the gel absorption water begins to lose, resulting in the tightening of the gel particles and the increasing of the drying shrinkage of concrete.

4. Gel water, which is firmly bonded with the gel particles by hydrogen bonds in the cement gel particles, is also called interlayer water. It is only lost when the environment is very dry (relative humidity less than 11%) and can cause concrete to shrink significantly.

When the concrete is in water or wet environment, the large pores and capillary pores are filled with water. When the external environment begins to dry, the free water in large pores evaporates firstly, and then the capillary pore water evaporates, which will make the negative capillary pore pressure increases and produce contraction force, resulting in the capillary pores being compressed and the shrinkage of the volume of concrete. If the water is continuously lost, the adsorption water film on the surface of the gel particles becomes thinner, resulting in the tightening between the particles and the further increase of the drying shrinkage of concrete. If the shrunk concrete re-absorb water, the pores in concrete can fill with water and the volume of concrete expand, which can restore most of the shrinkage deformation, but 30% to 50% of them cannot be restored.

ii. Deterioration effect of drying shrinkage

The drying shrinkage of concrete mainly occurs in the cement paste, and the drying shrinkage value of cement paste can usually reach 400 to 1000 micro-strain ($10^{-6}$ mm/mm). After hardening, concrete is a brittle material with very poor deformability and low tensile strength. When the shrinkage stress generated by drying shrinkage is greater than the tensile strength of concrete, cracks will occur in the concrete. Shrinkage cracking will not only reduce the strength of concrete, but also provide a channel for the invasion of erosion medium, reducing the durability of concrete, such as frost resistance, penetration resistance, erosion resistance, etc. As the strength of concrete grows gradually in the later ages, concrete is less likely to crack, so the deterioration effect of early-age drying shrinkage on concrete is greater than that of later-age drying shrinkage.

## IV. Temperature deformation

Due to the exothermic hydration of cement, the temperature difference between the inside and outside of concrete is formed. The temperature expansion coefficient of concrete is in a range of $0.6 \times 10^{-5} \sim 1.3 \times 10^{-5}/°C$, generally $1.0 \times 10^{-5}/°C$, that is, for every 1°C change in temperature, 1m long concrete will produce 0.01mm expansion or shrinkage deformation. The temperature difference between the inside and outside of the concrete leads to the internal expansion of concrete exceeding the external deformation; as the same time, the temperature difference between the inside and outside of the concrete causes temperature gradient inside the concrete and generates stress. Both of these two effects lead to temperature deformation and cracking in concrete.

In order to reduce the impact of temperature deformation on the structure, it should use less cement and use low-heat cement or admix slag and fly ash in the mix design of concrete, which can effectively reduce the hydration heat of concrete. The temperature is very high during the summer construction, and the raw materials should be cooled down. Mass concrete(refers to the minimum side size of concrete structure above 1m) should be cast in layers, and the next layer should be cast after the heat of the cast concrete has been roughly released. For long concrete pavement, large ground area, etc., in order to prevent cracking caused by temperature deformation, an expansion joint can also be set at intervals or reinforcement can be set in the structure to increase the tensile capacity of concrete.

## V. Elastoplastic deformation of concrete under short-term loading

ⅰ. Elastoplastic deformation of concrete

Concrete is a multiphase composite material, which is an elastoplastic body. The relationship between the stress $\sigma$ and the strain $\varepsilon$ of concrete under static compression is shown in Figure 4-21. As can be seen form this figure, when the load is removed at point A, the stress-strain curve is arc AC, and the elastic deformation $\varepsilon_e$ is recovered after the load is removed, leaving the plastic deformation $\varepsilon_p$.

ⅱ. Determination of modulus of elasticity

Since concrete is an elastoplastic body, it is not easy to determine its modulus of elasticity accurately. But its approximation can be sought indirectly. That is, at low stress (30% to 50% of the axial compressive strength $f_{cp}$), as the number of load repetitions (3 to 5 times) increases, the increase in the plastic deformation of concrete gradually decreases, and finally a stress-strain curve with a small curvature is obtained, which is almost parallel to the initial tangent line (the tangent line at the origin of the stress-strain curve when the concrete is initially compressed), see arc $A'C'$ in Figure 4-22. The secant modulus of elasticity of concrete can be obtained from the ratio of stress to strain corresponding to the arc $A'C'$, which is approximated as the static compression modulus of elasticity of concrete.

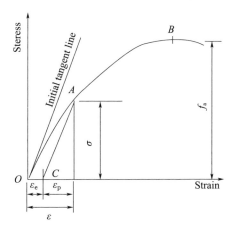
Figure 4-21  Stress-strain curve of concrete under compression

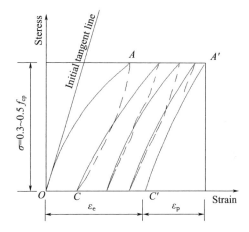
Figure 4-22  Stress-strain curve of concrete under low stress repeated loading

According to the provisions of China's standard, the determination of the modulus of elasticity of concrete uses 150mm × 150mm × 300mm prismatic specimens. The one-third of the axial compressive strength value $f_{cp}$ is set as the test control stress load value, the ratio of stress to strain measured after more than 3 repeated loading and unloading is the modulus of elasticity of concrete.

The modulus of elasticity of concrete is positively related to the compressive strength of concrete, and the empirical formula between then is shown in Equation (4-20). Generally, when the strength grade of concrete is in the range of C10 to C60, its modulus of elasticity is about 17.5GPa to 36.0GPa. The modulus of elasticity of concrete increases, then the deformation of concrete decreases.

$$E_c = \frac{10^5}{2.2 + \frac{34.74}{f_{cc}}} \tag{4-23}$$

## VI. Creep under the long-term load

Concrete under long-term load will occur the phenomenon of creep. The deformation of concrete along the loading direction under the long-term constant load increases with the extension of time, which generally develops for 2 to 3 years before gradually stabilizing. This nature of the development of deformation with time of concrete under long-term load is known as the creep of concrete. Concrete, whether under compression, tension or bending, will produce the phenomenon of creep. The relationship between the deformation of concrete under long-term load and the time of holding the load is shown in Figure 4-23.

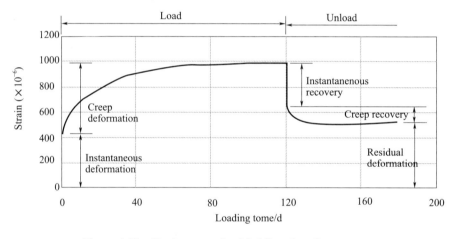

Figure 4-23   Strain versus load holding time for concrete

As can be seen from the figure, the instantaneous deformation is generated immediately when the concrete is loaded, which is mainly the elastic deformation. With the extension of the loading time, the creep deformation is produced, which is mainly the plastic deformation. The creep deformation of concrete is 1 to 3 times of the instantaneous elastic deformation. When the loading stress does not exceed a certain value, the creep deformation of concrete increases fast in the initial loading period, and then increases slowly, which finally stops increasing. When the load is removed after a certain period of time, part of the deformation can be recovered instantaneously, which is called the instantaneous elastic recovery. It is worth noting that the instantaneous elastic recovery is smaller than the instantaneous elastic deformation. There is a small part of the deformation that will be gradually recovered in a number of days, known as creep recovery, which is much smaller than the creep deformation. Finally, most of the deformation that cannot be recovered is called residual deformation.

The reason for the production of concrete creep is generally considered to be due to the tangential slip within the gel layered structure in cement stone under the long-term loading, as well as the migration and penetration of the adsorbed water or crystalline water in the gel into the capillary tubes. The creep deformation is consistent with the direction of loading, and the effects of creep on structures in structural design are as follows.

1. Increase the deformation of the structure

When designing bending members such as bridges and beams in buildings, not only the load-bearing capacity is considered, but also the maximum deflection at mid-span is needed to meet certain requirement. Creep makes the deflection increase with the extension of the loading time, which is unfavorable to the structural safety and normal use. Therefore, the effect of creep on the structure deformation should be fully considered in the design.

2. Cause prestress loss in prestressed reinforced concrete structures

Utilizing the high tensile strength of reinforcing steel, the pre-tension stress is first applied to the reinforcing steel in reinforced concrete. After the rebar is bonded or anchored to the concrete as a whole, the load is unloaded, which can apply compression stress on the concrete by using the properties of the reinforcing steel trying to recover the elastic deformation. In this case, the pre-added compression stress has been already generated in concrete before it is subjected to the external loading. When prestressed reinforced concrete is subjected to tensile stress, the pre-compression stress inside the concrete can offset some of the tensile stress, thus improving the tensile strength and cracking resistance of concrete. However, due to the nature of concrete creep, the deformation of concrete will increase in the direction of compression with the extension of time, which partially restores the tensile deformation of the reinforcement, resulting in the loss of prestress.

3. Reduce temperature stress and microcrack

Due to the hydration heat of concrete, large temperature stress often exists within the mass concrete, leading to temperature deformation cracking. Creep enables concrete to deform slowly in the direction of stress, thus reducing temperature stress and mitigating the risk of the formation of temperature deformation cracking.

4. Generate stress relaxation and relieve stress concentration

Creep can produce stress relaxation and relieve stress concentrations arising from cracks or other defective areas within concrete members, which is beneficial to the structure.

## Section VIII  Durability of concrete

Concrete used in buildings and structures should not only has the strength required by the design requirement to ensure that it can safely withstand the load, but also has the durability to meet the requirement of durability in the environment and employment conditions.

## I. Concept and meaning of durability

The durability of concrete refers to the performance of concrete structures that can maintain the original performance for a long time and resist deterioration and damage under the action of environmental effects. Environmental effects include physical, chemical and biological effects. The physical effect includes temperature change and freeze-thaw cycle, humidity change and wet-dry cycle, etc. ; the chemical effect includes the erosion of acid, alkali, salt substances in aqueous solution or other harmful substances, as well as the role of sunlight, ultraviolet light, etc. on the materials; the biological effect included the erosion of bacteria, insects, etc.

When it comes the safety of concrete structures, people often first think of the load-bearing capacity, i. e. , strength. However, engineering practice shows that there are not many cases of damage to concrete structures caused only by insufficient load-bearing capacity, and many concrete structures are damaged by water erosion, freeze-thaw cycles, corrosion of chemical substances, etc. , which results in the shortened service life and significant waste. Therefore, the design of concrete structures should consider not only strength, but also durability. Improving the durability of concrete is important for both the safety and economic performance of the structure.

## II. Common durability problems

The durability performance of concrete materials includes aspects such as permeability, freezing-thawing resistance, erosion resistance, carbonization, alkali-aggregate reaction. To directly examine these properties of materials requires long-term observation and testing. In practical engineering, accelerated tests are usually conducted under the simulated actual using conditions or enhanced testing conditions to evaluate the relevant durability performance of materials based on the fundamentals of these erosive factors.

i. Impermeability

The impermeability of concrete refers to the ability of concrete to resist the penetration of pressure liquids(water, oil, solution, etc. ). Impermeability is the most important factor in determining the durability of concrete. If the impermeability of concrete is poor, water and other liquid substances in the environment can easily penetrate into the interior of concrete, resulting in the concrete being susceptible to frost damage or erosion damage when encountering negative temperature conditions or water contains aggressive media. Poor impermeability can also cause corrosion of the reinforcement inside the reinforced concrete, which leads to cracking and spalling of the concrete protection layer. Therefore, for the engineering projects that subject to pressure water(or oil), such as underground buildings, pools, water towers, pressure water pipes, dams, oil tanks, as well as port engineering, marine engineering, etc. , must require a certain degree of impermeability of concrete.

The impermeability of concrete is expressed by the impermeability grade P. According to the gradual pressure loading in the *Standard for Test Methods of Long-term Performance and Durability of Ordinary Concrete* (GB/T 50082—2009), six specimens are used to conduct the impermeability test. The following formula is used to calculate the impermeability.

$$P = 10H - 1 \quad (4\text{-}24)$$

Where $H$——the water pressure when water has already seeped out of three specimens while the fourth specimen has not yet seeped (MPa).

Concrete is divided into P4, P6, P8, P10 and P12 five grades according to the impermeability, indicating that the concrete does not seep under the water pressure of 0.4MPa, 0.6MPa, 0.8MPa, 1.0MPa and 1.2MPa, respectively.

The main reason for concrete seepage is the existence of its internal connected seepage pores, which are mainly derived from the evaporation of excess water in the cement paste and the capillary channels left after water bleeding, as well as the water gap gathered at the lower edge of the coarse aggregate; in addition, it can also arise from the cracks caused by drying shrinkage, thermal expansion and other deformations generated in the process of casting non-densely and hardening. The impermeability of concrete is mainly affected by the following factors. Firstly, the amount of water seepage pores produced by cement paste is mainly related to the $W/C$. The smaller the $W/C$, the smaller the capillary porosity and pore size in concrete and the better the impermeability of concrete, and vice versa. Secondly, the use of mineral admixtures in concrete can improve the impermeability of concrete due to the secondary hydration effect of mineral admixtures, which fills some of the pores. In addition, the increase in the maximum particle size of aggregates in concrete could reduce its impermeability. Finally, strengthening the curing also has a role in improving the impermeability of concrete.

ii. Freezing-thawing resistance

Freezing-thawing resistance of concrete refers to the performance of hardened concrete to withstand the effects of multiple freezing-thawing cycles with damage and without serious reduction in strength in a water-saturated state. Concrete is subjected to freezing-thawing damage due to the volume expansion and cold water migration caused by water freezing in its internal voids and capillary pores. When the combined effect of expansion pressure and osmotic pressure exceeds the tensile strength of concrete, microcracks occur in the concrete, which gradually increase and expand under repeated freezing-thawing cycles, resulting in flaking and spalling of the concrete surface until complete destruction.

According to the *Standard for Test Methods of Long-term Performance and Durability of Ordinary Concrete* (GB/T 50082—2009), the freezing-thawing resistance of concrete can be determined by the resistance grade of the slow freezing-thawing method and the resistance class of the rapid freezing-thawing method. In the slow freezing-thawing method, the cubic specimen with side length of 100mm at the standard curing age of 28d is underwent freezing-thawing cycles (each of the air-freezing time and water-thawing time greater than 4h) in the case of $-18^\circ C$ and $20^\circ C$ after water-saturated. Finally, the freezing-thawing resistance of concrete is expressed by the maximum number of freezing-thawing cycles that the concrete can withstand when the compressive strength loss reaches 25% or the mass loss reaches 5%. The resistance grade to freezing-thawing of concrete is divided into D25, D50, D100, D150, D200, D250 and D300, where the number indicates the maximum number of freezing-thawing cycles that concrete can withstand.

For concrete with high freezing-thawing resistance requirements, the rapid freezing-thawing method can be used. In the rapid freezing-thawing method, the prismatic specimen with size of 100mm × 100mm × 400mm at the standard curing age of 28d is underwent freezing-thawing cycles (each cycle time of 2 ~ 4h) in the case of $-18℃$ and $5℃$ after water-saturated. Finally, the resistance class to freezing-thawing of concrete is expressed by the maximum number of freezing-thawing cycles that the concrete can withstand when the relative dynamic modulus of elasticity reduces to 60% or the mass loss reaches 5%, which can be divided into 9 classes, including F25, F50, F100, F150, F200, F250 and F300.

The freezing-thawing resistance of concrete is related to the number of pores inside the concrete, pore characteristics, the filling degree of water in pores, the reduction in ambient temperature and the number of repeated freezing and thawing. First, the reduced $W/C$ of concrete indicates an increased compactness of concrete, which is beneficial to its freezing-thawing resistance. Second, when there are many closed and small pores or the water is not filled in the open pores in concrete, the freezing-thawing resistance of concrete is good. In addition, the addition of water reducer can refine the pore structure of concrete and improve its freezing-thawing resistance. At last, the use of air-entraining agent can significantly improve the freezing-thawing resistance of concrete.

iii. Erosion resistance

The erosion resistance of concrete refers to the performance of concrete to resist damage caused by aggressive media, such as salt, acid and strong alkali in the environment in which it is located. The erosion suffered by concrete in the environment mainly includes sulfate erosion, chloride ion erosion and acidic erosion. The erosion resistance of concrete depends mainly on the type of cement and the compactness of concrete. Therefore, the measures to improve the erosion resistance of concrete are mainly the reasonable choice of cement type, the reduction of $W/C$, the improving the compactness of concrete, and minimizing the open pores in concrete.

iv. Carbonization

Carbonization of concrete refers to the chemical reaction between calcium hydroxide in the cement stone and carbon dioxide in the air, which occurs when the humidity is appropriate to produce calcium carbonate and water.

The adverse effect of carbonization on concrete is that it firstly weakens the protective effect on the reinforcement. The hydration of cement in concrete generates a large amount of calcium hydroxide, which makes it internal alkalinity of concrete high. The rebar in the alkaline environment will generate a passivation film on the surface to protect the rebar from rusting. When the carbonization penetrates the concrete protective layer to reach the surface of the rebar, it reduces the alkalinity of the environment in which the rebar is located and destroys the passivation film, which in turn induces corrosion of the rebar and produces volume expansion, resulting in cracking of the concrete protective layer. Once cracked, the carbonization of concrete will be further accelerated, and the corrosion of rebar will be more serious, finally leading to the destruction of concrete by cracking along the direction of rebar. In addition, the carbonization increases the shrinkage of concrete, which causes small cracks due to tensile stress on the concrete surface, thus reducing the impermeability of concrete.

Carbonization has some beneficial effects on concrete. Firstly, the calcium carbonate produced by carbonization fills the pores of cement stone, which can improve the compactness of the carbonated layer of concrete. In addition, the water released by carbonization helps cement hydration, which is beneficial to improve the compressive strength of concrete.

V. Alkali-aggregate reaction

Alkali-aggregate reaction refers to the chemical reaction between the alkaline oxides(sodium oxide and potassium oxide) in the cement and the active silica in the aggregate to produce an alkali-silicate gel, which will produce a large volume expansion (volume increase of up to three times or more) after water absorption, resulting in concrete expansion and cracking damage.

Three conditions must be present for the alkali-aggregate reaction to occur in concrete.

1. High alkali content in cement. $Na_2O$ equivalent(in $Na_2O + 0.658K_2O$) in cement needs to be greater than 0.6%;

2. The aggregate contains active silica components. Minerals containing active silica components include opal, chalcedony, scaly quartz, etc., which are often found in natural rocks such as rhyolite, andesite, tuff, etc.;

3. The presence of water. Alkali-aggregate reaction in concrete is not possible without the presence of water.

The alkali-aggregate reaction in concrete proceeds slowly and usually occurs only after a number of years, but the problem is difficult to repair once it occurs, so it must be prevented before it happens. To deal with the alkali-aggregate reaction, the following measures should be taken.

1. Use low alkali cement(alkali content less than 0.6%);

2. Use of non-alkaline reactive aggregates;

3. Add mineral admixtures, such as fly ash, slag and silica fume. Mineral admixture contains higher activity of silica, which reacts with the alkali in cement before the aggregate, reducing the alkali content in the concrete;

4. Admix alkali-aggregate reaction inhibitors, such as lithium carbonate and lithium chloride.

Scan the QR code to get the details of the *Standard for Test Methods of Long-term Performance and Durability of Ordinary Concrete*(GB/T 50082—2009).

## III. Durability design of concrete structure

The so-called durability design of concrete means that the concrete structure should be designed for durability according to the design service life and environmental category. The durability design of concrete structure mainly includes the following contents: determine the environmental category in which the structure is located; propose the basic requirements for durability of concrete; determine the thickness of concrete protective layer of rebar in members; propose the technical measures for durability under different environmental conditions; propose the requirements for testing and maintenance in the use phase of the structure, etc.

The environment categories to which concrete structures are exposed are classified according to the requirements of Table 4-17.

Table 4-17  Environmental categories of concrete structures

| Environment category | Conditions |
| --- | --- |
| I | Dry indoor environment;<br>Non-aggressive hydrostatic submerged environment |
| II a | Indoor humid environment;<br>Open-air environments in non-harsh and non-cold areas;<br>Environment in direct contact with non-erosive water or soil in non-harsh and non-cold areas;<br>Environment in direct contact with non-erosive water or soil below the freezing line in harsh and cold areas |
| II b | Wetting-drying cycle environment;<br>Environment with water level frequently changing;<br>Open-air environments in harsh and cold areas;<br>Environment in direct contact with non-erosive water or soil above the freezing line in harsh and cold areas |
| III a | Environment of winter water level changing zone in harsh and cold regions;<br>Environment affected by de-icing salt;<br>Sea breeze environment |
| III b | Saline soil environment;<br>Environment subjected to the action of de-icing salts;<br>Coastal environment |
| IV | Seawater environment |
| V | Environment affected by man-made or natural aggressive substances |

When the design service life is 50 years for the concrete structure that locating in categories I, II and III environments, the basic requirements of concrete should be in line with the provisions of Table 4-18. In these three categories of environments, it should take special effective measures for concrete structure with a designed service life of 100 years. For the concrete structures locating in the environment categories IV and V, their durability requirements should be in line with the provisions of the relevant standards.

**Table 4-18　Basic requirements for the durability of structural concrete**

| Environment category | Maximum water-to-binder ratio($W/B$) | Minimum strength grade |
| --- | --- | --- |
| I | 0.60 | C20 |
| II a | 0.55 | C25 |
| II b | 0.50(0.55) | C30(C25) |
| III a | 0.45(0.50) | C35(C30) |
| III b | 0.40 | C40 |

Note: The relevant parameters in parentheses can be used when using air-entraining agent in concrete.

## Section IX　Mix design of concrete

Mix proportion of concrete refers to mass proportional relationship between the constituent materials in a unit volume of concrete. Concrete mix proportion has two methods of expression, one is directly expressed by the amount of various material in $1m^3$ concrete; the other one is expressed by the mass proportional relationship between the constituent materials of concrete, in which the mass of binder is 1, such as binder : water : sand : rock = 1 : $W/B$ : $x$ : $y$.

The process of using the raw materials in the engineering projects to determine the proportional amount of each raw material in the concrete to obtain a specific performance of the concrete is known as mix proportion design of concrete.

### I. Basic requirements, basis and methods of mix proportion design of concrete

According to the *Specification for Mix Proportion Design of Ordinary Concrete* (JGJ 55—2011), the mix proportion of ordinary concrete should be calculated according to the performance of raw materials and the technical requirements of concrete and determined by the laboratory trial and adjustment.

i . Basic requirements of the mix proportion design

1. Meet the strength grade requirements of the structural design;
2. Meet the required workability of concrete construction;
3. Meet the durability requirements of the environment in which the project is located;
4. Comply with the economic principle, i.e., save cement to reduce the cost of concrete.

ii . Basis of the mix proportion design

1. The principle of determining the basic parameters of concrete mix proportion design

Water-to-cement ratio, unit water consumption and sand ratio are the three parameters of concrete mix proportion design. They have a very close relationship with the performance of concrete. Therefore, the concrete mix proportion design is mainly the correct determination of these three parameters to ensure the preparation of concrete which meets the requirements.

The principles for determining the three parameters in the concrete mix proportion design are: determine the $W/C$ of concrete on the basis of meeting the strength and durability of concrete; determine the unit water consumption of concrete according to the type and specification of coarse aggregate on the basis of meeting the workability of concrete; and determine the sand ratio based on the principle of slight surplus of sand after filling the rock voids.

2. The material calculation basis of mix proportion design

The amount of each material in $1m^3$ concrete is used as a benchmark in the concrete mix proportion design, in which the calculation of aggregate is based on its dry state. The so-called dry state of the aggregate refers to the moisture content of fine aggregate less than 0.5% and that of coarse aggregate less than 0.2%. If one needs the saturated surface dry aggregates as the basic for calculation, it should be modified accordingly.

The dosage of the chemical admixture in concrete is generally small. Therefore, the volume of the chemical admixture is negligible in the calculation of the volume of concrete. In the calculation of the apparent density of concrete, the mass of chemical admixture can also be negligible.

iii. Method and principle of mix proportion design

There are two methods for the mix proportion design of ordinary concrete, namely volume method and mass method.

1. Volume method

The basic principle of the volume method for the mix proportion design of concrete is that the volume of the concrete mixture is assumed to be equal to the sum of the absolute volume of the constituent materials and the volume of the small amount of air it contains. If using $V_h$, $V_c$, $V_f$, $V_s$, $V_g$, $V_w$ and $V_k$ represent the volume of concrete, cement, mineral admixtures, sand, rock, water and air, then the principle of the volume method can be expressed in the following formula.

$$V_h = V_c + V_f + V_s + V_g + V_w + V_k \tag{4-25}$$

If the amount of cement, mineral admixture, water, sand and rock are represented with $m_{c0}$, $m_{f0}$, $m_{s0}$, $m_{g0}$ and $m_{w0}$, respectively, and the density of cement, density of mineral admixture, apparent density of sand, apparent density of rock, and density of water are represented with $\rho_c$, $\rho_f$, $\rho_s$, $\rho_g$ and $\rho_w$, respectively, then the above formula can be rewritten as the following formula assuming that the air content percentage in the concrete mixture is $\alpha$.

$$\frac{m_{c0}}{\rho_c} + \frac{m_{f0}}{\rho_f} + \frac{m_{s0}}{\rho_s} + \frac{m_{g0}}{\rho_g} + \frac{m_{w0}}{\rho_w} + 0.01\alpha = 1 \tag{4-26}$$

Where $\alpha$ is the percentage of air content (%) in concrete, which is 1 in the concrete without air-entraining admixture.

2. Mass method

The basic principle of the mass method of concrete mix proportion design is that when the raw materials used in concrete are relatively stable, the apparent density of the prepared concrete is close to a constant value. The mass of $1m^3$ fresh concrete can be assumed in advance, and the following relationship can be established.

$$m_{c0} + m_{f0} + m_{s0} + m_{g0} + m_{w0} = m_{cp} \tag{4-27}$$

The assumed mass per cubic meter of concrete ($m_{cp}$) can be selected in the range of 2350 to 2450kg/m³.

Scan the QR code to get the details of *Specification for Mix Proportion Design of Ordinary Concrete* (JGJ 55—2011).

## II. The calculation of mix proportion design of concrete

i. Determination of the formulated strength of concrete ($f_{cc,0}$)

According to the concept of strength guarantee rate of concrete, it is known that if the average strength ($\bar{f}_{cc}$) of the formulated concrete is equal to the design strength grade value ($f_{cc,k}$), then its strength guarantee rate is only 50%. If a strength guarantee rate higher than 50% is to be achieved, the formulated strength of concrete must be higher than the design strength grade value. When the design strength grade of concrete is less than C60, the following equation can be obtained according to Equation(4-8) if the formulated strength of concrete is equal to the average strength.

$$f_{cc,0} = \bar{f}_{cc} = f_{cc,k} + t\sigma \tag{4-28}$$

China currently requires a strength guarantee rate of 95% for concrete, check Table 4-15 and get $t$ is equal to 1.645, which can be substituted into the above equation to obtain the formulated strength.

$$f_{cc,0} = f_{cc,k} + 1.645\sigma \tag{4-29}$$

Where the value of the $\sigma$ can be calculated from Equations(4-9) or (4-10) based on the historical statistical concrete strength information. If concrete strength information is not available, Table 4-19 can be referred to.

Table 4-19 Standard deviation $\sigma$ value (MPa)

| Concrete strength grade | Lower than C20 | C20 ~ C35 | Higher than C35 |
|---|---|---|---|
| $\sigma$ | 4.0 | 5.0 | 6.0 |

When the design strength grade of concrete is not less than C60, the formulated strength of concrete can be determined according to the following equation.

$$f_{cc,0} \geq 1.15 f_{cc,k} \tag{4-30}$$

ii. Determination of W/B

According to the known concrete formulated strength ($f_{cc,0}$) and the 28d compressive strength of the binder ($f_b$), then the W/B can be obtained from the concrete strength empirical formula, i.e.:

$$\frac{W}{B} = \frac{\alpha_a f_b}{f_{cc,0} + \alpha_a \alpha_b f_b} \tag{4-31}$$

Where the regression coefficients $\alpha_a$ and $\alpha_b$ can be determined through the relationship formula between the W/B and concrete strength established by testing using the raw materials used in the engineering project. When the above test statistics are not available, the value in Table 4-16 can be used.

When there is no measured value for the 28d compressive strength of the binder($f_b$), it can be calculated according to the following formula based on the 28d compressive strength of cement($f_{ce}$),

$$f_b = \gamma_f \gamma_s f_{ce} \qquad (4\text{-}32)$$

Where $\gamma_f$ and $\gamma_s$ are fly ash impact coefficient and slag impact coefficient, respectively, which can be selected according to Table 4-20.

Table 4-20  Impact coefficients of fly ash($\gamma_f$) and slag($\gamma_s$)

| Dosage/% | Fly ash impact coefficient($\gamma_f$) | Slag impact coefficient($\gamma_s$) |
|---|---|---|
| 0 | 1.00 | 1.00 |
| 10 | 0.85~0.95 | 1.00 |
| 20 | 0.75~0.85 | 0.95~1.00 |
| 30 | 0.65~0.75 | 0.90~1.00 |
| 40 | 0.55~0.65 | 0.80~0.90 |
| 50 | — | 0.75~0.85 |

When there is no measured 28d compressive strength value of cement($f_{ce}$), it can be calculated according to the following formula using the strength grade value of cement($f_{ce,g}$),

$$f_{ce} = \gamma_c f_{ce,g} \qquad (4\text{-}33)$$

Where $\gamma_c$ is the surplus coefficient of cement strength grade value, which can be determined according to the actual statistics; when it lacks actual statistics, the surplus coefficient can be selected according to Table 4-21.

Table 4-21  Surplus coefficient of cement strength grade value($\gamma_c$)

| Cement strength grade value | 32.5 | 42.5 | 52.5 |
|---|---|---|---|
| Surplus coefficient | 1.12 | 1.16 | 1.10 |

If the calculated $W/B$ based on Equation(4-31) is larger than the maximum $W/B$ specified in Table 4-18, the maximum $W/B$ specified in the table should be used.

iii. Selection of mixing water content($m_{w0}$)

Mixing water content($m_{w0}$) in concrete mixture can be selected from the value specified in Table 4-22 according to the category of coarse aggregate, the maximum particle size of coarse aggregate ad the slump value required for construction.

Table 4-22  Mixing water content in plastic and dry concrete(kg/m³)

| Item | Indicator | Maximum particle size of pebble/mm | | | | Maximum particle size of crushed stone/mm | | | |
|---|---|---|---|---|---|---|---|---|---|
| | | 10 | 20 | 31.5 | 40 | 16 | 20 | 31.5 | 40 |
| Slump/mm | 10~30 | 190 | 170 | 160 | 150 | 200 | 185 | 175 | 165 |
| | 35~50 | 200 | 180 | 170 | 160 | 210 | 195 | 185 | 175 |
| | 55~70 | 210 | 190 | 180 | 170 | 220 | 205 | 195 | 185 |
| | 75~90 | 215 | 195 | 185 | 175 | 230 | 215 | 205 | 195 |

continued

| Item | Indicator | Maximum particle size of pebble/mm | | | | Maximum particle size of crushed stone/mm | | | |
| --- | --- | --- | --- | --- | --- | --- | --- | --- | --- |
| | | 10 | 20 | 31.5 | 40 | 16 | 20 | 31.5 | 40 |
| VB consistency/s | 16~20 | 175 | 160 | — | 145 | 180 | 170 | — | 155 |
| | 11~15 | 180 | 165 | — | 150 | 185 | 175 | — | 160 |
| | 5~10 | 185 | 170 | — | 155 | 190 | 180 | — | 165 |

Note: 1. The mixing water in this table is the value when using medium sand. If fine sand or coarse sad is used, the mixing water content in $1m^3$ concrete should be increased or decreased by 5~10kg accordingly;
2. When mixed with various chemical admixtures or mineral admixtures, the mixing water content should be adjusted accordingly.

For concrete with a slump greater than 100mm, it is necessary to mix with water reducer. At this time, the determination of the mixing water content should be based on the mixing water content of 90mm slump in Table 4-22, and the mixing water content should be increased by 5kg for every 20mm increase in slump to calculate the mixing water content $m'_{w0}$ of $1m^3$ concrete without chemical admixture. Then the mixing water content $m_{w0}$ of $1m^3$ concrete containing water reducer according to the following equation.

$$m_{w0} = m'_{w0} \times (1 - \beta) \quad (4\text{-}34)$$

Where $\beta$——water reduction rate of the chemical admixture(%), determined by the test.

ⅳ. Calculation of the amount of binder($m_{b0}$)

The amount of binder in $1m^3$ concrete($m_{b0}$) is calculated according to the following equation.

$$m_{b0} = \frac{m_{w0}}{W/B} \quad (4\text{-}35)$$

Besides the preparation of C15 and the lower strength grades of concrete, the minimum amount of binder for concrete should in accordance with the provisions of Table 4-23. If the amount of binder calculated by Equation(4-32) is less than the minimum amount of binder specified in the following table, the minimum amount of binder specified in the table should be used.

Table 4-23 Minimum amount of binder for concrete($kg/m^3$)

| Maximum $W/B$ | Minimum amount of binder | | |
| --- | --- | --- | --- |
| | Plain concrete | Reinforced concrete | Prestressed concrete |
| 0.60 | 250 | 280 | 300 |
| 0.55 | 280 | 300 | 300 |
| 0.50 | 320 | | |
| ≤0.45 | 330 | | |

The amount of mineral admixture per cubic meter of concrete($m_{f0}$) is calculated according to the following formula.

$$m_{f0} = m_{b0} \beta_f \quad (4\text{-}36)$$

Where $\beta_f$——dosage of mineral admixture(%), determined by the test.

The amount of cement per cubic meter of concrete($m_{c0}$) is calculated according to the following formula.

$$m_{c0} = m_{b0} - m_{f0} \tag{4-37}$$

ⅴ. Calculation of the amount of chemical admixture ($m_{b0}$)

The amount of chemical admixture per cubic meter of concrete is calculated according to the following equation.

$$m_{a0} = m_{b0} \beta_a \tag{4-38}$$

Where $\beta_a$——dosage of chemical admixture (%), determined by the concrete test.

ⅵ. Determination of sand ratio ($\beta_s$)

The sand ratio of concrete should be determined according to the technical specifications of the aggregate, workability of concrete mixture, and construction requirements, with reference to the existing historical data. When it lacks information, the sand ratio of concrete should be determined accordance with the following provisions: sand ratio should be determined by test for concrete with a slump less than 10mm; sand ratio can be selected in Table 4-24 according to the type of coarse aggregate, the maximum nominal particle size, and $W/B$; sand ratio can be determined by test, and also can be adjusted on the basis of Table 4-24 b increasing the sand ratio by 1% for every 20mm increase in slump.

Table 4-24  Sand ratio of concrete (%)

| w/b | Maximum nominal particle size of pebble/mm | | | Maximum nominal particle size of crushed stone/mm | | |
| --- | --- | --- | --- | --- | --- | --- |
|  | 10 | 20 | 40 | 16 | 20 | 40 |
| 0.40 | 26~32 | 25~31 | 24~30 | 30~35 | 29~34 | 27~32 |
| 0.50 | 30~35 | 29~34 | 28~33 | 33~38 | 32~37 | 30~35 |
| 0.60 | 33~38 | 32~37 | 31~36 | 36~41 | 35~40 | 33~38 |
| 0.70 | 36~41 | 35~40 | 34~39 | 39~44 | 38~43 | 36~41 |

Note: 1. Sand ratio in the table the value when using medium sand. For fine sand or coarse sand, the sand ratio can be reduced or increased accordingly;

2. The sand ratio should be increased appropriately when using only a single grain coarse aggregate to prepare concrete;

3. When using manufactured sand to prepare concrete, the sand ratio can be increased appropriately.

ⅶ. Calculation of amount of sand and rock ($m_{s0}$, $m_{g0}$)

When the volume method is used to calculate the mix proportion design of concrete, the amount of the coarse and fine aggregates is calculated according to Equation (4-23), and the sand ratio is calculated according to Equation (4-39).

$$\frac{m_{s0}}{m_{s0} + m_{g0}} \times 100\% = \beta_s \tag{4-39}$$

The above two equations can be combined to find the amount of the coarse and fine aggregates.

When using the mass method to calculate the mix proportion design of concrete, the amount of the coarse and fine aggregates is calculated according to Equation (4-27), and the sand ratio is calculated according to Equation (4-39). The above two equations can be combined to find the amount of the coarse and fine aggregates.

ⅷ. Writing the mix proportion of concrete

At this point, the calculated mix proportion of concrete has been obtained.

## Ⅲ. Trial, adjustment and determination of mix proportion of concrete

In the actual construction, the trial batch of concrete should be prepared based on the calculated mix proportion using the actual materials used in the engineering project. Adjust the mix parameters to make the performance of concrete mixture being in line with the design and construction requirement, and put forward the trail mix proportion. On the basis of the trial mix proportion, the strength test of concrete and the adjustment of the apparent density of concrete are carried out to obtain the laboratory mix proportion. Finally, according to the moisture content of the aggregate used in the actual project, the laboratory mix proportion can be adjusted to obtain the construction mix proportion.

ⅰ. Trial mix proportion

The trial of concrete should use the actual raw materials used in the engineering project, and the concrete mixing method should also be the same as the method used in construction. When conducting the trial, the minimum mixing volume of each concrete trial shall conform to the provisions of Table 4-25 and shall not be less than one-fourth of the rated mixing volume of the mixer.

Table 4-25　Minimum mixing volume for concrete trial

| Maximum nominal particle size of coarse aggregate/mm | Volume of mixture/L |
| --- | --- |
| 31 and below | 20 |
| 40 | 25 |

On the basis of the calculated mix proportion, a trial is carried out to test the performance of the concrete mixture. If the slump of the test mixture is greater than the target slump, the $W/C$ should be kept unchanged while reducing the amount of binder and water simultaneously; if the slump is less than the target slump, the $W/B$ should be kept unchanged while increasing the amount of binder and water simultaneously; is the cohesion and water retention are not qualified, the sand ratio should be adjusted until it meets the requirements. Modify the calculated mix proportion and propose the trial mix proportion for the further concrete strength test.

ⅱ. Laboratory mix proportion

Concrete strength test is conducted on the basis of the trial mix proportion using three different mix proportion designs, one of which is the trial mix proportion, the other two feature an increased or reduced $W/B$ by 0.05, the same mixing water content, and an increased or reduced sand ratio by 1% compared with those of the trial mix proportion. The workability of the concrete mixture should be in line with the design and construction requirements during the concrete strength test.

According to the results of concrete strength test, the linear relationship between strength and binder-to-water ratio ($B/W$) is plotted, and the $B/W$ corresponding to the concrete formulated strength ($f_{cc,0}$) is determined. On the basis of the trial mix proportion, the amount of water used ($m_w$) and the amount of chemical admixture ($m_a$) are selected from the trial mix proportion design and adjusted appropriately according to the slump measured during the production of strength specimens; the amount of binder ($m_b$) is calculated by multiplying the amount of water used by the determined $B/W$; the amount of coarse and fine aggregates ($m_g$ and $m_s$) are selected from the trial mix proportion design and adjusted according to the amount of water used and the amount of binder.

After the adjustment on the trial mix proportion according to the strength test, the apparent density of concrete should also be corrected. The apparent density of the concrete mixture ($\rho_{c,c}$) after the adjustment is calculated as follows.

$$\rho_{c,c} = m_b + m_w + m_s + m_g \qquad (4\text{-}40)$$

After the adjustment on the trial mix proportion, conduct the trial batch of concrete, measure irs apparent density ($\rho_{c,t}$), and then calculated the correction factor $\delta$ according to the following formula.

$$\delta = \frac{\rho_{c,t}}{\rho_{c,c}} \qquad (4\text{-}41)$$

When the absolute value of the difference between the measured and calculated apparent density of concrete does not exceed 2% of the calculated value, then the above adjusted mix proportion can be determined as the laboratory mix proportion of concrete; if the difference between the two exceeds 2%, the amount of each material in the mix proportion should be multiplied by the correction factor to obtain the laboratory mix proportion.

iii. Construction mix proportion

The laboratory mix proportion of concrete is calculated based on dry aggregates, but the aggregates used on the actual site often contain a certain amount of moisture, so the laboratory mix proportion must be converted to the construction mix proportion.

Let the amount of binder, water, sand, rock and chemical admixture in $1m^3$ concrete of the construction mix proportion be $m'_b, m'_w, m'_s, m'_g$ and $m'_a$, respectively, and the moisture content in sand and rock on the site be $a\%$ and $b\%$, respectively, then the amount of each material in $1m^3$ concrete in the construction mix proportion should be:

$$m'_b = m_b \qquad (4\text{-}42)$$

$$m'_s = m_s(1 + a\%) \qquad (4\text{-}43)$$

$$m'_g = m_g(1 + b\%) \qquad (4\text{-}44)$$

$$m'_w = m_w - m_s \cdot a\% - m_g \cdot b\% \qquad (4\text{-}45)$$

$$m'_a = m_a \qquad (4\text{-}46)$$

So far, the concrete construction mix proportion is obtained which can be used for field construction.

## IV. Example of concrete mix proportion design

**Example 4-1** A cast-in-place reinforced concrete beam of a frame structure project in a dry indoor environment, the design strength grade of concrete is C30, which is constructed using machine mixing and machine vibration. The slump requirement of concrete slump is 30 to 50mm, while the standard deviation $\sigma$ of concrete strength is 5MPa according to the history statistics of the construction company. The raw materials used are as follows.

Cement: 42.5 grade ordinary Portland cement with a cement density of 3.00g/cm$^3$, and the surplus factor of the cement strength standard value is 1.16;

Fly ash: Grade II fly ash, with a density of 2.10g/cm$^3$ and a dosage of 10% of the binder;

Sand: medium sand with qualified gradation and an apparent density of 2650kg/m$^3$;

Rock: 5.0 ~ 31.5mm crushed stone with qualified gradation and an apparent density of 2700kg/m$^3$;

Chemical admixture: polycarboxylic acid high-range water reducer (non-air-entraining type) with a dosage of 0.2%, whose water reducing rate is 8% under this dosage.

Try to find:

1. Calculated mix proportion of concrete;
2. The workability and strength of the trial concrete are in line with the requirements, no needs to make adjustments. If the moisture content of sand is 3% and the moisture content of rock is 1%, try to calculate the construction mix proportion of concrete.

Solution: 1. Find the calculated mix proportion of concrete

(1) Determine the formulated strength of concrete ($f_{cc,0}$)

$$f_{cc,0} = f_{cc,k} + 1.645\sigma = 30 + 1.645 \times 5 = 38.23 \text{MPa}$$

(2) Determine the water-to-binder ratio ($W/B$)

Since the surplus coefficient of the cement strength grade is 1,16, the compressive strength value of cement is

$$f_{ce} = \gamma_c f_{ce,g} = 1.16 \times 42.5 = 49.3 \text{MPa}$$

The content of fly ash is 10%, in which the impact factor of fly ash is 0.9 from Table 4-20, then the compressive strength value of the binder is

$$f_b = \gamma_f \gamma_s f_{ce} = 0.9 \times 49.3 = 44.37 \text{MPa}$$

The regression coefficient $\alpha_a$ and $\alpha_b$ can be found in Table 4-16 as 0.53 and 0.20, respectively, and the $W/B$ is

$$\frac{W}{B} = \frac{\alpha_a f_b}{f_{cc,0} + \alpha_a \alpha_b f_b} = \frac{0.53 \times 44.37}{38.23 + 0.53 \times 0.20 \times 44.37} = 0.55$$

As the concrete beam of frame structure is in the indoor dry environment, the maximum $W/B$ is 0.60 through checking Table 4-18. The calculated value meets the requirement, and the $W/B$ of 0.55 is adopted.

(3) Determine the amount of water used ($m_{w0}$)

The amount of water used in 1m$^3$ concrete can be selected as 185kg when the required slump is 35~50mm for the concrete containing the crushed stone with the maximum particle size of 31.5mm. Since the water reduction rate of the water reducer is 8%, then the amount of water used after mixing with water reducer is

$$m_{w0} = m'_{w0} \times (1-\beta) = 185 \times (1-0.08) = 170.2 \text{kg}$$

(4) Calculate the amount of binder ($m_{b0}$)

The total amount of the binder in 1m³ concrete is

$$m_{b0} = \frac{m_{w0}}{W/B} = \frac{170.2}{0.55} = 309 \text{kg}$$

Since the fly ash content is 10%, then the amount of fly ash is

$$m_{f0} = m_{b0}\beta_f = 309 \times 0.1 = 30.9 \text{kg}$$

The amount of cemnet is

$$m_{c0} = m_{b0} - m_{f0} = 309 - 30.9 = 278.1 \text{kg}$$

Check Table 4-23, the minimum amout of bidner for reinfroced concrete with $W/B$ of 0.55 is 300kg/m³, and the calculated value is in accordance with the regulations. The amounts of cement and fly ash in 1m³ concrete are 278.1kg and 30.9kg, respectively.

(5) Calculate the amount of chemical admixture ($m_{a0}$)

The dosage of water reducer is 0.2%, then its amount is

$$m_{a0} = m_{b0}\beta_a = 309 \times 0.002 = 0.618 \text{kg}$$

(6) Determine the sand ratio ($\beta_s$)

Check Table 4-24, the sand ratio can be selected as 35% (using the interpolation method) for the concrete containing the crushed stone with the maximum particle size of 31.5mm and the $W/B$ of 0.55.

(7) Calculate the amount of sand and rock ($m_{s0}, m_{g0}$)

Using the volume method, i.e.,

$$\begin{cases} \dfrac{278.1}{3000} + \dfrac{30.9}{2100} + \dfrac{170.2}{1000} + \dfrac{m_{s0}}{2650} + \dfrac{m_{g0}}{2700} + 0.01 \times 1 = 1 \\ \dfrac{m_{s0}}{m_{s0} + m_{g0}} \times 100\% = 35\% \end{cases}$$

Solving the simultaneous equations gives: $m_{s0}$ is 668.8kg and $m_{g0}$ is 1241.9kg.

(8) Write the calculated mix proportion of concrete

The amount of each material in 1m³ concrete is: 278.1kg of cement, 30.9kg of fly ash, 170.2kg of water, 668.8kg of sand, 1241.9kg of crushed stone, and 0.618 kg of water reducer.

2. Convert to construction mix proportion

Let the amount of cement, fly ash, water, sand, rock and water reducer in 1m³ concrete of the construction mix proportion be $m'_c, m'_f, m'_w, m'_s, m'_g$ and $m'_a$, respectively, then we have

$$m'_c = m_c = 278.1 \text{kg}$$

$$m'_f = m_f = 30.9 \text{kg}$$

$$m'_s = m_s(1+a\%) = 668.8 \times (1+3\%) = 688.9 \text{kg}$$

$m'_g = m_g(1+b\%) = 1241.9 \times (1+1\%) = 1254.3\text{kg}$

$m'_w = m_w - m_s \cdot a\% - m_g \cdot b\% = 170.2 - 668.8 \times 3\% - 1241.9 \times 1\% = 137.7\text{kg}$

$m'_a = m_a = 0.618\text{kg}$

## Section X  Building Mortar

Building mortar is one of the building materials with large amount and wide use in building engineering. Building mortar is made up of cementitious materials, fine aggregate and water in accordance with a certain proportion. Building mortar can cement granular materials, bulk materials and sheet materials into the overall structure, and can also become the material of decoration and protection of the main body. It plays the role of bonding, liner and transferring stress in construction engineering, and is mainly used in masonry, plastering, repairing and decoration engineering.

According to the different cementing materials used in building mortar, it can be divided into cement mortar, lime mortar, cement clay mortar, polymer mortar and mixed mortar, etc. According to different uses, it can be divided into masonry mortar, plastering mortar, special mortar, etc. According to the production methods of mortar can be divided into two ways: site-mortar and factory-mortar, the latter is the developing trend of mortar at home and abroad, so that the construction of mortar is required to realize and spread of mortar in our country soon.

Masonry mortar is a mortar that bonds bricks, stones, blocks, etc. into masonry. Masonry mortar mainly plays the role of bonding, transferring load, making the distribution of stress more uniform and coordinating deformation. It is an important part of masonry and is the most widely used. This section mainly discusses the related contents of the masonry mortar.

### I. The composition of masonry mortar materials

i. Cementitious materials and admixtures

Common cements can be used to compound masonry mortar, cement varieties are selected in the same way as in concrete. Usually mortar for strength requirements are not very high, so the medium strength grade of concrete can meet the requirements. According to the requirements of *Design Regulations for Masonry Mortar Mix Ratio* (JGJ/T 98—2010), the selection of cement should conform to the current national standards *General Portland Cement* (GB 175) and *Masonry Cement* (GB/T 3183). Generally speaking, the cement strength grade of cement mortar should not be more than 32.5. The cement strength grade of cement mixed mortar should not be more than 42.5.

In order to improve the workability of mortar and save cement, appropriate amount of lime, fly ash and other admixtures are often added into mortar. The mortar prepared in this way is called cement mixed mortar. The admixtures such as fly ash and quicklime should conform to the relevant regulations.

ii. Fine aggregate

Natural sand is the most commonly used fine aggregate for preparing building mortars. Sand should first meet the technical properties of concrete sand requirements. Because the mortar layer is relatively thin, the maximum particle size of sand should be limited, theoretically it should not exceed 1/4 ~ 1/5 of the thickness of the mortar layer. For example, brick masonry mortar should choose medium sand, and the maximum particle size is not more than 2.5mm. Coarse sand should be selected as the mortar for stone masonry, and the maximum particle size should be no more than 5.0mm. Fine sand should be used for smooth surface and jointing mortar, and the maximum particle size is not more than 1.2mm. In order to ensure the quality of mortar, the content of sand should be limited. The mud content of sand for masonry mortar with strength grade above M2.5 shall not exceed 5%; The mud content of sand used for cement mixed mortar with strength grade M2.5 should not be more than 10%; The mud content of sand used for waterproof mortar should not be more than 3%.

iii. Water

The technical requirements of mortar mixing water are the same as those of concrete mixing water, and they all need to meet the provisions of *Concrete Mixing Water Standard*(JGJ 63—2006).

iv. The additive

In order to improve or increase some properties of fresh mortar and hardened mortar, appropriate additives are often added to the mortar. Various admixtures used in concrete also have corresponding effects on mortar. According to the standard *Masonry Mortar Plasticizer* (JG/T 164—2004), mortar plasticizer can be added to cement mortar for masonry, which can obviously improve its workability. But the variety of the admixture (air entraining agent, early strength agent, retarder, antifreeze, etc.) and dosage must be determined by test. When adding admixtures in cement mortar, besides the effect of admixture on the performance of mortar itself, the effect on the use function of mortar should also be considered. Scan the QR code to get the detailed information of *Masonry Mortar Plasticizer*(JG/T 164—2004).

## II. Main technical features of masonry mortar

i. The workability

1. Liquidity

Fluidity refers to the performance of mortar whether it is easy to flow under its own weight or external force. Mortar fluidity can be expressed in terms of consistency. The fluidity is measured by the mortar consistency meter, which is expressed by the number of millimeters of the cone of the mortar consistency meter into the depth of the mortar, called the degree of sink. Mortar with large degree of sinking has good fluidity.

The fluidity of mortar should be selected according to the types of substrate materials, construction conditions, weather conditions and other factors. However, it shall comply with the *Code for Masonry Construction and Quality Acceptance*(GB 50203—2011). The fluidity of mortar can be selected by referring to Table 4-26.

**Table 4-26  Mortar fluidity reference table (sink degree) (mm)**

| Types of masonry | Dry climate or Porous water absorbing material | Cold climates or dense materials | Rendering engineering | Construction by machinery | Manual operation |
|---|---|---|---|---|---|
| Sintered ordinary brick masonry | 80 ~ 90 | 70 ~ 80 | Layer of preparation | 80 ~ 90 | 110 ~ 120 |
| Sintered porous brick, hollow brick masonry | 70 ~ 80 | 60 ~ 70 | The underlying | 70 ~ 80 | 70 ~ 80 |
| Stone masonry | 40 ~ 50 | 30 ~ 40 | The surface layer | 70 ~ 80 | 90 ~ 100 |
| Ordinary concrete masonry | 60 ~ 70 | 50 ~ 60 | Mortar layer | | 90 ~ 120 |
| Lightweight aggregate concrete masonry | 70 ~ 90 | 60 ~ 80 | | | |

Scan the QR code to get the detailed information of *Code for Masonry Construction and Quality Acceptance* (GB 50203—2011).

2. Water retention

Water retention of mortar refers to the ability of fresh mortar to retain water, and also indicates whether the composition of the mortar is easy to segregate. The degree of stratification was used to represent the water retention of the mortar. In order to measure the stratification degree, the mortar which has been instrumented for consistency is filled with the stratification degree cylinder and divided into the upper and lower sections (The inner diameter of the delamination cylinder is 150mm, the height of the upper segment is 200mm, and the height of the lower segment is 100mm). Gently tap around the barrel three to five times to scrape off excess mortar and smooth. After placing it for 30min, remove the upper 200mm mortar, take out the remaining 100mm mortar and pour it into the stirring pot for 2min, then measure the consistency. The difference between the two measurements before and after is the stratification of the mortar (measured in mm). The reasonable stratification of mortar should be controlled in 10 ~ 30mm. The mortar with a delamination degree more than 30mm is easy to segregate, bleed water and delamination or water loss too fast, which is not convenient for construction. The mortar with a delamination degree less than 10mm is easy to produce dry shrinkage crack after hardening.

ii. Compressive strength and strength grade

According to *Test Methods for Basic Properties of Building Mortar* (JGJ 70—2009), the strength grade of mortar is the average compressive strength (MPa) measured by the standard test method after curing to 28 days under standard conditions of 70.7mm × 70.7mm × 70.7mm cubic standard test blocks (3 blocks in each group), which is indicated by $f_z$. The strength grade of masonry mortar is divided into 7 grades: M30, M25, M20, M15, M10, M7.5 and M5.0. Scan the QR code to get the detailed information of *Test Methods for Basic Properties of Building Mortar* (JGJ 70—2009).

According to the regulations of *Design Regulations for the Proportion of Masonry Mortar*

(JGJ 98—2010), the actual strength of mortar is not only related to the strength and dosage of cement, but also related to the water absorption of the material at the bottom of mortar.

1. non-absorbent base material (such as dense stone)

The factors affecting the strength of mortar are basically the same as those of concrete, which mainly depend on the strength of cement and the ratio of water to cement. In a word, the strength of mortar is proportional to the strength of cement and the ratio of grey water. The relationship is as follows:

$$f_{m,cu} = \alpha f_{ce}\left(\frac{C}{W} - \beta\right) \tag{4-47}$$

Where  $f_{m,cu}$——28 day compressive strength of mortar (MPa);

$f_{ce}$——the measured strength value of cement (MPa);

$C/W$——cement-to-water ratio;

$\alpha$ and $\beta$——coefficient, which can be determined from the test data statistics. Generally, $\alpha$ is about 0.29, $\beta$ is about 0.40.

2. Water-absorbing base material (such as brick or other porous material)

When the base level absorbs water, the amount of water retained in the mortar depends on its own water retention. Therefore, even if the water consumption of mortar is different, the water retained in the mortar is almost the same because it has certain water retention and water absorption of the base level. Therefore, the compressive strength of mortar mainly depends on the cement strength and the amount of cement, but has little relationship with the water cement ratio. Mortar strength calculation formula is as follows:

$$f_{m,cu} = A f_{ce} \frac{Q_c}{1000} + B \tag{4-48}$$

Where  $f_{m,cu}$——28 day compressive strength of mortar (MPa);

$f_{ce}$——the measured strength value of cement (MPa);

$Q_c$——amount of cement per cubic meter of mortar (kg/m$^3$);

$A$ and $B$——characteristic coefficient of mortar, $A = 3.03$, $B = -15.09$.

Scan the QR code to get the related information of the testing for masonry mortar.

### III. Mix ratio design of masonry mortar

Different from concrete, the design of mortar mix ratio follows the principle of incompact filling. According to the regulations of *Design Regulations for Masonry Mortar Mix Ratio* (JGJ 98—2010), the design process of masonry mortar mix ratio is briefly described as follows:

i. Mix ratio design of cement mixed mortar

1. Determine the mortar trial strength

Masonry mortar shall have a 95% guarantee rate, and the mortar trial strength should be calculated according to the following formula:

$$f_{m,0} = k f_2 \tag{4-49}$$

Where  $f_{m,0}$——the mortar trial strength, which should be accurate to 0.1 MPa;

   $f_2$——the mortar trial strength grade value, which should be accurate to 0.1 MPa;

   $k$——coefficient. The value is listed in Table 4-27.

Table 4-27  Selection values of coefficient of the mortar trial strength

| Construction level | Standard deviation of intensity $\sigma$/MPa | | | | | | | k |
| --- | --- | --- | --- | --- | --- | --- | --- | --- |
| | M5 | M7.5 | M10 | M15 | M20 | M25 | M30 | |
| Good | 1.00 | 1.50 | 2.00 | 3.00 | 4.00 | 5.00 | 6.00 | 1.15 |
| General | 1.25 | 1.88 | 2.50 | 3.75 | 5.00 | 6.25 | 7.50 | 1.20 |
| Poor | 1.50 | 2.25 | 3.00 | 4.50 | 6.00 | 7.50 | 9.00 | 1.25 |

The standard deviation of strength of mortar on site shall be determined in accordance with the following provisions:

(1) When statistical information is available, it should be calculated as follows:

$$\sigma = \sqrt{\frac{\sum_{i=1}^{n} f_{m,i} - n\overline{f}_m}{n-1}} \quad (4\text{-}50)$$

Where  $f_{m,i}$——the strength of group $i$ of the same mortar in the statistical period(MPa);

   $\overline{f}_m$——the average strength of group n specimens of the same type of mortar in the statistical period(MPa);

   $n$——the total number of mortar specimens of the same variety within the statistical period, $n \geq 25$.

(2) When there is no recent unified capital and material, the standard deviation $\sigma$ of mortar on-site strength can be obtained according to Table 4-28.

Table 4-28  Selection values of standard deviation of mortar strength/MPa

| Construction level | Mortar strength grade | | | | | |
| --- | --- | --- | --- | --- | --- | --- |
| | M2.5 | M5 | M7.5 | M10 | M15 | M20 |
| Good | 0.50 | 1.00 | 1.50 | 2.00 | 3.00 | 4.00 |
| General | 0.62 | 1.25 | 1.88 | 2.50 | 3.75 | 5.00 |
| Poor | 0.75 | 1.50 | 2.25 | 3.00 | 4.50 | 6.00 |

2. Calculation of cement consumption

The amount of cement per cubic meter should be calculated according to the following formula:

$$Q_c = \frac{1000(f_{m,0} - B)}{A f_{ce}} \quad (4\text{-}51)$$

Where  $Q_c$——amount of cement per cubic meter of mortar(kg/m³);

   $f_{m,0}$——the mortar trial strength(MPa);

   $f_{ce}$——the measured strength value of cement(MPa);

   $A$ and $B$——characteristic coefficient of mortar, $A = 3.03$, $B = -15.09$.

When the measured strength value of cement cannot be obtained, the following formula can be used to calculate:

$$f_{ce} = \gamma_c f_{ce,k} \tag{4-52}$$

Where  $f_{ce,k}$——cement dosage per cubic meter of mortar($kg/m^3$);

$\gamma_c$——the surplus coefficient of the cement strength grade value, which should be determined according to the actual statistical data. If there is no statistical data, it can be 1.00

3. The amount of admixture of cement mixed mortar

The amount of admixture of cement mortar should be calculated according to the following formula:

$$Q_D = Q_A - Q_c \tag{4-53}$$

Where  $Q_D$——the dosage of admixture per cubic meter of mortar is accurate to 1kg. The consistency of stone plaster and clay paste is $(120 \pm 5)$mm;

$Q_c$——the amount of cement per cubic meter of mortar, which is accurate to 1kg;

$Q_A$——the total amount of cement and admixture per cubic meter of mortar, accurate to 1kg. It should be between 300 and 350kg.

When the lime paste is of other consistencies, conversion shall be carried out according to Table 4-29。

Table 4-29  Conversion coefficient of different consistency of plaster

| Lime paste consistency (mm) | 120 | 110 | 100 | 90 | 80 | 70 | 60 | 50 | 40 | 30 |
|---|---|---|---|---|---|---|---|---|---|---|
| Conversion factor | 1.00 | 0.99 | 0.97 | 0.95 | 0.93 | 0.92 | 0.90 | 0.88 | 0.87 | 0.86 |

4. Determine the amount of sand

The addition of water, cementitious material and admixture in the mortar basically fills the void in the sand. Therefore, the amount of sand in each cubic meter of mortar $Q_s$($kg/m^3$) should be calculated according to the bulk density value of dry state(moisture content is less than 0.50%).

5. Determination of water consumption

The water consumptionfor per cubic meter of mortar $Q_w$($kg/m^3$) can be selected according to the mortar consistency and other requirements. Mixed mortar can be 240~310kg.

Note: (1) water consumption in mixed mortar does not include water that in limestone paste or clay paste; (2) when using fine sand or coarse sand, the upper or lower limits of water consumption are taken respectively; (3) when the consistency is less than 70mm, the water consumption can be less than the lower limit; (4) in the hot or dry season of the construction site, the water consumption can be increased.

ii. Choose the mix proportion of cement mortar

Due to the amount of cement is generally less in calculating the mix proportion of cement mortar. The main reason for this unreasonable situation is that the strength of cement is too high, but the strength of mortar is too low in calculation. Therefore, the specification stipulates that the proportion of cement mortar materials can refer to American ASTM and British BS standards, using direct check Table 4-30 selection.

Table 4-30  Cement mortar material consumption per cubic meter, kg/m³

| Strength grade | Cement consumption/kg | Sand consumption/kg | Water consumption/kg |
|---|---|---|---|
| M5 | 200 ~ 230 | Bulk density value of 1m³ sand | 270 ~ 330 |
| M7.5 | 230 ~ 260 | | |
| M10 | 260 ~ 290 | | |
| M15 | 290 ~ 330 | | |
| M20 | 340 ~ 400 | | |
| M25 | 360 ~ 410 | | |
| M30 | 430 ~ 480 | | |

Note: 1. The cement strength grade in this table is 32.5, and the lower limit of cement dosage is appropriate if it is more than 32.5;
2. Choose cement dosage reasonably according to the construction level;
3. The upper or lower limit of water consumption shall be taken respectively, when fine sand or coarse sand is used;
4. When the consistency is less than 70mm, the water consumption can be less than the lower limit;
5. In the hot or dry season of the construction site, water consumption can be increased;
6. The mortar trial strength should be calculated according to the following formula: $f_{m,0} = kf_2$.

The dosage of cement should be chosen reasonably according to the strength grade and construction level of cement. Generally, when the strength level of cement is higher or the construction management level is higher, the dosage of cement is selected as the lowest value. The water consumption is selected according to the thickness of sand, mortar consistency and climate conditions. When the sand is coarser, the consistency is smaller or the climate is humid, the water consumption is selected as the low value. And vice versa.

## III. Trial matching, adjustment and determination of the mix proportion

The mortar used in calculation or trial test should be the material actually used in the project. The mixing time is from the end of feeding, cement mortar and cement mixed mortar shall not be less than 120s, cement fly ash mortar and mortar with admixture shall not be less than 180s. The consistency and delamination of the mixture shall be measured when the mixture ratio is calculated or looked up in the table. When the mortar cannot meet the requirements, the amount of material should be adjusted until it meets the requirements. The mix proportion at this time is the benchmark mix proportion of mortar at the trial test.

In order to measure the mortar strength within the range of design requirements, at least 3 different mix proportion should be used in the trial mixing, one of which is the benchmark mix proportion, and the cement dosage of the other two mix proportion should be increased and decreased by 10% respectively according to the benchmark mix proportion. Under the condition of ensuring consistency and layered degree qualified, water or mixed feed amount can be adjusted correspondingly. According to the *Basic Performance Test Method for Building Mortar* (JGJ/T 70—2009), specimens are made to measure the strength of mortar. The mix proportion that meets the requirements of mortar trial strength and the least amount of cement is selected as the mix proportion of mortar.

The mortar mix proportion is expressed as the proportion of material consumption:

Cement : admixture : sand : water $= Q_c : Q_D : Q_S : Q_w$

Or, cement : admixture : sand : water $= 1 : \dfrac{Q_D}{Q_c} : \dfrac{Q_S}{Q_c} : \dfrac{Q_w}{Q_c}$

## Questions for review

1. How to select cement when formulating concrete?

2. What is the gradation of aggregate? How is it expressed? What is the chatacteristics of a good gradation of aggregate?

3. If the finesness moduli of two sands are same, then whether they have the same gradation or not? If they have the same gradation, then whether they have the same fineness moduli?

4. What are the properties of slag and fly ash, respectively?

5. Please describe the mechanism of water reducer, and describe the technical and economical effects of the admixing of water reducer in concrete.

6. Please dexcribe the concept and the measured methods of workability of concrete mixture.

7. What is the slump loss over time of concrete? What is the reason of it?

8. What are the main factors that affect the workability of concrete mixture? What are the measures to improve the workability of concrete mixture?

9. What is the fixed water demand law?

10. What is the sand ratio of concrete? What is the reasonable sand ratio of concrete? What is the meaning of adopting the reasonable sand ratio to concrete?

11. How is the strength grade of concrete determined?

12. What are the main factors that affect the strength of concrete? What are the main measures that can increase the strength of concrete?

13. What is the classification of shrinkage of concrete? What is the reason for each type of shrinkage?

14. What is the mechanism of the cracking damage caused by temperature deformation during the early-age of hardening of concrete?

15. What is the creep of concrete? What are the effects of creep on the concrete structure?

16. What is the impermeability of concrete? What are the measures that can improve the impermeability of concrete?

17. What is the resistance of concrete to freezing-thawing cycles? What are the main factors that influence the freezing-thawing resistance of concrete? Which type of chemical admixture can significantly improve the freezing-thawing resistance of concrete?

18. What is the carbonization of concrete? What is the influence of carbonization on the performance of reinforced concrete?

Chapter IV  Ordinary Concrete and Construction Mortar

19. What are the basic pricinples and bases of the mix proportion design of concrete?

20. If the workability of the trial concrete using the calculated mix proportion fails to meet the requirements, what measures can be taken to adjust the workability of concrete?

21. What are the differences and similarities between the technical requirements for fresh mortar and concrete mixtures?

22. What aspects does the workability of fresh mortar include? How is it measured? What impact does poor workability of mortar have on engineering application?

23. What are the requirements of component materials for masonry mortar? Why add admixtures or plasticizers?

24. What are the differences in the factors affecting the strength of masonry mortar used for water absorbing base and non-water absorbing base? How to calculate?

25. What are the requirements for the composition materials and technical properties of plastering mortar and masonry mortar? Why?

26. Cement-lime mortar is used in a project to build sintered ordinary clay bricks, and the strength grade of mortar is required to be M10. Slag Portland cement of strength grade 32.5 and 42.5 is available on site. It is known that the bulk density of the cement used is 1280kg/$m^3$; The water content of medium sand is 1% ~ 3%, and the bulk density is 1550kg/$m^3$; The apparent density of lime paste is 1350kg/$m^3$. The construction level is good. Try to calculate the mortar mix proportion.

27. What are the main finishes of decorative mortars?

28. Try to describe the significance of using ready-mixed mortar.

# Chapter V  Asphalt and Asphalt Mixture

Asphalt is a black or dark brown amorphous organic material composed of polymer hydrocarbons and their derivatives, which is insoluble in water but almost completely soluble in carbon disulfide. Asphalt mixture is a material that uses asphalt as a cementing meterial to bond coarse aggregate, fine aggregate, filler and so on.

Asphalt and asphalt mixture are indispensable building materials in civil engineering construction. They are widely used in construction, highway, bridge and waterproof engineering. Asphalt mixture using asphalt as a cementing meterial has become the main material of high-grade pavement. Not only has it good mechanical properties, skid resistance and waterproof, but also it is smooth, comfortable quieter and so on, and it can be layered thickening and easy to repair; but there are also shortcomings such as easy aging and poor temperature sensitivity.

## Section I  Asphalt Material

### I. Classification of asphalt

For the naming and classification of asphalt material, the world has not yet reached a unified understanding. The naming and classification of asphalt in China are briefly described as follows:

Asphalt can be divided into two categories: land-bitumen and tar asphalt according to the way it is obtained in nature. Land-bitumen is a kind of asphalt material that exists naturally or is processed by petroleum refining. According to its source, it can be divided into natural asphalt and petroleum asphalt. Tar asphalt is a product obtained by reprocessing tar obtained by retorting various organic matter, including coal pitch, shale pitch, etc.

The above types of asphalt can be summarized as follows:

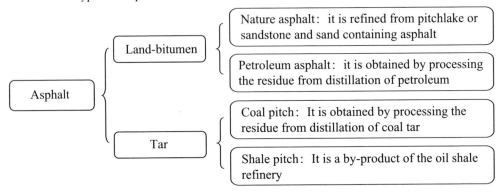

## II. Petroleum asphalt

i. Component and Structure

Petroleum asphalt is a residue produced by distillation, oxygen blowing and blending of petroleum. It is mainly semi-solid viscous substance of hydrocarbons soluble in carbon disulfide. Petroleum asphalt is a mixture of a variety of hydrocarbons and their non-metallic (oxygen, sulfur, nitrogen) derivatives. Its molecular expression formula is $C_n H_{2n+a} O_b S_c N_d$. The chemical composition is mainly carbon (80% ~ 87%), hydrogen (10% ~ 15%), followed by non-hydrocarbon elements, such as oxygen, sulfur, nitrogen ( < 3% ). In addition, it also contains some trace metal elements, such as nickel, vanadium, iron, manganese, magnesium, sodium, etc. But the content of metal elements is very small, about a few million to tens of millions.

Due to the complexity of asphalt chemical composition and structure, although many chemists have been devoted to this research for many years, the relationship between asphalt element content and engineering performance is still not directly obtained. At present, the research on asphalt composition and structure mainly focuses on component theory, colloid theory and polymer solution theory.

1. Component analysis

Chemical composition analysis is to separate asphalt into several chemical components with similar chemical properties and certain connection with its engineering performance, which are called components. The current *Highway Engineering Asphalt and Asphalt Mixture Test Procedures* (JTG E20—2011) in China stipulates two methods of three-component analysis and four-component analysis (scanning the QR code to learn more).

(1) Three-component analysis

The three-component analysis method of petroleum asphalt is to separate petroleum asphalt into three components: oil, resin and asphaltene (Table 5-1). Because China is rich in paraffin base and intermediate base asphalt, and wax is often contained in oil, so the separation of oil and wax should also be carried out in the analysis. Since this component analysis method combines selective dissolution and selective adsorption, it is also called dissolution-adsorption method. The advantage of three-component analysis is that the component boundary is very clear, and the component content can explain its engineering performance to a certain extent. The main disadvantage is that the analysis process is complex and the analysis time is long.

Oil gives asphalt fluidity, and the amount of oil content directly affects the flexibility, crack resistance and construction difficulty of asphalt. Oil can be converted into resin or even asphaltenes under certain conditions.

The resin is divided into neutral resin and acidic resin. The neutral resin makes the asphalt have certain plasticity, fluidity and adhesion. The rise of its content increases the adhesion and extensibility of the asphalt. In addition to the neutral tree finger, asphalt resin also contains a small amount of acid resin, namely asphalt acid and asphalt anhydride, which are resin like dark brown viscous substances and are the products of oil oxidation. It is solid or semi-solid, acidic, capable of alkali saponification, easily soluble in alcohol and chloroform, but hardly soluble in petroleum ether and benzene. Acid resin is the most active component in asphalt. It can improve the wettability of asphalt to mineral materials, especially the adhesion to carbonate rocks, and increase the emulsification of asphalt.

Asphaltenes determines the cohesion, viscosity and temperature stability of asphalt, as well as the hardness and softening point of asphalt. With the increase of asphaltenes content, the viscosity and cohesion of asphalt increase, and the hardness and temperature stability increase.

Table 5-1  Component properties of petroleum asphalt by three-component analysis

| Component | Apperance feature | Average relative molecular mass | C/H | Content (%) | Physical and chemical characteristics |
|---|---|---|---|---|---|
| Oil | Pale yellow transparent liquid | 200 ~ 700 | 0.5 ~ 0.7 | 45 ~ 60 | It is almost soluble in most organic solvents and has optical activity. It is often found to emit fluorescence. Its relative density is about 0.7 ~ 1.00 |
| Resin | Reddish-brown viscous semisolid | 800 ~ 3000 | 0.7 ~ 0.8 | 15 ~ 30 | It has high temperature sensitivity. Its melting point is lower than 100℃, and its relative density is greater than 1.0 ~ 1.1 |
| Asphaltenes | Dark brown solid particles | 1000 ~ 5000 | 0.8 ~ 1.0 | 5 ~ 30 | It is not melted but carbonized when heated, and its relative density is 1.1 ~ 1.5 |

(2) Four-component analysis

The four-component analysis method was first proposed by Colbert, which separated asphalt into saturate, naphthene-aromatics, polar-aromatics and asphaltenes. Later, the above four components were called saturate, aromatics, resin and asphaltenes, so this method is also called SARA method. The current four-component analysis in China separates asphalt into asphaltenes (At), saturate(S), aromatics(A) and resin(R). There are the properties of each component obtained by the four-component analysis method of petroleum asphalt in Table 5-2.

Table 5-2  Component properties of petroleum asphalt by four-component analysis

| Component | Apperance feature | Average relative density | Average relative molecular mass | Main chemical composition |
|---|---|---|---|---|
| Saturate | Colorless liquid | 0.89 | 625 | Alkanes, cycloalkanes |
| Aromatics | Yellow to red liquid | 0.99 | 730 | Aromatic hydrocarbons, S-containing derivatives |
| Resin | Brown viscous liquid | 1.09 | 970 | Polycyclic structure, containing S, O, N derivatives |
| Asphaltenes | Dark brown solid particles | 1.15 | 3400 | Condensation structure, containing S, O, N derivatives |

We study the influence of each component on the properties of asphalt by using four-component analysis method. According to Colbert's research, the increase of saturated content can reduce the consistency of asphalt, that is, the penetration increases; the increase of resin content can increase the ductility of asphalt. With the presence of saturates, the increase of asphaltene content can make asphalt obtain low temperature sensitivity; the increase of resin and asphaltene content can increase the viscosity of asphalt, that is, the penetration is reduced.

2. Colloidal structure

Because the composition of asphalt can not fully reflect the properties of asphalt materials, the properties of asphalt are also closely related to its structure. Modern colloidal theory suggests that the colloidal structure of asphalt is solid ultra-fine particles of asphalt as a dispersed phase, usually a number of asphalt together, they adsorbed a polar semi-solid colloid, and the formation of micelles. Due to the peptization of the peptizer-gum, the micelles are peptized and dispersed in a dispersion medium composed of liquid aromatics and saturated components to form a stable colloid.

According to the chemical composition and relative content of each component in asphalt, it can be divided into the following three types.

(1) Sol-type structure

When the asphaltene molecular weight and content in asphalt is lower (e. g. less than 10%) and there is a certain amount of rein with a higher aromaticity, the micelle can be completely peptised and dispersed in the media of aromatic and saturated phenols. In this case, the micelles are far apart, the attraction between them is small (even no attraction), and the micelles can move freely within the permitted range of viscosity of the dispersion medium. This colloidal structure of asphalt is called sol-type asphalt [Figure 5-1(a)]. The characteristics of sol-type asphalt are good fluidity and plasticity, strong self-healing ability after cracking, strong sensitivity to temperature, poor stability to temperature, and excessive temperature will flow. Generally, most of the straight-run asphalt belongs to sol-type asphalt.

(2) Solution-gel structure

The asphaltene content in the asphalt is appropriate (e. g. between 15% and 25%), and there are a large number of resin with higher aromaticity. The number of micelles formed in this way increases, the concentration of micelles in the colloid increases, and the distance between the micelles is relatively close [Figure 5-1(b)], and there is a certain attraction between them. This is a structure between solution and gel called a solution-gel structure. The asphalt of this structure is called solution-gel asphalt. The asphalt used to build modern high-grade asphalt road should belong to this type of colloidal structure. Typically, naphthenic heavy oil straight-run asphalt or semi-oxidized asphalt and restructured (newly prepared) solvent asphalt according to the requirements, etc. , are often able to meet such colloidal structure. This kind of asphalt has low temperature sensitivity at high temperature and good deformation ability at low temperature.

(3) Gel-type structure

The asphaltene content in asphalt is very high (e. g. >30%), and there is a considerable amount of aromatic resin to form micelles. In this way, the concentration of micelles in asphalt increases to a large extent, and the mutual attraction between them is enhanced, so that the micelles are close together to form a spatial network structure. At this time, the liquid aromatic aroma and saturated aroma become dispersed phases in the network of micelles, and the continuous micelles become dispersion media Figure 5-1c. The asphalt with colloidal structure is called gel asphalt. It is characterized by high elasticity and viscosity, low temperature sensitivity, poor self healing ability after cracking, and low fluidity and plasticity. In engineering performance, it has good temperature stability and poor low-temperature deformation capacity.

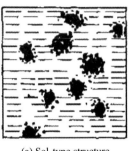    

(a) Sol-type structure  (b) Solution-gel structure  (c) Gel-type structure

**Figure 5-1  Colloidal structure of asphalt**

ii. Technical property

1. Needle penetration

The relative viscosity of viscous petroleum asphalt is represented by the needle penetration measured by the penetrometer, as shown in Figure 5-2. It reflects the viscosity of petroleum asphalt. The smaller the penetration value, the greater the viscosity. The penetration of viscous petroleum asphalt is the depth of penetration of a standard needle with a specified weight of 100g into the sample at a specified temperature of 25℃ for 5s after a specified time, represented in 1/10mm, with the symbol $P_{(25℃,100g,5s)}$.

2. Softening point

The softening point of asphalt is an important index reflecting the temperature sensitivity of asphalt. Since there is a certain change interval from the solid to the liquid of asphalt materials, it is specified that a certain state is the starting point for the transition from the solid to the viscous flow state(or a specified state), and the corresponding temperature is called the asphalt softening point.

The current Highway Engineering Asphalt and Asphalt Mixture Test Procedures(JTG E20—2011) in China adopt the global softening point method( Figure 5-3). This method is to inject the viscous asphalt sample into a steel ring with an inner diameter of 18.9mm, place a 3.5g steel ball on the ring, and heat the steel ring at the specified heating speed(5℃/min). The asphalt sample gradually softens until the temperature at which the asphalt falls 25.4mm because of the load of the steel ball is called the softening point. According to the existing research, the viscosity of asphalt at softening point is about 1200Pa. s, or equivalent to that the softening point is 800(0.1mm). Therefore, the softening point can be considered as an artificial "isoviscosity temperature".

## 3. Ductility

Ductility is the property that petroleum asphalt deforms without destroying (cracks or breaks) under external force, and remains the same shape after deformation except for external force. It reflects the ability of asphalt to withstand plastic deformation under force.

The ductility of petroleum asphalt is related to its components. When the content of resin in petroleum asphalt is more and the content of other components is appropriate, the plasticity is larger. The factors affecting the plasticity of asphalt are temperature and asphalt film thickness. The higher the temperature, the greater the ductility, and the thicker the film, the higher the plasticity. On the contrary, the thinner the film, the worse the plasticity. When the film is very thin, the plasticity is close to disappear, that is, it is close to elasticity.

At room temperature, asphalt with good ductility can also rely on its viscoplasticity to heal itself when cracks occur. Therefore, the ductility also reflects the self-healing ability of asphalt after cracking. Asphalt can be used to produce flexible waterproof materials with good performance largely because of the ductility of asphalt. The ductility of asphalt can absorb impact and vibration loads to a certain extent, and reduce the noise during friction, so asphalt is an excellent pavement material.

Ductility is usually used as the ductility index. The ductility test method is to make the asphalt sample into an 8-shaped standard test piece (the minimum sectional area is $1 cm^2$), and the length (cm) of the asphalt sample when it is broken at the specified tensile speed and temperature is called ductility, as shown in Figure 5-4. The test temperatures commonly used are 15℃ and 10℃.

**Figure 5-2   Asphalt penetrometer**

**Figure 5-3   Asphalt softening point meter**

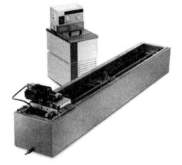

**Figure 5-4   Ductility meter and ductility test die**

The needle penetration, softening point and ductility mentioned above are the most commonly used empirical indicators for evaluating the engineering performance of viscous petroleum asphalt. Therefore, they are collectively referred to as the "three major indicators". Scan the QR code to learn the relevant experimental content of petroleum asphalt performance indicators.

4. Adhesion

Adhesion is the the interface bonding performance and anti-stripping performance of asphalt and other materials (mainly aggregate). The adhesion of asphalt and aggregate directly affects the service quality and durability of asphalt pavement, so it is an important indicator to evaluate the technical performance of road asphalt.

The most commonly used methods to evaluate the adhesion of asphalt and aggregate are boiling method and water immersion method. According to the current *Highway Engineering Asphalt and Asphalt Mixture Test Procedures* (JTG E20—2011) in China the adhesion test of asphalt and coarse aggregate is determined by the maximum particle size of asphalt mixture. The water boiling method is used for those larger than 13.2mm. Water immersion method shall be adopted for those less than (or equal to) 13.2mm. The water boiling method is to select five regular aggregates with a particle size of 13.2~19mm whose shape close to the cube. After being coated with asphalt, they are boiled in distilled water for 3min. The adhesion between asphalt and aggregate is evaluated by dividing them into five grades according to the peeling of asphalt film. The water immersion method is to mix 100g 9.5~13.2mm aggregate and 5.5g asphalt under specified temperature to prepare asphalt aggregate mixture. After cooling, it is immersed in 80℃ distilled water for 30min, and then the adhesion between asphalt and aggregate is evaluated according to the percentage of spalling area.

5. Atmospheric stability

Under the combined effect of sunlight, air and heat, the components of asphalt will constantly change. Low molecular compounds will gradually change into high molecular substances, that is, oil and resin will gradually decrease, while asphaltene will gradually increase. Therefore, with the development of time, the fluidity and plasticity of petroleum asphalt gradually decrease, and the brittleness gradually increases until brittle fracture. This process is called aging of petroleum asphalt. Consequently, the atmospheric stability of asphalt can be explained by its anti-aging performance.

According to the current *Highway Engineering Asphalt and Asphalt Mixture Test Procedures* (JTG E20—2011) in China, the anti-aging performance of petroleum asphalt is evaluated by the percentage of mass loss of asphalt samples before and after heating and evaporation, the needle penetration ratio and the ductility after aging. The measurement method is as follows: first we measure the mass and penetration of asphalt sample, then place the sample in the oven to heat and evaporate at 163℃ for 5 hours. We measure the mass and penetration after cooling, then calculate the percentage of evaporation loss mass in the original mass, which is called evaporation loss percentage. The ratio of the measured penetration after aging to the original penetration is called the penetration ratio. Meanwhile, the ductility after aging is measured. After asphalt aging, if the percentage of mass loss of asphalt is less, and the penetration ratio and ductility of asphalt is greater, atmospheric stability of asphalt is better.

6. Construction safety

Viscous asphalt must be heated when it is used. When it is heated to a certain temperature, the volatile oil vapor in the asphalt material forms a mixed gas with the surrounding air. This mixed gas is easy to flash when it meets the flame. If heating continues, the oil vapor and saturation will increase. Because the mixed gas composed of oil vapor and air is very easy to burn when encountering the flame and cause a fire, it is necessary to measure the asphalt heating flash and combustion temperature, namely the so-called flash point and ignition point.

The flash point and ignition point indicate the possibility of fire or explosion caused by asphalt, which is related to the safety of transportation, storage, heating and use. Generally, the temperature of petroleum asphalt during cooking is 150℃. Therefore, the flash point of asphalt shall be generally controlled to be greater than 230℃. However, for the sake of safety, the asphalt shall be isolated from the flame when heated.

iii. Technical standard

According to the diffirent properties of petroleum asphalt, the asphalt is divided into diffirent types and grades by select appropriate technical standard, so as to facilitate the selection of asphalt materials. At present, petroleum asphalt is mainly divided into three categories: road petroleum asphalt, construction petroleum asphalt and ordinary petroleum asphalt, among which road petroleum asphalt is the main type.

1. Classification of road petroleum asphalt

Road petroleum asphalt is divided into three grades : Grade A, Grade B and Grade C, and their respective application scope complies with the provisions in Table 5-3.

Table 5-3 The application scopes of road petroleum asphalt

| Grade of asphalt | Application scope |
| --- | --- |
| A | It is applicable to all classes of highways, and applicable to all occasions and layers. |
| B | 1. It is applicable to the asphalt lower surface layer and lower layers of expressways and Class I highways, and all layers of Class II and lower highways.<br>2. It is used as matrix asphalt of modified asphalt, emulsified asphalt, modified emulsified asphalt and diluted asphalt. |
| C | It is applicable to all layers of Class III and below highway. |

2. The numbers of road petroleum asphalt

According to penetration, road petroleum asphalt is divided into sevennumbers: 160, 130, 110, 90, 70, 50 and 30. At the same time, corresponding requirements are put forward for the technical indexes of each grade asphalt, such as ductility, softening point, flash point, wax content, film heating test, etc. Specific requirements are shown in Table 5-4.

Civil engineering materials

**Table 5-4　The technical requirements of road petroleum asphalt**

| Index | Unit | Grade | 160[4] | 130[4] | 110 | 90 | 70[3] | 50 | 30[4] | Test method[1] |
|---|---|---|---|---|---|---|---|---|---|---|
| Needle penetration (25°C, 5s, 100g) | 0.1mm | | 140~200 | 120~140 | 100~120 | 80~100 | 60~80 | 40~60 | 20~40 | T 0604 |
| Applicable climate zones[6] | | | Note[4] | Note[4] | 2-1, 2-2, 3-2 | 1-1, 1-2, 1-3, 2-2, 2-3 | 1-3, 1-4, 2-2, 2-3, 2-4 | 1-4 | Note[4] | Appendix A[5] |
| Needle penetration PI[2] | | A | | | | −1.5 ~ +1.0 | | | | T 0604 |
| | | B | | | | −1.8 ~ +1.0 | | | | |
| Softening point (R&B) no less than | °C | A | 38 | 40 | 43 | 45 | 46 | 49 | 55 | T 0606 |
| | | B | 36 | 39 | 42 | 43 | 44 | 46 | 53 | |
| | | C | 35 | 37 | 41 | 42 | 43 | 45 | 50 | |
| 60°C dynamic viscosity[2] no less than | Pa·s | A | — | 60 | 120 | 160 | 180 | 200 | 260 | T 0620 |
| 10°C ductility[2] no less than | cm | A | 50 | 50 | 40 | 45/30/20/20/20 | 25/20/15/15/15 | 15 | 10 | T 0605 |
| | | B | 30 | 30 | 30 | 30/20/15/15/15 | 20/15/10/10/10 | 10 | 8 | |
| 15°C ductility no less than | cm | A, B | 80 | 80 | 60 | 100 | 40 | 80 | 50 | |
| | | C | | | | | | 30 | 20 | |
| Wax content (distillation method) no greater than | % | A | | | | 2.2 | | | | T 0615 |
| | | B | | | | 3.0 | | | | |
| | | C | | | | 4.5 | | | | |
| Flash point no less than | °C | | 230 | | 260 | 245 | | | | T 0611 |
| Solubility no less than | % | | | | | 99.5 | | | | T 0607 |
| Density (15°C) | g/cm³ | | | | | Measured records | | | | T 0603 |

## Chapter V  Asphalt and Asphalt Mixture

continued

| Index | Unit | Grade | The number of asphalt | | | | | | | Test method[1] |
|---|---|---|---|---|---|---|---|---|---|---|
| | | | 160[4] | 130[4] | 110 | 90 | 70[3] | 50 | 30[4] | |
| | | | After TFOT (or TFOT)[5] | | | | | | | |
| Mass variation no greater than | % | | ±0.8 | | | | | | | T 0610 or T 0609 |
| Residual penetration ratio no less than | % | A | 48 | 54 | 55 | 57 | 61 | 63 | 65 | T 0604 |
| | | B | 45 | 50 | 52 | 54 | 58 | 60 | 62 | |
| | | C | 40 | 45 | 48 | 50 | 54 | 58 | 60 | |
| Residual ductility (10℃) no less than | cm | A | 12 | 12 | 10 | 8 | 6 | 4 | — | T 0605 |
| | | B | 10 | 10 | 8 | 6 | 4 | 2 | — | |
| Residual ductility (15℃) no less than | cm | C | 40 | 35 | 30 | 20 | 15 | 10 | — | T 0605 |

Note:

[1] The test method shall be in accordance with the current *Test Procedures for Asphalt and Asphalt Mixture of Highway Engineering* (JTG E20—2011). The correlation coefficient of penetration relationship of 5 temperatures in arbitration test, which are used for calculating PI, shall not be less than 0.997.

[2] With the consent of the owner, the PI value, 60℃ dynamic viscosity and 10℃ ductility in the table can be used as selective indicators or not as construction quality inspection indexes.

[3] For No. 70 asphalt, the supplier can be required to provide asphalt with a penetration range of 60~70 or 70~80 as required, and for No. 50 asphalt, the supplier can be required to provide asphalt with a penetration range of 40~50 or 50~60.

[4] No. 30 asphalt is only applicable to asphalt stabilized base course. No. 130 and No. 160 asphalt are generally used as emulsified asphalt, diluted asphalt and matrix asphalt of modified asphalt, except that they can be directly used on medium and low-grade highways in cold regions.

[5] The aging test shall be subject to TFOT, which can also be replaced by RTFOT.

[6] See Appendix A of *Technical Specification for Construction of Highway Asphalt Pavement* (JTG F40—2004) for climate zoning. Scan the following QR code for more information.

## III. Other types of asphlt

ⅰ. Emulsified asphalt

Emulsified asphalt is a uniform and stable emulsion formed by heating viscous asphalt to flow state and forming microdroplets (particle size of $2 \sim 5 \mu m$) by mechanical force, which disperses in water containing emulsifier-stabilizer. It is also known as asphalt emulsion and referred to as emulsion. It is mainly used for road pavement construction, and its main components are asphalt, emulsifier, stabilizer and water.

1. Asphalt

Asphalt is the main material of emulsified asphalt. The quality of asphalt directly determines the performance of emulsified asphalt. When selecting asphalt for emulsified asphalt, its easy emulsification should be considered first. The emulsifiability of asphalt is closely related to its chemical structure. For the purpose of engineering application, it can be considered that the emulsifiability is related to the content of asphaltic acid in asphalt. It is generally believed that the asphalt with asphaltic acid whose total amount is greater than 1% is very easy to form emulsified asphalt with general emulsifier and general process. In addition, the asphalt with the same oil source and process is easy to form lotion with a larger penetration, but the choice of penetration should be determined by the use of emulsified asphalt in pavement engineering.

2. Emulsifier

Emulsifier is the key material for the formation of emulsified asphalt. From the perspective of chemical structure, it is a kind of surfactant. Its molecular structure contains an "amphiphilic" molecule, one part of which is hydrophilic, while the other part is lipophilic. The lipophilic part is generally composed of hydrocarbon atomic groups, especially long-chain alkyl groups, with little structural difference. There are many kinds of hydrophilic groups, and their structures are quite different. Therefore, the classification of emulsifiers is based on the structure of hydrophilic groups, which can be divided into ionic and non-ionic types according to whether their hydrophilic groups are ionized in water. Ionic emulsifiers are divided into anionic emulsifiers, cationic emulsifiers and zwitterionic emulsifiers in the light of their ionic electrical properties. In addition, with the development of emulsified asphalt, in order to meet various special requirements, many compound emulsifiers with more complex chemical structures have been derived.

3. Stabilizer

In order to make the emulsion have good storage stability and stability under the mechanical action of spraying or mixing during construction, an appropriate amount of stabilizer can be added when necessary. General stabilizers can be divided into organic stabilizers and inorganic stabilizers. Common organic stabilizers include polyvinyl alcohol, polyacrylamide, sodium methylcellulose, dextrin, MF waste liquid, etc. This kind of stabilizer can improve the storage stability and construction stability of lotion. Common inorganic stabilizers include calcium chloride, magnesium chloride, ammonium chloride, chromium chloride and so on.

ⅱ. Mixing of asphalt

In addition to the modification of asphalt, the required specific properties(such as the required softening point conditions) can also be obtained by mixing two or three kinds of asphalt. It should be noted that the blending should follow the principle of homology, that is, only the same petroleum asphalt or coal tar asphalt can be blended.

The mixing proportion of the two asphalts can be calculated by using the following formula:

$$Q_1 = \frac{T_2 - T_1}{T - T_1} \times 100\% \tag{5-1}$$

$$Q_2 = 100 - Q_1 \tag{5-2}$$

Where   $Q_1$—the amount of softer asphalt, %;

$Q_2$—the amount of softer asphalt, %;

$T_1$—the softening point of softer asphalt, ℃;

$T_2$—the softening point of softer asphalt, ℃;

$T$—the softening point of asphalt required, ℃.

The estimated mixing proportion and its adjacent proportion ($\pm 5\% \sim \pm 10\%$) shall be used for trial mixing(mixing and boiling uniformly). The softening point of asphalt after mixing shall be measured, and then the mixing proportion-softening point relationship curve shall be drawn to determine the required mixing proportion from the curve. Similarly, the penetration index can also be estimated by the above method.

## Ⅳ. Asphalt aging and modification

ⅰ. Asphalt aging

In the process of asphalt use and production, a series of changes occur to the internal chemical structure and colloidal structure of asphalt because of the function of temperature, UV, oxygen, and so on, which will lead to changes in macro performance. The above process is called asphalt aging.

Due to the aging of asphalt, the physical and chemical properties of asphalt change. The general rule is that the penetration and the ductility decreases, and the softening point and brittleness point increase, which shows that the asphalt becomes hard, brittle, and the extensibility decreases, resulting in cracks, looseness and other damages to the asphalt. The changes of physical mechanical properties of asphalt after aging are shown in Table 5-5.

Table 5-5   Examples of technical properties of aging asphalt and recycled asphalt

| The name of asphalt | Technical properties | | | |
|---|---|---|---|---|
| | Needle penetration(1/10mm) | Ductility(cm) | Softening point(℃) | Brittle point(℃) |
| Original asphalt | 106 | 73 | 48 | -6 |
| Aging asphalt | 39 | 23 | 55 | -4 |
| Recycle asphalt | 80 | 78 | 49 | -10 |

In practical application, people require that asphalt should have as good durability as possible, and the aging speed should be as small as possible. In this reason, the requirements for asphalt durability are put forward. Durability is a very important comprehensive index of asphalt performance. Due to the deterioration of performance caused by aging, the projects using asphalt materials will undergo large-scale renovation after a certain service life. Therefore, how to improve the durability of asphalt and extend the service life of asphalt materials has a very important position in the national economy, which is also a very urgent topic in the professional research and production of asphalt science.

Contrary to the aging process, certain processes and materials are used to improve the composition, colloidal structure and macro performance of asphalt, which makes restore engineering-behavior to a certain extent. This process is called regeneration.

At present, the recycling of asphalt materials is a hot spot in the theoretical and engineering fields. However, due to the complexity of asphalt materials, these works are still in the initial stage.

ii. Modified petroleum asphalt

Asphalt materials are directly exposed to the natural environment whether they are used as roofing waterproof materials or as pavement bonding materials, and the performance of asphalt is greatly affected by environmental factors. At the same time, in the modern civil engineering asphalt is not only required to have good performance, but also a long service life. It is difficult to achieve the requirements of modern civil engineering for asphalt in many aspects simply depending on its own properties. Therefore, in modern civil engineering, other materials are often added to asphalt to further improve its performance, which is called modified asphalt.

At present, the commonly used modified asphalt at home and abroad includes the following four categories:

1. Mineral filler modified asphalt

Adding a certain amount of mineral filler to asphalt can improve the viscosity and heat resistance of asphalt and reduce the temperature sensitivity of asphalt, which is mainly applicable to the production of asphalt adhesive.

There are two kinds of mineral fillers: powder and fiber. Common fillers include talc powder, limestone powder, diatomite, asbestos wool, mica powder, ground sand, fly ash, cement, Shangling soil, chalk, etc.

2. Resin modified asphalt

The cold resistance, heat resistance, adhesion and air impermeability of asphalt can be improved by using resin to modify asphalt. It shall be used in the production of coiled materials, sealing materials, waterproof coatings and other products.

The resin used for asphalt modification is mainly thermoplastic resins, which are usually polyethylene and polypropylene. The performance of modified asphalt composed of them is mainly to improve the viscosity, the high temperature anti-fluidity and the toughness of asphalt, but sometimes the improvement of low-temperature performance is not obvious.

3. Rubber modified asphalt

Rubber is an important modified material of petroleum asphalt. It has good compatibility with petroleum asphalt and can make asphalt have many advantages of rubber, such as small deformation at high temperature, good flexibility at low temperature. Meanwhile, the strength, elongation and aging resistance of modified asphalt improve. Due to different rubber varieties and mixing methods, the properties of various rubber asphalt are different.

The performance of rubber modified asphalt mainly depends on the performance of asphalt, the type and preparation of rubber, technology and other factors. At present, the rubber materials commonly used in synthetic rubber modified asphalt include neoprene, butyl rubber, styrene butadiene rubber, etc.

4. Resin and rubber blend modified asphalt

Using rubber and resin together to improve the properties of petroleum asphalt can make asphalt have the characteristics of both rubber and resin. Because resin is cheaper than rubber and rubber and resin have better miscibility, it can achieve satisfactory comprehensive effect.

In the heating and melting state of rubber, resin and petroleum asphalt, mutual invasion and diffusion occur between asphalt and polymer. Meanwhile, asphalt molecules are filled in the gaps of polymer macromolecules, and some chain links of polymer molecules are diffused into asphalt molecules. In the end, a cohesive network mixed structure is formed and then the modified asphalt obtains better performance.

# Section Ⅱ  Asphalt mixture

Asphalt mixture is a uniform mixture formed by mixing mineral mixture and a proper amount of asphalt materials, in which the mineral mixture is made of coarse aggregate, fine aggregate and filler by selecting reasonable grading with people. Asphalt concrete and asphalt macadam, as well as open graded or discontinuous graded asphalt mixture belong to asphalt mixture.

Because asphalt concrete pavement has good flatness, smooth and comfortable driving, and low noise, many countries give priority to the use of asphalt concrete in the construction of expressways. The semi-rigid base has the characteristics of high strength, good stability and high stiffness, and it is widely used to build the base or subbase of high-grade highway asphalt pavement. Many expressway pavements under construction or completed in China use semi-rigid base asphalt pavement. Because asphalt mixture can best adapt to the characteristics of modern traffic, it is the most important pavement material of modern highways, and it is widely used in trunk highways and urban roads.

## Ⅰ. Classification of asphalt mixture

Hot mix asphalt mixture is a mixture of manually prepared mineral mixture and easily thickened asphalt, which is heated and mixed in special equipment. The mixture is transported to the construction site by thermal insulation transportation tools, and then paved and compacted under hot mixing. It is generally called hot mix hot paving asphalt mixture, or hot mix asphalt mixture for short. Hot mix asphalt mixture is the most typical variety of asphalt mixture, from which other kinds of asphalt mixtures are developed. This section mainly describes its classification, composition, technical properties, composition materials and design methods.

i. Classification by the grading type of mineral aggregate

1. Continuous graded asphalt mixture. The mineral aggregate in asphalt mixture is a mixture composed of particles of all levels from large to small in proportion according to the grading principle, which is called continuous graded mixture.

2. Discontinuous graded asphalt mixture. The asphalt mixture lacking one or several grades of particle size for the mineral aggregate of continuous graded asphalt mixture is called discontinuous graded asphalt mixture.

ii. Classification by the density of asphalt mixture

1. Dense graded asphalt concrete mixture. The continuous dense graded asphalt mixture, which is designed according to the dense grading principle, has a small particle size decline coefficient, and the residual porosity after compaction is less than 10%. Dense graded asphalt concrete mixture can also be divided into Type I dense asphalt concrete mixture according to its residual void ratio, whose residual void ratio is 3% ~ 6%. Type II semi-dense asphalt concrete mixture has a residual void ratio of 4% ~ 10%.

2. Open graded asphalt mixture. The continuous graded mixture, which is designed according to the open grading principle, has a large particle size decline coefficient, and the residual void ratio after compaction is greater than 15%. It is also called semi-open graded asphalt mixture, also known as asphalt macadam mixture, if the residual void content is between dense grading and open grading (namely, the residual void content is 10% ~ 15%).

iii. Classification by nominal maximum particle size

Asphalt concrete can be divided into the following four categories according to the nominal maximum particle size of aggregate:

1. Coarse grained asphalt mixture: it is the asphalt mixture with the maximum aggregate size of 26.5mm or 31.5mm.

2. Medium grained asphalt mixture: it is the asphalt mixture with the maximum aggregate size of 16mm or 19mm.

3. Fine grained asphalt mixture: it is the asphalt mixture with the maximum aggregate size of 9.5mm or 13.2mm.

4. Sand asphalt mixture: it is an asphalt mixture with the maximum aggregate size equal to or less than 4.75mm, also known as asphalt chips or asphalt sand.

In addition to the above four types of asphalt macadam mixture, there is also extra coarse asphalt macadam mixture, with the maximum aggregate size of 37.5mm or more.

## II. Composition and structure of asphalt mixture

ⅰ. The theory of asphalt composition and structure

As far as the existing research results are concerned, there are mainly two kinds of asphalt mixture composition and structure theory.

According to the surface theory, asphalt mixture is a dense graded mineral skeleton artificially composed of coarse aggregate, fine aggregate and filler. This asphalt mixture is distributed on the surface of mineral skeleton with thinner consistency which is cemented into a whole with strength.

In the mortar theory, asphalt mixture is regarded as a multi-level spatial network structure dispersion system, and namely it is a coarse dispersion system that takes coarse aggregate as dispersion phase and disperses in the asphalt mortar.

ⅱ. Composition and structure type of asphalt mixture

The asphalt mixture formed according to the grading principle can be generally composed in the following three ways, as shown in Figure 5-5.

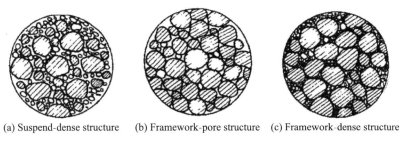

(a) Suspend-dense structure　　(b) Framework-pore structure　　(c) Framework-dense structure

**Figure 5-5　Types of asphalt mixture mineral aggregate skeleton**

1. Suspend-dense structure

For the dense mixture composed of continuously graded mineral mixture, the aggregate continuously exists from large to small. In fact, the larger particles of the same grade are squeezed by the smaller ones, and the larger particles are suspended in the smaller ones. This structure is usually designed according to the principle of optimum gradation, so its compactness and strength are high. However, it is greatly affected by the properties and physical state of asphalt materials and its stability is poor.

2. Framework-pore structure

In framework-pore structure, the coarser stones are closely connected with each other, and the number of finer particles is small, which is not enough to fully fill the gap. The gap of the mixture is large, and the stone can fully form a skeleton. In this structure, the internal friction between coarse aggregates plays an important role, and its stability is better because its structural strength is less affected by the nature and physical state of asphalt.

3. Framework-dense structure

It is a structure composed of the above two ways. In the mixture, A certain amount of coarse aggregate forms a skeleton, and fine aggregate is added according to the amount of coarse aggregate voids to form a higher compactness. The discontinuous gradation is formed in the light of this principle.

## III. Technical properties of asphalt mixture

The asphalt mixture is directly affected by vehicle load and atmospheric factors in the pavement. At the same time, the physical and mechanical properties of the asphalt mixture are greatly affected by climate factors and time factors. Therefore, in order to enable the pavement to provide stable and durable services for vehicles, the asphalt pavement must have a certain stability and durability, which include high-temperature stability, low-temperature crack resistance, durability, skid resistance, construction workability and so on.

ⅰ. High-temperature stability

The high-temperature stability of asphalt mixture is traditionally the ability of asphalt mixture to resist permanent deformation under load. As the strength and stiffness (modulus) of asphalt mixture decrease significantly with the increase of temperature, in order to ensure that the asphalt pavement will not produce such diseases as waves, pushing, rutting and bumps under the repeated action of driving load in high temperature seasons, the asphalt pavement should have good high-temperature stability.

According to the current JTG F40—2004 *Technical Code for Construction of Highway Asphalt Pavement*, Marshall test is used to evaluate the high-temperature stability of asphalt mixture; For expressways, Class I highways, urban expressways and main roads with asphalt mixture, their anti-rutting ability shall be tested by dynamic stability test, and their water stability shall be tested by immersion Marshall test and freeze-thaw split test. Scan the two-dimensional code to learn more.

1. Marshall test

The test piece prepared according to the standard method shall be kept in a constant temperature water bath at 60℃ for 45min, and then placed on the Marshall stability meter. The load shall beincreased at the deformation rate of (50 ± 5) mm/min until the maximum load (kN), i. e. Marshall stability ($MS$), when the test piece is damaged. Flow value ($FL$) is the vertical deformation (in 0.1m) of the test piece when the maximum failure load is reached. The relationship curve between load ($P$) and deformation ($F$) can be automatically drawn on the Marshall stability meter with X-Y recorder.

Marshall modulus is the quotient of stability divided by flow value. The calculation formula is as follows:

$$T = \frac{MS \times 10}{FL} \qquad (5\text{-}3)$$

Where  $T$——Marshall modules, kN/mm;

$MS$——Marshall stability, kN;

$FL$——flow value, 0.1mm.

2. Wheel tracking test

The wheel tracking test is to adopt the following standard forming method. The first step is to make the 300mm × 300mm × 50mm asphalt mixture test piece, and then a wheel under a certain load walks repeatedly on the same track for a certain time to form a certain rut depth when the temperature is 60℃. After the test, the experimenter calculates the number of times that the test wheel needs to travel when the test piece deforms by 1mm, which is the dynamic stability. The calculation formula is as follows:

$$DS = \frac{N \times (t_2 - t_1)}{d_2 - d_1} \times c_1 \times c_2 \quad (5\text{-}4)$$

Where  $DS$——dynamic stability of asphalt mixture, times/mm;

$N$——round trip rolling speed of the test wheel, usually is 42 times/min;

$c_1$——correction factor of testing machine;

$c_2$——coefficient of test piece;

$d_2$——rut depth at test time $t_2$, mm;

$d_1$——rut depth at test time $t_1$, mm.

The dynamic stability of asphalt mixture for the surface course and middle course of expressway shall be greater than 1 200times/mm. For the surface course and middle course of Class I highway, the dynamic stability of asphalt mixture shall be more than 800times/mm. If modified asphalt is used, its dynamic stability standard is related to the type of modified asphalt.

ii. Water stability

The water stability of asphalt mixture is tested by Immersion Marshall test and freeze-thaw split test.

1. Immersion Marshall test

The immersion Marshall stability test is used to test the ability of asphalt mixture to resist spalling when it is damaged by water. The feasibility of mix design is tested by testing its water stability.

The difference between the immersion Marshall test method and the standard Marshall test method is that the holding time of the test piece in the constant temperature water tank that has reached the specified temperature is 48h, and the rest are the same as the standard Marshall test method.

The residual stability is calculated according to the following formula:

$$MS_0 = \frac{MS_1}{MS} \times 100 \quad (5\text{-}5)$$

Where  $MS_0$——residual stability of test piece after immersion in water, %;

$MS_1$——stability of test piece after 48h immersion, kN.

2. Freeze-thaw split test

The freeze-thaw split test is to conduct freeze-thaw cycles on asphalt mixture under specified conditions and measure the strength ratio of the mixture specimen before and after being damaged by water, which is used to evaluate the water stabilized asphalt mixture. The standard test temperature is 25℃, and the loading rate is 50mm/min.

The residual strength ratio of freeze-thaw split test is calculated according to the following formula:

$$TSR = \frac{R_{T2}}{R_{T1}} \times 100 \qquad (5\text{-}6)$$

Where  $TSR$——residual strength ratio of freeze-thaw split, %;

$R_{T2}$——splitting tensile strength of the second group of specimens after freeze-thaw cycles, MPa;

$R_{T1}$——splitting tensile strength of the first group of specimens without freeze-thaw cycles, MPa.

iii. Low-temperature crack resistance

In addition to the high-temperature stability, the asphalt mixture should also have good low-temperature crack resistance to ensure that the pavement will not crack when the temperature is low in winter.

In the current standard *Technical Specifications for Construction of Highway Asphalt Pavement*, it is required to conduct bending test on dense graded asphalt mixture at a temperature of $-10℃$ and a loading rate of 50mm/min. The experimenter measures the failure strength, failure strain and failure stiffness modulus, and comprehensively evaluate the low-temperature crack resistance of asphalt mixture according to the shape of the stress-strain curve.

iv. Durability

The aging problem exists in the mixing, paving, rolling and later use of asphalt mixture. The aging process is generally divided into two stages: short-term aging during construction and long-term aging during pavement use. The aging degree of asphalt mixture in the mixing process is related to the mixing temperature, asphalt storage temperature, asphalt storage time, etc. The long-term aging of asphalt mixture in the use process is related to the shape of asphalt materials and asphalt in the mixture, such as the size of mixture porosity, asphalt consumption, light, oxygen and other natural climatic conditions.

At present, the air void ratio, saturation, residual stability and other indicators are used to characterize the durability of asphalt mixture in china.

v. Skid resistance

The current national standard *Code for Construction and Acceptance of Asphalt Pavement* (GB 50092—96) has proposed three indicators, namely, the polishing value, Dow Rayleigh wear value and impact value, for antiskid aggregate. Scan the QR code for more information.

The skid resistance of asphalt mixture pavement is related to the micro surface property of mineral aggregate, the gradation composition of mixture, the amount of asphalt, wax content and other factors. In order to ensure the safety of long-term high-speed driving, special attention shall be paid to the abrasion resistance of coarse aggregate when batching.

The skid resistance is very sensitive to the amount of asphalt. If the amount of asphalt exceeds 0.5% of the optimal amount, the skid resistance coefficient will be significantly reduced. Wax content has obvious influence on skid resistance of asphalt mixture, and the wax content of heavy traffic road asphalt shall not be greater than 3%.

ⅵ. Construction workbility

To ensure the smooth realization of indoor batching under on-site construction conditions, the asphalt mixture should not only meet the aforementioned technical requirements, but also have appropriate construction workability. There are many factors affecting the workability of asphalt mixture, such as local temperature, construction conditions and mixture properties.

From the perspective of mixture material properties, the first factor affecting the workability of asphalt mixture construction is the gradation of the mixture. If the particle sizes of coarse and fine aggregates are too large apart and lack of intermediate size, the mixture is easy to be layered (the coarse aggregates are concentrated on the surface and the fine aggregates are concentrated on the bottom). If the fine aggregate is too little, the asphalt layer is not easy to be evenly distributed on the surface of coarse particles. If there is too much fine aggregate, it will be difficult to mix the asphalt mixture. In addition, the mixture is easy to loose and not easy to compact when the amount of asphalt is too small or the amount of mineral powder is too large. On the contrary, when the amount of asphalt is too much or the quality of mineral powder is poor, it is easy to bond the mixture into a lump and not easy to pave.

## Ⅳ. Mix proportion design of asphalt mixture

The mix proportion design of asphalt mixture aims to determine the material variety and proportion, mineral aggregate gradation, and the optimal asphalt dosage of asphalt mixture. The design includes three stages: target proportioning design, production proportioning design and production proportioning verification. The relevant specifications in China clearly stipulate that Marshall test proportioning design method is adopted, and the proportioning design flow chart is shown in Figure 5-6 below. If other methods are used to design asphalt mixture, Marshall test and design inspection of various mix proportions shall be carried out according to the specifications.

ⅰ. Target proportioning Design

The target proportioning design is carried out in the laboratory, which is divided into two parts: the design of mineral mixture composition and the determination of the optimal asphalt dosage.

1. Composition design of mineral mixture

The purpose of mineral mixture composition design is to mix all kinds of mineral aggregates in the best proportion, so that after asphalt is added, asphalt concrete is dense and has certain gaps to adapt to asphalt expansion in summer

(1) Selection of asphalt mixture type

The asphalt mixture is applicable to the asphalt pavement of various highways. The type shall be selected by taking into account the nominal maximum particle size of aggregate, mineral aggregate gradation, porosity and other factors, as shown in Table 5-6.

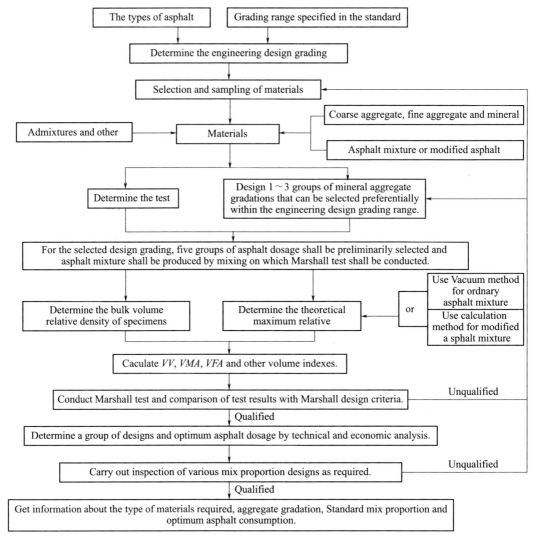

Figure 5-6  Flow chart of asphalt mixture proportion design

Table 5-6  The type of asphalt mixture

| Layer of structure | Expressway, First-class highway, urban expressway and main-road | | Other highways | | General urban roads and other roads | |
|---|---|---|---|---|---|---|
| | Three-layer asphalt concrete pavement | Two-layer asphalt concrete pavement | Asphalt concrete pavement | Asphalt macadam pavement | Asphalt concrete pavement | Asphalt macadam pavement |
| Upper layer | AC-13<br>AC-16<br>AC-20 | AC-13<br>AC-16 | AC-13<br>AC-16 | AC-13 | AC-5<br>AC-10<br>AC-13 | AM-5<br>AM-10 |
| Middle layer | AC-20<br>AC-25 | — | — | — | — | — |

continued

| Layer of structure | Expressway, First-class highway, urban expressway and main-road | | Other highways | | General urban roads and other roads | |
|---|---|---|---|---|---|---|
| | Three-layer asphalt concrete pavement | Two-layer asphalt concrete pavement | Asphalt concrete pavement | Asphalt macadam pavement | Asphalt concrete pavement | Asphalt macadam pavement |
| Lower layer | AC-25<br>AC-30 | AC-20<br>AC-30 | AC-20<br>AC-25<br>AC-AM-25<br>AM-30 | AM-25<br>AM-30 | AC-20<br>AC-25<br>AM-25<br>AM-30 | AC-25<br>AM-30<br>AM-40 |

(2) Determine the engineering design grading range

The design grading range of mixture for asphalt pavement works is specified in the engineering design documents or bidding documents. The design grading of different asphalt mixtures should be within the grading range specified in the specifications, as shown in Table 5-7 below. According to the highway grade, project nature, climate conditions, traffic conditions and material varieties, the design gradation is adjusted and determined after investigating the use of projects with similar conditions. If necessary, it is allowed to exceed the specification gradation range. The determined engineering design grading range is the basis for mix proportion design and shall not be changed at will.

(3) Calculation of mix proportion of mineral mixture

① Selection and preparation of mineral aggregate. According to the methods specified in the current standard *Specifications for Aggregate Test of Highway Engineering* (JTG E42—2005, scan the QR code for more information) various mineral aggregates designed for mix proportion must be taken representative samples from the materials actually used in the project. All materials must meet the needs of climate and traffic conditions, and their quality shall meet the technical requirements specified in the specification. Record the screening test results of various materials for use in mix design and calculation.

② Mineral aggregate mix proportion design. The design of mineral aggregate mix proportion for asphalt pavement of expressway and Class I highway should be carried out with the help of electronic computer spreadsheet and trial mix method. The asphalt pavement of other highways can also refer to this method.

③ For expressways and Class I highways, it is advisable to calculate 1~3 groups of different mix proportions within the engineering design grading range and draw design grading curves, which are respectively located above, at the middle value and below the engineering design grading range. The designed composite grading shall not have too many zigzag intersections, and there shall be no "hump" within the range of 0.3~0.6mm. When repeated adjustment is not satisfactory, the material design should be replaced.

Table 5-7  Aggregate grading ranges of asphalt mixture

| Gradation types | | | | Mass percentage passing the following sieve openings(mm)/% | | | | | | | | | | | | | | | Asphalt consumption for reference /% |
|---|---|---|---|---|---|---|---|---|---|---|---|---|---|---|---|---|---|---|---|
| | | | | 53 | 37.5 | 31.5 | 26.5 | 19 | 16 | 13.2 | 9.5 | 4.75 | 2.36 | 1.18 | 0.6 | 0.3 | 0.15 | 0.075 | |
| Asphalt concrete | Coarse grain | AC-30 | I | | 100 | 90~100 | 79~92 | 66~82 | 59~77 | 52~72 | 49~63 | 32~52 | 25~42 | 18~32 | 13~25 | 8~18 | 5~13 | 3~7 | 4.0~6.0 |
| | | | II | | 100 | 90~100 | 66~85 | 52~70 | 45~85 | 38~58 | 30~50 | 18~38 | 12~28 | 8~20 | 4~14 | 3~11 | 2~7 | 1~5 | 3.5~5.0 |
| | | AC-25 | I | | | 100 | 95~100 | 75~90 | 62~80 | 53~73 | 43~63 | 32~52 | 25~42 | 18~32 | 13~25 | 8~18 | 5~13 | 3~7 | 4.0~6.0 |
| | | | II | | | 100 | 90~100 | 65~85 | 52~70 | 42~62 | 32~52 | 20~40 | 13~30 | 9~23 | 6~16 | 4~12 | 3~8 | 2~5 | 3.5~5.0 |
| | Medium grain | AC-20 | I | | | | 100 | 95~100 | 75~90 | 62~80 | 52~72 | 38~58 | 28~46 | 20~34 | 15~27 | 10~20 | 6~14 | 4~8 | 4.0~6.0 |
| | | | II | | | | 100 | 90~100 | 65~85 | 52~70 | 40~60 | 26~45 | 16~33 | 11~25 | 7~18 | 4~13 | 3~9 | 2~5 | 4.0~5.5 |
| | Fine grain | AC-16 | I | | | | | 100 | 95~100 | 75~90 | 58~78 | 42~63 | 32~50 | 22~37 | 16~28 | 11~21 | 7~15 | 4~8 | 4.0~6.0 |
| | | | II | | | | | 100 | 90~100 | 65~85 | 50~70 | 30~50 | 18~35 | 12~26 | 7~19 | 4~14 | 3~9 | 2~5 | 4.0~5.5 |
| | | AC-13 | I | | | | | | 100 | 95~100 | 70~88 | 48~68 | 36~53 | 24~41 | 18~30 | 12~22 | 8~16 | 4~8 | 5.0~7.0 |
| | | | II | | | | | | 100 | 90~100 | 60~80 | 34~52 | 22~38 | 14~28 | 8~20 | 5~14 | 3~10 | 2~8 | 4.5~6.5 |
| | Sand | AC-10 | I | | | | | | | 100 | 95~100 | 55~75 | 35~58 | 26~43 | 17~33 | 10~24 | 6~16 | 4~9 | 5.0~7.0 |
| | | | II | | | | | | | 100 | 90~100 | 40~60 | 24~42 | 15~30 | 9~22 | 6~15 | 4~10 | 2~6 | 4.5~6.5 |
| | | AC-5 | I | | | | | | | | 100 | 95~100 | 55~75 | 35~55 | 20~40 | 12~28 | 7~18 | 5~10 | 6.0~8.0 |

continued

| Gradation types | | | 53 | 37.5 | 31.5 | 26.5 | 19 | 16 | 13.2 | 9.5 | 4.75 | 2.36 | 1.18 | 0.6 | 0.3 | 0.15 | 0.075 | Asphalt consumption for reference /% |
|---|---|---|---|---|---|---|---|---|---|---|---|---|---|---|---|---|---|---|
| | | | | | | | | | Mass percentage passing the following sieve openings (mm)/% | | | | | | | | | |
| Asphalt macadam | Coarse grain | AM-40 | 100 | 90~100 | 50~80 | 40~65 | 30~54 | 25~30 | 20~45 | 13~28 | 5~25 | 2~15 | 0~10 | 0~8 | 0~6 | 0~5 | 0~4 | 2.5~3.5 |
| | | AM-30 | | 100 | 90~100 | 50~80 | 38~65 | 32~57 | 25~50 | 17~42 | 8~30 | 2~20 | 0~15 | 0~10 | 0~8 | 0~5 | 0~4 | 3.0~4.0 |
| | | AM-25 | | | 100 | 90~100 | 50~80 | 43~73 | 38~65 | 25~55 | 10~32 | 2~20 | 0~14 | 0~10 | 0~8 | 0~6 | 0~5 | 3.0~4.5 |
| | Medium grain | AM-20 | | | | 100 | 90~100 | 60~85 | 50~75 | 40~65 | 15~40 | 5~22 | 2~16 | 1~12 | 0~10 | 0~8 | 0~5 | 3.0~4.5 |
| | | AM-16 | | | | | 100 | 90~100 | 60~85 | 45~68 | 18~42 | 6~25 | 3~18 | 1~14 | 0~10 | 0~8 | 0~5 | 3.0~4.5 |
| | Fine grain | AM-13 | | | | | | 100 | 90~100 | 50~80 | 20~40 | 8~20 | 4~20 | 2~16 | 0~10 | 0~8 | 0~6 | 3.0~4.5 |
| | | AM-10 | | | | | | | 100 | 85~100 | 35~65 | 10~35 | 5~22 | 2~16 | 0~12 | 0~9 | 0~6 | 3.0~4.5 |
| Anti-skid surface | | AK-13A | | | | | | 100 | 90~100 | 60~80 | 30~53 | 20~40 | 15~30 | 10~23 | 7~18 | 5~12 | 4~8 | 4.0~5.5 |
| | | AK-13B | | | | | | 100 | 85~100 | 50~70 | 18~40 | 10~30 | 8~22 | 5~15 | 3~12 | 3~9 | 2~6 | 4.0~5.5 |
| | | AK-16 | | | | | 100 | 90~100 | 60~82 | 45~70 | 85~45 | 15~35 | 10~25 | 8~18 | 6~13 | 4·10 | 3~7 | 4.0~5.5 |

313

2. Determination of the optimum asphalt consumption

For the current specification, the optimal asphalt content of asphalt mixture is determined by marshall test, which is expressed in $OAC$. The asphalt content can be expressed in two ways: asphalt-aggregate ratio and asphalt content. The asphalt-aggregate ratio is the percentage of asphalt in total mineral aggregate. The amount of asphalt is the percentage of asphalt in the total amount of asphalt mixture.

To determine the best asphalt dosage, first select the appropriate asphalt dosage according to the local practical experience, and then make several groups of graded Marshall specimens. Determine the $VMA$ of the test piece, and initially select a group of gradings that meet or are close to the design requirements as the design gradations. Then Marshall test is carried out to determine the optimal asphalt content.

Take the estimated asphalt aggregate ratio as the median, and take 5 or more different asphalt-aggregate ratios to form Marshall specimens at certain intervals (0.5% for dense graded asphalt mixture, 0.3 ~ 0.4% for asphalt macadam mixture). Various Marshall test indexes related to different oil stone ratios are determined by tests:

$VV$——porosity of test pieces(%);

$VMA$——mineral aggregate gap probability of test pieces(%);

$VFA$——effective asphalt saturation of test pieces(That is, the volume ratio of effective asphalt content to $VMA$,%);

$\gamma s$——bulk volume relative density of test pieces;

$FL$——flow value(mm);

$MS$——stability(kN);

The bulk volume relative density takes the asphalt aggregate ratio or asphalt content as the abscissa, and the indicators of Marshall test as the ordinate. The test results are plotted to form a smooth curve, as shown in Figure 5-7. Determine the asphalt dosage range $OAC_{min}$ ~ $OAC_{max}$ in accordance with the technical standard of asphalt mixture specified. The selected asphalt dosage range must cover the whole range of design void fraction and the required range of asphalt saturation as far as possible, and make the density and stability curve peak.

(1) According to the trend of the test curve, determine the optimal asphalt content $OAC_1$ of asphalt mixture according to the following methods.

Calculate the asphalt dosage $a_1, a_2, a_3, a_4$ corresponding to the maximum density, maximum stability, target porosity (or median), and median value of asphalt saturation range. Take the average value as $OAC_1$ according to Equation(5-7).

$$OAC_1 = (a_1 + a_2 + a_3 + a_4)/4 \qquad (5\text{-}7)$$

If the selected asphalt dosage range fails to cover the required range of asphalt saturation, the average value of the three is taken as $OAC_1$ according to the following formula.

$$OAC_1 = (a_1 + a_2 + a_3)/3 \qquad (5\text{-}8)$$

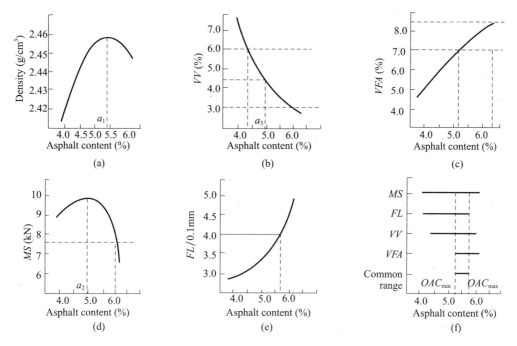

Figure 5-7 Relation curve between asphalt consumption and various indexes

For the asphalt consumption range of the selected test, if there is no peak in density or stability (the maximum value is often at both ends of the curve), the asphalt consumption $a_3$ corresponding to the target void fraction can be directly used as $OAC_1$, but $OAC_1$ must be within the range of $OAC_{min} \sim OAC_{max}$. Otherwise, the mix proportion design shall be carried out again.

(2) The median value of asphalt consumption range $OAC_{min} \sim OAC_{max}$ in which all indicators meet the technical standards is taken as $OAC_2$.

$$OAC_2 = (OAC_{min} + OAC_{max})/2 \quad (5\text{-}9)$$

(3) Generally, the median value of $OAC_1$ and $OAC_2$ is taken as the optimal asphalt consumption OAC.

$$OAC = (OAC_1 + OAC_2)/2 \quad (5\text{-}10)$$

Check whether all indicators corresponding to $OAC$ conform to Marshall test technical standards. In addition, the best asphalt consumption $OAC$ shall be adjusted and determined according to practical experience, highway grade, climate and traffic conditions.

For asphalt mixture, various performance tests shall be conducted on the basis of mix design. For asphalt mixture that does not meet the requirements, materials must be replaced or mix design must be conducted again. The inspection of performance mainly includes high temperature stability inspection, water stability inspection, low temperature crack resistance inspection, water permeability coefficient inspection, steel slag activity inspection, etc.

ii. Design of production mix proportion

After the target mix proportion is determined, the production mix proportion shall be designed. In the production of asphalt mixture, although the materials used are the same as the target mix design, the actual situation is still different from the laboratory. In addition, during production, the sand and stone may be different from the cold material screening in the laboratory after being heated and screened. For batch mixers, samples and sieves shall be taken from the materials that enter each hot bin after twice screening to determine the material proportion of each hot bin, so that the gradation formed is consistent with or basically close to the gradation of the target mix proportion design, which is used by the mixer control room. At the same time, the feeding proportion of the cold silo shall be adjusted repeatedly to make the material supply balanced, and the Marshall test shall be carried out with three asphalt consumption, including the optimal asphalt consumption designed for the target mix proportion and the optimal asphalt consumption of ± 0.3%, to determine the optimal asphalt consumption for the production mix proportion for trial mixing and paving.

iii. Verification of production mix proportion

After the production mix proportion is determined, it is also necessary to pave the test section and conduct Marshall test with the mixed asphalt mixture. At the same time, core drilling and sampling shall be carried out, and production mix proportion shall be inspected. If the standard requirements are met, the whole mix proportion design shall be completed to determine the standard mix proportion for production. Otherwise, adjustment is required.

The standard mix proportion is the basis for production control and quality inspection. For the mineral aggregate composite gradation of standard mix proportion, the passing rate of the three sieve openings of 0.075mm, 2.36mm and 4.75mm shall be close to the median value of the required grading.

# Questions for review

1. Try to describe three components of petroleum asphalt and their characteristics.
2. What are the main technical properties of petroleum asphalt? What is the index of each property?
3. What is asphalt mixture? How to classify asphalt mixture?
4. Try to describe the test conditions, specimen size and evaluation indexes of Marshall test.
5. How to determine the optimum amount of asphalt in asphalt concrete mix design?

# Chapter VI  Construction steel

Metallic materials are generally divided into ferrous metals and non-ferrous metals. The main component of ferrous metals such as iron, steel and alloy steel are iron. Non-ferrous metals refer to metals that are mainly composed of other metal elements. The metal materials commonly used in civil engineering are mainly steel and aluminum alloy.

Steel has high tensile strength, the ability to form sheet material and wire, weldability and easy welding with other metals. With the rapid development of high-rise and long-span structures, steel is more and more widely used in civil engineering.

## Section I  Chemical composition of steel and its influence on steel properties

The main chemical composition of steel is iron(Fe), but also contains a small amount of carbon(C), silicon(Si), manganese(Mn), phosphorus(P), sulfur(S), oxygen(O), nitrogen(N), titanium(Ti), vanadium(V), niobium(Ni) and other elements. Although the content of these elements is small, they have great influence on the properties of steel. These components can be divided into two categories. One can improve the performance of the optimization of steel called alloy elements, mainly Si, Mn, Ti, V, Ni, etc.. Another type of deterioration of the performance of steel called steel impurities, mainly aerobic, sulfur, nitrogen, phosphorus, etc.

### I. Carbon

Carbon is the most important element that determines the properties of steel. The main components of steel and pig iron are both iron and carbon, and the difference lies in the carbon content. Iron carbon alloy with carbon content less than 2% is called steel, and iron carbon alloy with carbon content more than 2% is called pig iron.

Figure 6-1 shows the influence of carbon content on mechanical properties and technological properties of steel, such as strength, plasticity and toughness. It can be seen from the figure that the strength and hardness of steel increase with the increase of carbon content. Plasticity, toughness and cold bending decreased with the increase of carbon content. As the carbon content increases, the machinability of the steel decreases. When the carbon content increases to about 0.8%, the strength is maximum, but when the carbon content exceeds 0.8%, the strength decreases instead, which is caused by the brittle steel. With the increase of carbon content in steel, the weldability of steel will become worse (When the carbon content of steel is more than 0.3%, the weldability of steel decreases significantly.), cold brittleness and aging sensitivity will increase, and atmospheric corrosion resistance of steel will decrease. Carbon steel for general engineering is low carbon steel. In other words, the carbon content is less than 0.25%, and the carbon content of low alloy steel for engineering is less than 0.52%.

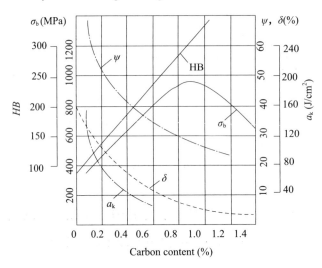

Figure 6-1  The effect of the carbon content on the performance of steel

## II. Silicon

Silicon is a beneficial element in steel, which deoxygenates when steel-making. Silicon is the main alloying element of steel reinforcement in China, and its functions are mainly to improve the strength, fatigue limit, corrosion resistance and oxidation resistance of steel, but it has little influence on plasticity and toughness. However, since the content of silicon in steel is very low, this effect is not obvious. When the content is increased to 1.0% ~ 1.2%, the tensile strength of the steel can be increased by 15% ~ 20%, but the plasticity and toughness are obviously decreased, the welding performance is poor, and the cold brittleness of the steel is increased. Generally, the silicon content of carbon steel is less than 0.3%, and low alloy steel is less than 1.8%.

## III. Manganese

Manganese is a beneficial element, which is the main alloying element of low alloy steel in China. When making steel, it can deoxidize and remove sulfur, which can reduce the thermal brittleness caused by sulfur, improve the thermal processing performance of steel, and improve the strength and hardness of steel. Manganese content is generally in the range of 1.0% ~ 2.0%. When manganese content is less than 1.0%, it has little effect on the plasticity and toughness of steel. When the manganese content reaches 11% ~ 14%, it is called high manganese steel, which has high wear resistance.

## IV. Phosphorus

Phosphorus is one of the harmful elements in steel. With the increase of phosphorus content at normal temperature, the strength and hardness of steel can be improved, but the plasticity and toughness can be significantly reduced. In particular, the lower the temperature, the greater the impact on plasticity and toughness, which significantly increase the cold brittleness of steel.

Phosphorus also significantly reduces the weldability of steel, but phosphorus can improve the wear resistance and corrosion resistance of steel. Therefore, other elements such as copper can be used as alloying elements in low alloy steel. The phosphorus content of building steel is generally required to be less than 0.045%.

## V. Sulfur

Sulfur is one of the harmful elements, which exists in steel as non-metallic sulfide (FES). It will increase the thermal brittleness of steel, reduce various mechanical properties of steel, and reduce the weldability, impact toughness, fatigue resistance and corrosion resistance of steel. The sulfur content of building steel shall be less than 0.045%.

## VI. Nitrogen

The effect of nitrogen on steel properties is similar to that of carbon and phosphorus. It can improve the strength of steel, but the plasticity will be reduced, especially the toughness decreased significantly. Nitrogen can also aggravate the aging sensitivity and cold brittleness of steel, and make the weldability worse. In steel, if nitrogen reacts with aluminum or titanium, the resulting compounds can refine the grains and improve the properties of the steel. Therefore, with aluminum, vanadium and other elements, nitrogen can be used as an alloying element of low alloy steel. The nitrogen content in steel is generally less than 0.008%.

## VII. Oxygen

Oxygen is a harmful impurity in steel. With the increase of oxygen content, the mechanical properties of steel are reduced, plasticity and toughness are reduced, aging is promoted, thermal brittleness is increased, and welding performance is poor. The oxygen content in steel is usually required to be less than 0.03%.

# Section II  Technical properties of steel

As the main force structure material in civil engineering, steel mainly bears the action of tensile force, pressure, bending, impact and other external forces. Therefore, it is required to have good mechanical properties and easy machining properties.

## I. Tensile performance

Tensile property is the most important property of steel, which is widely used in design and construction. Through tensile test, yield strength, tensile strength and elongation after fracture can be measured, which are important technical performance indexes of steel. The tensile properties of steel and low carbon steel can be illustrated by a stress-strain diagram during tension(as shown in Figure 6-2). The curve in the figure can be obviously divided into elastic stage($O\rightarrow A$), yield stage($A\rightarrow B$), reinforcement stage($B\rightarrow C$) and neck contraction stage($C\rightarrow D$).

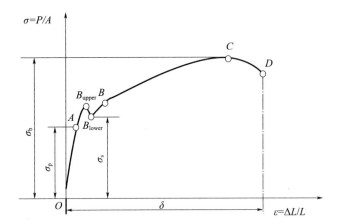

Figure 6-2   Stress-strain diagram for low carbon steel in tension

1. Elastic stage(section $OA$)

In the $OA$ range, with the increase of load, the stress becomes proportional to the stress. If the load is removed, it shall be restored to the original state. This property is called elasticity. $OA$ is a straight line, and the deformation in this range is called elastic deformation. The stress at point $A$ is called the elastic limit and is denoted by $\sigma_p$. In this range, the ratio of stress to strain is a constant, called the elastic modulus, denoted by $E$, that is, $E=\sigma/\varepsilon$. Elastic modulus reflects the stiffness of steel and is an important index to calculate the structural deformation of steel under stress. The elastic modulus of carbon structural steel Q235 commonly used in civil engineering is $E=(2.0\text{-}2.1)\times10^5\text{MPa}$, and the elastic limit $\sigma_p=180\sim200\text{MPa}$.

2. Yield stage(section $AB$)

When the stress exceeds the elastic limit, the stress and strain change in non-direct proportion in the range of AB curve. When the stress exceeds point $A$, plastic deformation begins. After the stress reaches $B_{\text{lower}}$, the deformation increases sharply, and the stress fluctuates in a small range until point $B$. This stage is called the yield stage. In the yield phase, the external force does not increase, but the deformation continues to increase. $B_{\text{upper}}$ is the upper limit of yield strength, $B_{\text{lower}}$ is the lower limit of yield strength, also known as yield limit or yield strength $\sigma_s$. When the stress reaches the upper point of $B_{\text{upper}}$, the ability of steel to resist the external force decreases, and the "yield" phenomenon occurs. $\sigma_s$ is the lowest value of stress fluctuation in the yield stage, which represents the allowable stress value of the steel in the working state, that is, the steel will not undergo large plastic deformation before $\sigma_s$. Therefore, the lower yield strength is generally used as the basis of strength value in the design. The $\sigma_s$ of commonly used carbon structural steel Q235 should not be less than 235MPa.

For the hard steel with no obvious yield phenomenon under the action of external force. Therefore, the stress when the residual deformation is 0.2% of the original standard distance length is set as the yield strength of the steel, which is represented by $\sigma_{0.2}$ (As shown in Figure 6-3).

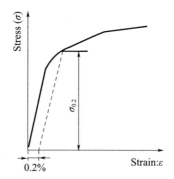

**Figure 6-3 Stress-strain diagram for hard steel in tension**

3. Reinforcement stage (secton $BC$)

After the steel yield to a certain degree, the further development of plastic deformation is prevented due to the internal lattice distortion and grain breakage. The ability of the steel to resist the external force is improved again. In the stress-strain diagram, the curve rises from point $B$ to the highest point $C$. This process is called the reinforcement phase. At this stage, the deformation develops rapidly and increases with the increase of stress. The stress corresponding to the highest point, $C$, is called the tensile strength and is denoted by $\sigma_b$. The tensile strength cannot be used directly, but the ratio of the down strength to the tensile strength ($\sigma_s/\sigma_b$) can reflect the safety, reliability and utilization of the steel. The smaller the strength ratio, the higher the safety and reliability of the material, and the material is not prone to dangerous brittle fracture. If the yield ratio is too small, the utilization rate is low, resulting in waste of steel. The $\sigma_b$ of the commonly used carbon structural steel Q235 should not be less than 375MPa, and the strength ratio should be between 0.58 and 0.63.

4. Necking stage (section $CD$)

When the steel tensile strengthening reached the highest point (point $C$), the deformation of the specimen began to concentrate in the weak section, which resulted in the significant reduction of the section and the "neck shrinkage phenomenon" (as shown in Figure 6-4). Due to the sharp reduction of the cross-sectional area of the specimen, the plasticity increases rapidly, and the tensile force decreases with it. Finally, fracture occurs.

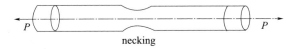

Figure 6-4 Schematic disgram of necking shrinkage phenomenon of steel under tensile

Splice the test pieces after pulling together at the broken place (as shown in Figure 6-5), and measure the gauge distance $l_1$ after breaking. The percentage of the extension of the scale $\Delta l$ to the original scale $l$ is called the elongation ($\delta$).

$$\delta = \frac{l_1 - l_0}{l_0} \times 100\% \qquad (6\text{-}1)$$

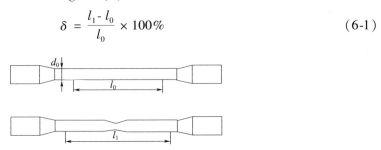

Figure 6-5 Specimen before and after broken

The distribution of plastic deformation is not uniform within the standard distance of the specimen. The deformation at the neck contraction is the largest, and the further away from the neck contraction, the smaller it becomes. Therefore, the smaller the ratio of the original standard distance to the diameter, the larger the proportion of the elongation value at the neck contraction in the whole elongation value, and the calculated elongation will be larger. Usually, steel tensile specimens are taken as $l_0 = 5d$ or $l_0 = 10d$, and their elongation is expressed as $\delta_5$ and $\delta_{10}$, respectively. For the same steel, $\delta_5$ is greater than $\delta_{10}$.

The elongation is an important technical index to measure the plasticity of steel. The larger the elongation, the better the plasticity of steel. Although the structure is used within the elastic range, the stress at the stress concentration may exceed the yield point. Good plastic deformation ability can redistribute stress and avoid premature failure of structure. The elongation of common carbon structural steels is generally 20% ~ 30%.

## II. Cold bending performance

Cold bending property refers to the ability of steel to bear bending deformation at room temperature, which is an important technological property of steel used in civil engineering.

The cold bending performance of steel is expressed by bending angle ($\alpha$) and bending core diameter ($d$). The steel cold-bending test adopts the specimen with diameter (or thickness) $a$, and the inner diameter of the bending center $d$ ($d = na$, $n$ is an integer) specified in the standard. When the steel is bent to the specified Angle (180° or 90°), if there is no crack, fracture and layer formation at the bending point, it is considered that the cold-bending performance is qualified. The greater the bending Angle and the smaller the diameter of the bending center, the better the cold bending performance. Figure 6-6 shows the cold bending test and the bending situation under the same bending Angle and different $d/a$.

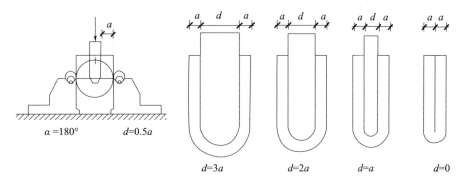

Figure 6-6  Bending diagrams of different $d/a$ flexion at $\alpha = 180°$ in cold bending test

The cold bending property and elongation of steel reflect the plastic deformation ability. Elongation reflects the plastic deformation ability of steel under uniform deformation conditions. The cold bending property is the plastic deformation ability of the steel under the condition of local deformation. Cold bending properties reveal whether the steel structure is uniform, whether there are defects such as internal stress and inclusions. In civil engineering, cold bending test is often used to check the welding quality of steel welded joints.

### III. High temperature performance

Steel is a material that is heat resistant but not fire resistant. When the temperature is less than 200℃, the mechanical properties of the steel are basically unchanged. However, when the temperature exceeds 300℃, the elastic modulus, yield strength and ultimate strength decrease significantly, and the deformation increases sharply. When the temperature exceeds 400℃, the strength and elastic modulus decrease sharply. When the temperature reaches 600℃, the elastic modulus, yield strength and ultimate strength are close to zero, and the bearing capacity has been lost. Therefore, according to the operating environment of the steel structure, fire protection maintenance should be carried out when necessary.

## Section III  Cold working and aging of steel

### I. Cold processing

Cold processing refers to the processing of steel at room temperature. The common cold working methods of steel used in civil engineering are: cold drawing, cold drawing, cold rolling, etc. The cold processing causes certain plastic deformation, significantly improve the strength, and reduce the plasticity and toughness. This process is called cold processing strengthening or "three cold treatments" of steel. After cold working, the yield strength of steel bar is increased, while the plasticity, toughness and elastic modulus are decreased.

The law of performance change of reinforcement after cold drawing can be seen from the tensile curve of low carbon steel sample (Figure 6-7), in which $OBCD$ is the deformation curve of the specimen without cold drawing aging. When the specimen was pulled to any point K beyond the yield point and the load was removed, the deformation $OO'$ of the specimen was generated, and the curve decreased along $KO'$, which was roughly parallel to $OB$. If it is restretched immediately, it can be found that the yield point increases to K, and the subsequent development curve is similar to that of

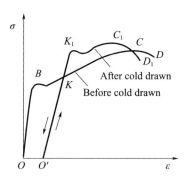

**Figure 6-7 Stress-strain diagram of steel bar after cold drawing and aging**

$KCD$. This phenomenon shows that when the steel is subjected to external force, it will produce plastic deformation. As the deformation increases, the resistance of the metal itself to deformation increases. This can be explained by the mechanism of "lattice slip": in the elastic deformation stage of steel, the arrangement of crystal atoms does not change, only in the direction of force, the distance between atoms increases or shortens (stretching or compression). Until the steel in plastic deformation stage, the crystal to produce the crystal-melt interface slip along the binding force of the worst. After sliding, the crystal is broken into small grains, resulting in bending and twisting, which is not easy to slip again and improve the ability of resistance to deformation. Therefore, greater external force is required to make it continue to produce plastic deformation, which is called "cold work hardening" or "work hardening".

## II. The aging

The phenomenon that the strength and hardness of steel increase with time, but the plasticity and toughness decrease is called aging. The aging of steel under natural conditions is very slow. If it is cold processed or often subjected to vibration and impact loads in use, the aging will develop rapidly.

As shown in Figure 6-7, if the sample is pulled to point K, and the load is not added immediately after removal, but after aging treatment (stored at room temperature for 15 ~ 20 days, or heated to 100 ~ 200℃ and kept for a certain time) and then stretched, it can be found that the yield point of the sample increases to the $K$-point, and the curve develops along $K_1 C_1 D_1$. This process is called aging treatment, the former is called natural aging, and the heating method is called artificial aging. At present, it is believed that the reason for the aging effect of steel after cold processing is that carbon (supersaturated) dissolved in ferrite slowly precipitates from ferrite to form cementite with the increase of time, which is distributed on the sliding surface of crystal to prevent sliding and enhance its ability to resist external forces, thus producing a strengthening effect.

There are two methods of aging treatment: storage at room temperature for 15~20 days, which is called natural aging, and is suitable for low-strength reinforcement; After heating to 100~200℃ and keeping for a certain time (2~3h), it is called artificial aging and is suitable for high-strength reinforcement.

Aging sensitivity refers to the degree of change in steel properties due to aging. The toughness and plasticity of steel with high aging sensitivity change greatly after aging treatment. Therefore, for important structures (such as crane beams, bridges, etc.) bearing vibration and impact loads, steel with low time sensitivity should be selected. Cold processing and aging are often used to improve the strength, increase the varieties and specifications of steel, and save steel.

## III. Application of cold processing and aging treatment of steel in engineering

There are obvious economic benefits in cold working of steel bar. After cold drawing, the yield point of steel bar can be increased by 20%~25%, and the yield point of cold drawn steel wire can be increased by 40%~90%. Thus, the design section of reinforced concrete structure can be appropriately reduced, or the number of steel wire in concrete can be reduced, and steel can be saved. Cold drawing of reinforcement is also conducive to simplifying the construction process. For example, the opening and straightening processes of wire rod reinforcement can be omitted. When cold drawing the bar reinforcement, it can be completed together with straightening, rust removal and other processes.

In civil engineering, cold working and aging treatment are often used at the same time. In actual construction, the cold drawing control parameters and aging treatment method should be determined by test. The control of cold drawing parameters is directly related to cold drawing effect and steel quality. Generally, the cold drawing of steel bar only controls the cold drawing rate, which is called single control. For the reinforcement used for prestressing, it is necessary to adopt dual control. Dual control not only controls the cold drawing stress, but also controls the cold drawing rate. When cold drawing to control the stress, the steel bar does not have to reach the control cold drawing rate. On the contrary, when the controlled cold drawing rate is reached but not the controlled stress, the reinforcement should be degraded to use.

It should be noted that the cold processed steel is not suitable for seismic structures. Although the tensile strength of the cold processed steel is improved, the elasticity and plasticity have decreased in the processing process, which easily leads to excessive structural stiffness and reduced seismic resistance. Seismic reinforcement shall be selected in the seismic structure. In addition to meeting all performance indexes of ordinary reinforcement specified in the standard, the seismic reinforcement shall also meet the following requirements: (1) the ratio of the measured tensile strength to the measured yield strength characteristic value of the seismic reinforcement shall not be less than 1.25; (2) The ratio between the measured yield strength of reinforcement and the characteristic value of yield strength specified in the standard shall not be greater than 1.30;(3)The maximum total stress elongation of reinforcement shall not be less than 9%.

## Section IV  Common category of steel in civil engineering

Steel commonly used in civil engineering can be divided into steel structure steel and reinforced concrete structure steel two categories. All kinds of steel sections no matter used in steel structures or reinforcement bars, steel wires and anchors used in reinforced concrete structures are basically carbon structural steels and low-alloy structural steels which processed by hot rolling or cold rolling, cold drawing and heat treatment.

Ordinary carbon structural steel, high quality carbon structural steel and low alloy structural steel arecommonly used in civil engineering.

### I. Carbon structural steel

According to the provisions of *Carbon Structural Steel*(GB/T 700—2006), carbon structural steel is divided into Q195, Q215, Q235, Q275 four grades. The steel grade is composed of four parts in sequence, the letter Q representing the yield strength of the steel, the yield strength value, the quality grade symbol and the deoxidation method symbol. Scan the QR code to get the detailed information of *Carbon Structural Steel*(GB/T 700—2006).

Symbol meaning: Q-represents the yield strength of steel, because the word "yield" is the first letter of Chinese Pinyin; A, B, C and D-are quality grades respectively; F-the first letter of the Chinese Pinyin of the word "boil" in boiling steel; Z-The first letter of the Chinese phonetic alphabet of the word "Zhen" in sedation steel; TZ-special killed steel "Tezhen" is the first letter of Chinese Pinyin. The symbols "Z" and "TZ" can be omitted in the brand composition representation method.

For example, Q235 AF refers to boiling carbon structural steel with yield strength not less than 235MPa and quality grade A; Q235 A refers to killed or specially killed carbon structural steel with yield strength not less than 235MPa and quality grade A.

The chemical composition and mechanical properties of carbon structural steel shall comply with the provisions in Table 6-1, Table 6-2 and Table 6-3 respectively.

Table 6-1  Grade and chemical composition of carbon structural steel(GB/T 700—2006)

| Brand | Uniform numeric code | Level | Thickness (or diameter)/mm | Deoxidizing method | Chemical composition (mass fraction)/%, not more than | | | | |
|---|---|---|---|---|---|---|---|---|---|
| | | | | | C | Si | Mn | P | S |
| Q195 | U11952 | — | — | F,Z | 0.12 | 0.30 | 0.50 | 0.035 | 0.040 |
| Q215 | U12152 | A | — | F,Z | 0.15 | 0.35 | 1.20 | 0.045 | 0.050 |
| | U12155 | B | | | | | | | 0.045 |

continued

| Brand | Uniform numeric code | Level | Thickness (or diameter)/mm | Deoxidizing method | Chemical composition (mass fraction)/%, not more than | | | | |
|---|---|---|---|---|---|---|---|---|---|
| | | | | | C | Si | Mn | P | S |
| Q235 | U12352 | A | — | F,Z | 0.22 | 0.35 | 1.40 | 0.045 | 0.050 |
| | U12355 | B | | | 0.20 | | | | 0.045 |
| | U12358 | C | | Z | 0.17 | | | 0.040 | 0.040 |
| | U12359 | D | | TZ | | | | 0.035 | 0.035 |
| Q275 | U12752 | A | — | F,Z | | 0.35 | 1.50 | 0.045 | 0.050 |
| | U12755 | B | ≤40 | Z | 0.20 | | | 0.045 | 0.045 |
| | | | >40 | | 0.22 | | | | |
| | U12758 | C | — | Z | 0.20 | | | 0.040 | 0.040 |
| | U12759 | D | | TZ | | | | 0.035 | 0.035 |

Note: 1. The unified numbers of killed steel and special killed steel grades are shown in the table. The unified numbers of boiling steel grades are as follows:
Q195F-U11950; Q215AF-U12150; Q215BF-U12153; Q235AF-U12350; Q235BF-U12353; Q275AF-U12750.

2. With the consent of the buyer, the carbon content of Q235B is not more than 0.22%.

**Table 6-2  Mechanical properties of carbon structural steels (GB/T 700—2006)**

| Brand | Grade | Yield strength[a] $R_{eH}$ (N/mm²), not less than | | | | | | Tensile strength[b] $R_M$/ (N/mm²) | Elongation after breakage A/%, not less than | | | | | Impact test | |
|---|---|---|---|---|---|---|---|---|---|---|---|---|---|---|---|
| | | Thickness (or diameter)/mm | | | | | | | Thickness (or diameter)/mm | | | | | Temperature /°C | Impact absorption (longitudinal) /J, not less than |
| | | 16 | 16~40 | >40~60 | >60~100 | >100~150 | >150~200 | | ≤40 | >40~60 | >60~100 | >100~150 | >150~200 | | |
| Q195 | — | 195 | 185 | — | — | — | — | 315~430 | 33 | — | — | — | — | — | — |
| Q215 | A | 15 | 205 | 195 | 185 | 175 | 165 | 335~450 | 31 | 30 | 29 | 27 | 26 | — | — |
| | B | | | | | | | | | | | | | +20 | 27 |
| Q235 | A | 35 | 225 | 215 | 215 | 195 | 185 | 370~500 | 26 | 25 | 24 | 22 | 21 | — | — |
| | B | | | | | | | | | | | | | +20 | 27[c] |
| | C | | | | | | | | | | | | | 0 | |
| | D | | | | | | | | | | | | | -20 | |
| Q275 | A | 75 | 265 | 255 | 245 | 225 | 215 | 410~540 | 22 | 21 | 20 | 18 | 17 | — | — |
| | B | | | | | | | | | | | | | +20 | 27 |
| | C | | | | | | | | | | | | | 0 | |
| | D | | | | | | | | | | | | | -20 | |

Note: a. The yield strength of Q195 is only for reference and is not a delivery condition.

b. When the steel thickness is greater than 100mm, the lower limit of tensile strength is allowed to be reduced by 20N/mm². Upper limit of tensile strength of wide strip steel (including shears) is not a delivery condition.

c. For Q235B steel with thickness less than 25mm, if the supplier can guarantee that the impact absorption work value is qualified, the inspection is not required with the consent of the buyer.

Table 6-3 Cold bending properties of carbon structural steels (GB/T 700—2006)

| Brand | Sample direction | Cold bending test 180° $B = 2a$ [a] | |
|---|---|---|---|
| | | Steel thickness (or diameter) [b] /mm | |
| | | ≤60 | >60~120 |
| | | Heart bending diameter d | |
| Q195 | Longitudinal | 0 | — |
| Q195 | Transverse | 0.5a | |
| Q215 | Longitudinal | 0.5a | 1.5a |
| Q215 | Transverse | a | 2a |
| Q235 | Longitudinal | a | 2a |
| Q235 | Transverse | 1.5a | 2.5a |
| Q275 | Longitudinal | 1.5a | 2.5a |
| Q275 | Transverse | 2a | 2a |

Note: a. $B$ is the width of the sample, and $a$ is the thickness (or diameter) of the sample.
  b. When the thickness (or diameter) of steel is greater than 100mm, the bending test shall be determined by both parties through negotiation.

With the increase of grade, the carbon content and manganese content of carbon structural steel increase, and the strength and hardness increase. However, the plasticity and toughness decrease, and the cold bending performance gradually deteriorates.

## II. High quality carbon structural steel

According to the provisions of *High Quality Carbon Structural Steel* (GB/T 699—1999), it can be divided into two groups according to its different manganese content: ordinary manganese content steel (manganese content less than 0.8%, a total of 20 steel grades) and higher manganese content steel (manganese content 0.7%~1.2%, a total of 11 steel grades). Scan the QR code to get the detailed information of *High Quality Carbon Structural Steel* (GB/T 699—1999).

High quality carbon structural steel is generally supplied by hot rolled steel. The content of sulfur, phosphorus and other impurities in steel is less than that in common carbon steel, and other defects are more restrictive. Thus, it has good performance and stable quality.

The steel grade of high-quality carbon structural steel is represented by two digits, which represents the average carbon content in the steel in ten thousand. For example, No. 45 steel means that the average carbon content in the steel is 0.45%. If there is "manganese" or "Mn" after the number, it is a higher manganese content of steel, otherwise it is ordinary manganese content of steel. For example, 35Mn steel, the average carbon content is 0.35%, manganese content is 0.7%~1.2%.

In case of boiling steel, "boiling" (F) should also be added after the steel number.

High quality carbon structural steel has high cost and is not widely used in construction. It is only used for steel castings and high strength bolts of important structures. Such as 30, 35, 40 and 45 steel for high strength bolts, 45 steel is also commonly used as an anchor for prestressed steel. Steels of 65,70,75 and 80 can be used to produce carbon steel wire, notched steel wire and stranded steel wire for prestressed concrete.

### III. Low alloy high strength structural steel

Low alloy high strength structural steel is formed by adding less than 5% alloying elements on the basis of carbon structural steel. The purpose of adding alloying elements is to improve the strength and properties of steel. Common alloying elements are silicon, manganese, titanium, vanadium, chromium, nickel and copper. According to the national standard GB/T 1591—2018 *Low Alloy High Strength Structural Steel*, there are 8 grades of low alloy high strength structural steel: Q345, Q390, Q420, Q460, Q500, Q550, Q620, Q690. The grade of low alloy high strength structural steel is composed of three parts: the yield point which is represented by the Chinese phonetic alphabet, the yield point value, and the quality grade symbol(grade A, B, C, D, E). For example, Q345D where:

Q—steel yield strength of the "flexion" word Chinese pinyin the first word;

345—yield strength value(MPa);

D—Grade D of quality.

If there is a requirement for the performance of the steel plate in the thickness direction, a symbol representing the thickness direction(Z direction) performance level shall be appended to the above specified grade, for example: Q345DZ15.

Low alloy steel not only has higher strength, but also has better plasticity, toughness, weldability and corrosion resistance. Therefore, it is a civil engineering steel with better comprehensive performance. Low alloy high-strength structural steel is mainly used for rolling various sections(angle steel, channel steel, I-beam), steel plates, steel pipes and reinforcement, and is widely used in steel structures and reinforced concrete structures. Especially in long-span structures, large structures, heavy structures, high-rise buildings, bridge projects, and structures bearing dynamic and impact loads. Scan the QR code to get the detailed information of GB/T 1591— 2018 *Low Alloy High Strength Structural Steel*.

# Questions for review

1. How to express the elongation of steel? How to evaluate the cold bending performance?
2. What is the yield ratio of steel? What is the effect of strength ratio on service performance?
3. What is the relationship between the elongation of steel and the gauge length of specimen? Why is that?
4. How does the degree of deoxidation affect the properties of steel?
5. What are the main harmful chemical elements in steel? What effect do they have on the properties of steel?
6. How does cold working of steel affect mechanical properties?
7. What does MnV stand for? Which structural steels are Q295-B and Q345-E?
8. What steels are mainly used in civil engineering? What are the principles of selection?

# Chapter VII  Other engineering materials

## Section I  Natural stone

The nature stone which is used in civil engineering and mechanically processed (or not) to varying degrees, are collectively called natural stones after natural rocks are mechanically processed (or not processed) to different degrees. Natural stone has high compressive strength, durability, wear resistance and other characteristics. Some rock varieties can also obtain unique decorative effect after processing, so it is widely used in civil engineering.

Stone is the oldest civil engineering structure and decorative materials. Many ancient buildings in the world are made of natural stone. For example, the Colosseum in Italy, the pyramids in ancient Egypt, the Zhaozhou Bridge in Hebei Province, the Luoyang Bridge in Quanzhou, Fujian Province, the base of the Palace Museum of the Ming and Qing Dynasties, the Great Hall of the People and the Monument to the People's Heroes all use a lot of natural stones. Natural stone has been gradually replaced by concrete and other materials as structural materials due to its high brittleness, low tensile strength, large self-weight, poor seismic performance of stone structures, difficult mining and processing of rocks, and high prices. In modern times, with the improvement of stone processing level, stone material unique decorative effect gets people's favor. Therefore, in the field of modern architectural decoration, the application prospect of stone is very broad. In addition, natural rocks can be naturally weathered or artificially broken into pebbles, gravels, sand and other materials, which can be used in large quantities as aggregates for concrete and are the main components of concrete. Some rocks are also the main raw materials for the production of artificial building materials, such as limestone for the production of Portland cement and lime, and quartzite for the production of ceramics and glass. Scan the code to learn about Zhaozhou Bridge and the Monument to the People's Heroes.

### I. Formation and classification of natural stone materials

Rocks are aggregates of natural solid minerals formed by different geological processes. The minerals that make up rocks are called rock forming minerals. Different rock forming minerals form rocks with different properties under different geological conditions.

Natural rocks can be divided into magmatic rocks, sedimentary rocks and metamorphic rocks according to their different geological conditions.

## Chapter VII  Other engineering materials

### i. Magmatic rock

#### 1. Formation and types of magmatic rocks

Magmatic rock, also known as igneous rock, is the rock formed by the condensation of molten magma in the deep crust after being underground or ejected from the ground. According to different formation conditions, magmatic rocks can be divided into plutonic rocks, extrusive rocks and volcanic rocks.

(1) Plutonic rock

Plutonic rock is the rock formed by the slow condensation of magma in the deep crust under the pressure of the upper overburden. It is crystal integrity, coarse grain, compact structure, with high compressive strength, small water absorption, high apparent density, good freezing resistance and other characteristics. Civil engineering commonly used deep diagenesis granite, syenite, peridotite, diorite and so on.

(2) Extrusive rock

The extrusive rock is the rock formed under the condition of pressure reduction and rapid cooling when magma is ejected from the surface. Because most magmas are too late to crystallize completely, they are usually fine crystalline(cryptocrystalline) or vitreous(pyrocrystalline) structures. When the magma is ejected into a thick rock layer, the structure and properties of the rock are similar to those of a plutonic rock. When a thin rock layer is formed, a porous rock is easily formed due to rapid cooling and atmospheric pressure, and its properties are similar to volcanic rocks. The ejecta rocks commonly used in civil engineering are diabase, basalt, andesite, and so on.

(3) The volcanic rock

Volcanic rock is rock that cools rapidly when magma is ejected into the air during a volcanic eruption. There are porous glass structure of scattered granular volcanic rocks, such as volcanic ash, cinder, pumice, and so on. There are also bulk cemented volcanic rocks, such as volcanic tuff, which are condensed by overburden pressure due to the accumulation of scattered granular volcanic rocks.

#### 2. Magmatic rock commonly used in civil engineering

(1) Granite

Granite is a widely distributed rock in magmatic rock, mainly composed of feldspar, quartz and a small amount of mica(or amphibolite, etc.). Granite has dense crystalline structure and massive structure, whose color generally gray black, gray white, light yellow, light red, and so on. In most cases, the same granite will present a combination of different colors. Due to the compact structure, its porosity and water absorption rate is small. The apparent density is large(2600 ~ 2800$kg/m^3$), compressive strength of 120 ~ 250MPa, good frost resistance (F100 ~ F200), weathering resistance and durability, service life of about 75 ~ 200 years. And granite has strong resistance for sulfuric acid and nitric acid corrosion. It is an excellent decorative material, the surface of granite is glossy and beautiful after grinding and polishing. In civil engineering, granite is commonly used as foundation, gate dam, bridge pier, step, road surface, wall stone, choke and memorial buildings. However, under the action of high temperature, the expansion of quartz in granite will cause damage to the stone, so its fire resistance is not good.

(2) Basalt

Basalt color is dark, and often glassy or cryptic structure, sometimes porous or spot structure. It has high hardness, fragility, weathering resistance, apparent density of $2900 \sim 3500 kg/m^3$, compressive strength of $100 \sim 500 MPa$.

(3) Pozzolanic ash

Pozzolanic ash is a powdery volcanic rock with particle size less than 5mm. Pozzolanic ash has pozzolanic activity, which can react with lime [$CaO$ or $Ca(OH)_2$] at room temperature and in the presence of water to produce hydrates with hydraulic cementing ability. Therefore, it can be used as mixing material of cement and admixture of concrete.

ii. Sedimentary rocks

1. Formation and types of sedimentary rocks

Sedimentary rocks, also known as hydromorphic rocks, are rocks formed by the re-deposition of various types of rocks on the surface by natural weathering, denudation, transport, deposition, and washing by running water, etc., and by compaction, mutual cementation and recrystallization, which mainly exist on the surface and not too deep underground. It is characterized by laminated structure, multi-layered appearance, low apparent density, high porosity and water absorption, low strength and poor durability. Sedimentary rock is the most widely distributed rock on the surface of the earth's crust, covering about 75% of the land surface area. It is widely used in civil engineering because it is mainly found on the surface, easy to mine and easy to process. According to the generation conditions of sedimentary rocks, they can be divided into mechanical sedimentary rocks, chemical sedimentary rocks, and bio-organic sedimentary rocks.

(1) Mechanical sedimentary rocks

Rocks and sands that have been gradually broken and loosened by natural weathering, transported by wind, water and glacial movement, and recompacted or cemented by mechanical forces such as sedimentation, commonly sandstones and shales.

(2) Chemical sedimentary rocks

Rocks formed by the aggregation, deposition, recrystallisation and chemical reaction of minerals dissolved in water, commonly limestone, gypsum and dolomite.

(3) Bio-organic sedimentary rocks

Rocks formed by the deposition of the remains of various organic bodies, such as diatomaceous earth.

2. Sedimentary rocks commonly used in civil engineering

Limestone is commonly known as chert or lapis lazuli, the main chemical composition is $CaCO_3$, the main mineral composition is calcite, but often contains dolomite, magnesite, quartz, opal, iron-bearing minerals and clay. The chemical composition, mineral composition, degree of denseness and physical properties of limestone vary greatly and are not as strong and durable as granite. Its apparent density is $2600\text{-}2800 kg/m^3$, its compressive strength is 80-160MPa, and its water absorption is 2%-10%. If the clay content of the rock does not exceed 3-4%, its water resistance and frost resistance are better. Limestone is widely sourced, low hardness, easy to split, has a certain strength and durability, and can be used in civil engineering as foundations, walls, steps, pavements, etc. Crushed stone can be used as concrete aggregate. Limestone is also a major raw material for the production of cement.

iii. Metamorphic rocks

1. Formation and types of metamorphic rocks

Metamorphic rocks are new rocks formed from magmatic or sedimentary rocks that were originally present in the earth's crust and have undergone recrystallisation in the solid state under the pressure or temperature of the strata, resulting in partial or complete changes in their mineral composition, structural structure and even chemical composition.

The mineral composition of metamorphic rocks retains the mineral composition of the original rocks, such as quartz, feldspar, mica, hornblende, pyroxene, calcite and dolomite, but also produces new metamorphic minerals, such as chlorite, argillite and serpentine. After the metamorphic process, the magmatic rocks will produce lamellar formations, their strength will decrease and the sedimentary rocks will become more dense.

2. Metamorphic rocks commonly used in civil engineering

(1) Marble

Marble, also known as marble and dolomite, is made from limestone or dolomite that has been re-crystallised and metamorphosed by high temperature and pressure. The main mineral composition of marble is calcite or dolomite, the chemical composition is mainly $CaO$, $MgO$, $CO_2$ and a small amount of $SiO_2$, after metamorphism, the crystalline particles directly combined in a whole block structure, high compressive strength (100~150MPa), dense texture, apparent density of 2500~2700kg/m$^3$, Mohs hardness is 3~4, than granite easy to carve. Pure marble is white, often called Chinese white jade in China, less distribution. General marble often contains iron oxide, silica, mica, graphite, serpentine and other impurities, so that the stone appears red, yellow, brown, black, green and other color mottled texture, stone is delicate, glossy, gorgeous. Marble has good decorative properties after grinding and polishing, so it is an excellent interior decoration material. Scan code to understand the knowledge of white marble.

(2) Quartzite

Quartzite is formed by the metamorphosis of siliceous sandstone and has a crystalline structure. It has a uniform and dense structure, high compressive strength (250~400MPa) and good durability, but its hardness makes it difficult to process. It is commonly used as the veneer material for important buildings and the wear-resistant and acid-resistant veneer material.

## II. Commonly used decorative stones

i. Natural marble slabs

1. Classification and grade of marble slabs

Marble slabs are slabs made of marble blocks (i.e. natural marble blocks with regular shapes extracted from mines) processed by sawing, grinding and polishing.

Natural marble slabs are classified according to their shape as ordinary type slabs(PX), rounded type slabs(HM) and shaped type slabs(XX) according to *Natural Marble Building Slabs*(GB/T 19766—2016). Ordinary type of plate for the square, rectangular; the arc plate is the plate with the same curvature radius of the decorative surface contour; other shapes of plate for the shaped plate. According to its appearance quality, mirror gloss, etc. is divided into three grades of superior quality(A), first grade(B) and qualified product(C). Scan the QR code to get the detailed information of *Natural Marble Building Slabs*(GB/T 19766—2016).

2. Technical requirements of marble slabs

The technical requirements of marble slabs are executed according to GB/T 19766—2016.

3. Application of marble

Marble slabs are mainly used for interior finishes, such as walls, floors, columns, countertops, balustrades, stepping stones, etc. Marble processed into building slabs, used for the decoration of buildings with higher grade requirements. After grinding, polishing the marble plate polished and delicate, white marble(Chinese white jade) as white as jade, crystal pure; pure black marble right solemn and elegant, beautiful and generous; colourful marble colorful, pattern strange. The durability of marble is generally several decades to several hundred years.

Marble should not be used as exterior finishing material for urban buildings because the air in the city often contains sulphur dioxide, which generates sulphurous acid when it meets water, and then becomes sulphuric acid, which reacts with the calcium carbonate in marble to produce $CaSO_4 \cdot 2H_2O$, which is easily soluble in water, causing the surface to lose its lustre and become rough and porous, thus reducing the decorative effect of the building. Only a few dense, pure varieties of sandstone, quartzite and alabaster, which contain mainly quartz, can be used outdoors.

ii. Natural granite slabs

Granite is very dense and hard. The minerals are all crystalline and coarse grained, in blocky formations or patchy formations with coarse crystals embedded in a glassy structure. They are ground and polished to form a mirror-like surface with a mottled pattern.

1. Classification and grade of granite slab products

According to the national standard *Natural Granite Building Slabs*(GB/T 18601—2009), granite slabs can be classified into three shapes: (1) general slabs(PX), (2) rounded slabs(HM), (3) shaped slabs(YX). There are three types of surface finish: (1) matt(YG), (2) mirror(JM) and (3) rough(CM). According to the general type plate specification size deviation, flatness tolerance, Angle tolerance, appearance quality and so on, the plate is divided into three grades: excellent product(A), first-class product(B) and qualified product(C). Scan the QR code to get the detailed information of *Natural Granite Building Slabs*(GB/T 18601—2009).

2. Technical requirements of granite slabs

The technical requirements of marble slabs, according to the national standard GB/T 18601—2009 implementation.

3. Application of granite

Granite slabs are rich in texture, with gorgeous and noble decorative effect, and hard texture, good durability, is a common material for indoor and outdoor advanced decoration. It is mainly used for the surface decoration of walls, columns, floors, stairs, steps and railings of buildings. In addition, granite can also be used as the foundation of important large buildings, dykes, bridges, road surfaces, street stones, and so on.

Polished granite plate used for decoration which characteristic is gorgeous and solemn, rough granite plate used for decoration which characteristic gritty and rough, mirror sense, with a mirror-like surface, bright colours and a moving lustre. Different physical properties and surface decorative effects of granite should be selected according to different use occasions. Among them, chopped axe plate, machine planed plate, rough grinding plate is used for external wall surface, column surface, step, plinth and other parts. Polished plates are mainly used for interior and exterior wall, column, ground and other decoration.

In recent years, the granite exterior decoration tends to be mainly rough granite, polished granite only for some lines or local lining. This kind of rough granite production process is the first granite grinding flat. Then burned with high temperature flame, no colour difference can be seen, uniform colour, no reflection. Give people a sense of visual comfort, the enjoyment of natural beauty.

4. The radioactivity of natural stone

Radioactivity in natural stone is a common concern. The test proves that the vast majority of natural stone contains only a very small amount of radioactive material, will not cause any harm to the human body. However, some of the stone products exceed the radioactivity index, which can cause pollution to the environment during long-term use. Radium, thorium and other radioactive elements in natural stone products with excessive radioactivity levels will produce the natural radioactive gas radon during the decay process. Radon is a colourless, tasteless and imperceptible gas that easily gathers in poorly ventilated places, leading to pathological changes in the lungs, blood and respiratory tract. Therefore, products that have been tested for radioactivity and issued with a certificate of conformity for radioactive products in decoration works. In addition, the doors and windows of the living room should be opened frequently during use to promote indoor air circulation.

# Section II  Timber

The use of wood in civil engineering has a long history and was a major building material in ancient China, especially during the Tang, Song and Liao periods when wooden structures were very popular and the most representative building preserved to this day is the Yingxian wooden towers in Shanxi. Wood is still an important building material in modern engineering applications because of its light weight, high strength, easy processing, low electrical and thermal conductivity, better flexibility and toughness, diverse patterns and good decorative qualities. Scan the code to learn about wood towers.

## I. Classification of timber

According to the species, timber is divided into two main categories: coniferous and broad-leaved trees.

i. Coniferous trees

Coniferous trees are mostly evergreen trees with long, thin leaves like needles, straight and tall trunks, smooth textures, uniform materials, soft wood and easy processing, so they are also known as softwood. Although the conifer material is soft, it has high strength, small dry and wet deformation and strong corrosion resistance. So common conifer species, such as red pine, larch, spruce, fir, cypress and so on, are widely used in construction projects, mostly used as load-bearing components and doors and Windows, floor panels and decorative materials.

ii. Broad-leaved trees

Broad-leaved trees are mostly deciduous trees with broad leaves, reticulated veins, curved trunks and short straight parts. Broad-leaved trees are dense, strong and easily cracked and warped in the wet, making them suitable for smaller timber components. Broadleaf trees are mostly deciduous trees, with broad leaves, reticular veins, curved trunks, and short straight parts. The material is heavy and hard, which is difficult to process, so it is also called hard wood. Broad-leaved trees have high apparent density, high strength, large wet deformation and easy to crack and warping, which is suitable for smaller wood components. Common broad-leaved tree species are: ash, birch, beech, oak, elm and so on. Some of these trees have beautiful grain and color after processing, and are very decorative. They are often used as architectural decoration materials.

## II. The structure of timber

The structure of timber is a major factor in determining its properties. Due to the type of timber and the environment in which it grows, the structure of various types of timber varies considerably. The structure of timber is usually viewed from both a macroscopic and microscopic.

i. Macrostructure of timber

Macrostructure of timber refers to the characteristics of timber tissue that can be seen with the naked eye or a magnifying glass. The three main parts of the tree, the pith center, xylem and bark, as well as the growth rings and pith lines, can be seen from the transverse section.

The pith center is the soft part in the center of the trunk, and its timber strength is low and easy to decay, which is the defect of the timber. Therefore, sawn boards should not have a pith core part. The radiating line outward from the heart of the pulp is called the medullary line, which is poorly connected to the surrounding area and tends to crack along this line when the timber is dry.

Xylem is the main body of wood, which is the part from the bark to the heart of the pith. According to the growth stage is divided into cambium, sapwood, heartwood and other parts. The forming layer is a thin layer of tree growth cells close to the bark, and growth is achieved by the continuous expansion of the forming layer, which grows year by year in the outermost layer and forms the "annual rings". The annual rings are concentric rings of varying depths. Within the same annual rings, wood grown in spring is lighter in colour and softer in quality, known as spring wood(early wood), while wood grown in summer and autumn is darker in colour and harder in quality, known as summer wood(late wood). For the same species, the denser and more uniform the annual rings, the better the material; the more summer wood, the higher the strength of the wood. The wood is darker near the centre of the trunk and is called heartwood; the outer parts are lighter in colour and heartwood is more valuable than sapwood.

The bark is the outer structural layer of the trunk, which is the protective layer for the tree's growth. Generally the bark has no use, and only rarely can it be processed into high grade insulation material. Scan the code to learn more about the macro structure of wood.

ii. Microstructure of wood

The structure of wood as observed under the microscope is called microstructure. Under the microscope, the wood can be seen to be made up of numerous tubular cells, most of which are arranged longitudinally to form a fibrous structure, with a few arranged horizontally to form a medullary line. Each cell is divided into two parts: the cell wall, which consists of fine fibres with very small spaces between them, and the cell lumen, which attracts and permeates water. The thicker the cell wall and the smaller the cavity, the denser the wood and the greater the apparent density and strength, but also the greater the expansion and contraction.

The microstructure of coniferous and broad-leaved trees differs considerably. The microstructure of coniferous trees is simple and regular, consisting mainly of canals and pith lines. In coniferous trees, the pith lines are thin and inconspicuous, and some species, such as pine, have resinous channels between the canals and brains to store resin. Broad-leaved trees are mainly composed of ducts, wood fibres and medullae. The pith line is thick and conspicuous, and the ducts have thin walls and large lumens. Thus, the presence or absence of ducts and the thickness of the medullary line are therefore distinctive features in identifying broad-leaved or coniferous trees.

## III. Properties of wood

The main physical and mechanical properties of wood are moisture content, wet and dry shrinkage, strength.

i. Water content

The moisture content is the mass of water contained in the wood as a percentage of the dry mass of the wood.

The water contained in wood consists of adsorbed water in the cell walls and free water in the cell cavities, as well as chemically bound water in the wood.

The adsorbed water is present in the individual wood fibres within the cell walls and this part of the water has a significant effect on the dry and wet deformation and mechanical strength of the wood. When dry wood absorbs water from the atmosphere, it is usually first absorbed by the cell walls as adsorbed water; after the cell walls are saturated, the water enters the cell cavities and cell interstices as free water. Free water is water that is present in the cell cavities and intercellular spaces and is poorly adsorbed to the wood cells. Changes in free water only affect the wood's performance density, thermal conductivity, resistance to decay and combustibility, and have little effect on deformation and strength. Chemically bound water is the water absorbed during the formation of organic polymers in the wood fibre and is an essential component of the wood. The bonded water in wood in its normal state should be saturated and has little effect on the wood at room temperature.

When damp wood is dried and evaporated, it is first stripped of free water and then of sorbed water. When the adsorbed water in the wood cell wall reaches saturation, but there is no free water in the lumen and intercellular space, the wood moisture content is called the fiber saturation point of wood. The fibre saturation point of wood is often the turning point in the pattern of change of wood properties, and is mostly 25% to 35% for general wood.

Moist wood can lose moisture in drier air, and dry wood can also absorb moisture from the surrounding air. When wood is in a certain temperature and humidity of the air for a long time, it will reach a relatively stable moisture content, which means the evaporation and absorption of water tend to balance, the moisture content of wood at this time is called the equilibrium moisture content. The equilibrium moisture content varies with the temperature and relative humidity of the atmosphere.

ii. Wet and dry expansion and contraction of wood

Wood has significant wet swelling and dry shrinkage properties. When wood is dried from a moist state to the saturation point of the fibres, free water evaporates, its dimensions do not change and it continues to dry, while volume contraction occurs when the adsorbed water in the cell walls evaporates. Conversely, when dry wood absorbs moisture, volume expansion will occur until the water content reaches the fibre saturation point, after that the water content of the wood continues to increase, but the volume no longer expands.

The wet and dry condition of timber has a serious impact on the use of timber, with dry shrinkage causing gaps in the joints of components resulting in loose bonds, and wet swelling causing bulges. Due to the significant wet expansion and dry shrinkage of timber, it should be dried as much as possible before processing. The moisture content should be brought up to the equilibrium moisture content corresponding to the average annual local temperature and humidity in order to reduce the dry shrinkage and deformation in the wood product during use. In addition, the storage time of wood also affects the wet expansion and dry shrinkage deformation. Longer storage time results in the ageing of the wood cells and correspondingly less deformation.

### iii. Strength

#### 1. Strength of wood

Wood is divided into tensile, compressive, bending and shear strengths depending on the state of stress. As wood is a non-homogeneous material with anisotropic properties, the strength of wood is highly directional and is therefore differentiated between smooth-grained and cross-grained strength. There is a significant difference between the smooth-grain and cross-grain strengths of wood and the proportional relationship between them is shown in Table 7-1.

**Table 7-1　Relationships between the various strengths of wood**

| compressive strength | | tensile strength | | bending strength | shear strength | |
|---|---|---|---|---|---|---|
| parallel grain | horizontal grain | parallel grain | horizontal grain | | parallel grain | shear across the grain |
| 1 | 1/10 ~ 1/3 | 2 ~ 3 | 1/20 ~ 1/3 | 3/2 ~ 2 | 1/7 ~ 1/3 | 1/2 ~ 1 |

The parallel to grain compressive strength of wood is the compressive strength when the direction of the force acts parallel to the direction of the wood fibres. The compression damage is the result of the destabilisation of the tubular cells under pressure, rather than the fracture of the fibres. This strength is the most widely used in engineering, as it is often used for load-bearing elements such as columns, piles, diagonal braces and trusses. The horizontal grain compressive strength of timber is the compressive strength when the direction of the force is perpendicular to the direction of the timber fibres. When this pressure is applied, the deformation is initially proportional to the external force, and when the proportional limit is exceeded, the cell wall is destabilised and the cell cavity is compressed, resulting in a large amount of deformation. The cross-grain compressive strength of wood is therefore much lower than the smooth-grain compressive strength.

The tensile strength of wood in the parallel grain is the tensile strength when the direction of tension is the same as the direction of the wood fibres. This tensile damage often occurs before the wood fibres are broken and the fibres are first torn apart. The tensile strength of wood in the grain is the greatest of all the strengths of wood and is usually between 70 and 170MPa. Faults in the wood such as knots and twills have a significant effect on the tensile strength of the wood, and the wood is more or less defective, so the actual tensile strength of the wood is lower than the tensile strength of the wood, making it difficult to make full use of the tensile strength of the wood.

#### 2. The main factors affecting the strength of wood

In addition to its own structural factors, the strength of wood is also strongly related to moisture content, load duration, temperature factors and defects in the wood.

(1) Water content. When the moisture content of wood is below the saturation point of the fibres, the lower the moisture content, the less water is adsorbed, the closer the cell walls are bonded and the strength of the wood increases. Conversely, the strength decreases. When the

moisture content exceeds the fibre saturation point, only the free water changes (apparent density increases) and the strength of the wood remains unchanged. The degree of influence of moisture content on the strength of each wood varies, with the greater influence being on the compressive strength and bending strength of the parallel grain, with little influence on the shear strength of the parallel grain, and almost no influence on the tensile strength of the parallel grain.

(2) Durable strength. The maximum strength of wood under long-term loading without causing damage is called the enduring strength. The enduring strength of timber is much smaller than its ultimate strength, which is generally 50-60% of the ultimate strength. In the design of timber structures, the design is generally based on the enduring strength.

(3) Ambient temperature. The strength of wood decreases as the ambient temperature rises. When the temperature rises from 25℃ to 50℃, the tensile strength of coniferous trees is reduced by 10% to 15% and the compressive strength by 20% to 24%. When the wood is at 60 ~ 100℃ for a long time, it will cause the evaporation of moisture and volatiles, and dark brown, so that the strength decreases and deformation increases. Therefore, buildings that are under high temperature for a long time should not use wood structures.

(4) Defects in timber. Some defects are generated during the growth, harvesting, storage, processing and use of wood, such as cracks, knots, decay, insect infestation, slanting and swirling, also known as blemishes. These defects cause discontinuities and unevenness in the structure of the timber, resulting in a reduction in the strength and mechanical properties of the timber, and even loss of value.

## Section III  Waterproof material

Waterproofing materials are functional materials that prevent rainwater, groundwater and other water from penetrating into buildings or structures. Waterproofing materials are widely used in construction projects, but also in highway and bridge projects, water conservancy projects.

Waterproofing materials are indispensable functional materials for construction projects, which play an important role in improving the quality of building components and ensuring that buildings play a normal engineering role. The cost of waterproofing materials currently accounts for about 15% of the total cost of the project, basement buildings are up to 25% to 30%. Although more investment but the effect is not good, according to statistics in recent years, new projects in the year leakage repair costs of about a billion, so the research and improvement of waterproofing materials is the construction sector is an urgent issue to be resolved.

The traditional waterproof material is the petroleum asphalt felt of paper tire, which has poor anti-aging ability, low elongation of paper tire and easy to rot. The asphalt on the surface of the linoleum has poor heat resistance. When the temperature changes, the joint between the linoleum and the base and the linoleum is easy to break off and crack, forming the water connection and leakage. The new waterproof material, a large number of polymer modified asphalt material to improve the mechanical properties of the carcass and anti-aging property. The application of synthetic materials and composite materials can enhance the low temperature flexibility, temperature sensitivity and durability of waterproof materials, and greatly improve the physical and chemical properties of waterproof materials.

## I. Waterproof coil

Waterproof coil is one of the most important waterproof materials in civil engineering. Before 1980s asphalt waterproof material is the mainstream product. After the 1980s, it gradually developed to rubber, resin base and modified asphalt series, forming three types of asphalt waterproof rolling material, polymer modified asphalt rolling material and synthetic polymer waterproof rolling material.

### i. Asphalt waterproofing roll-roofing

Since the 1950s to 1960s, our waterproofing materials have been represented by paper base asphalt felt. Due to poor durability, paper tyres are now largely obsolete. At present, the modified body of fiber fabric, fiber felt and asphalt roll modified by high polymer have become the development direction of asphalt waterproof roll.

Asphalt waterproof coil can be divided into reinforced asphalt waterproofing and no-reinforced asphalt waterproofing according to its matrix material. Reinforced asphalt waterproofing is a kind of glass cloth, asbestos cloth, cotton fabric, thick paper as the body, impregnated with petroleum asphalt, the surface is sprinkled with powdery, granular or sheet anti-stick material made of coil, also known as impregnated material. No-reinforced asphalt waterproofing is a waterproof material made of rubber powder and asbestos powder mixed into asphalt material through mixing and calendering, also known as roll press coil. Asphalt waterproof coil is a flexible waterproof material commonly used in civil construction.

1. Petroleum asphalt glass fiber matrix waterproof coil

Petroleum asphalt glass fiber matrix waterproof coil, also known as glass fiber asphalt waterproofing roll-roofing or fiberglass linoleum, belonging to the "elastomeric asphalt waterproofing roll-roofing". It uses thin glass fibre felt as the base material, dipping coated with petroleum asphalt, and in the surface coated with mineral powder or covered with polyethylene film and other isolation materials, to make a crimp sheet waterproof material.

Glass fibre linoleum has a high tensile strength and good leak-proof performance, and can reach the A-class waterproof standard. The material is anti-ageing, corrosion resistant and weather resistant. After compounding with modified asphalt, the elasticity, flexibility and shock resistance are greatly improved. For example, by SBS (styrene-butadiene-styrene) modified products, can be in $-25℃ \sim -15℃$ low temperature to maintain good flexibility. Scan the code to understand the petroleum asphalt glass fibre waterproofing roll-roofing grade and application related knowledge.

2. Aluminium foil-faced asphalt waterproofing roll-roofing material

Aluminum foil surface asphalt waterproofing roll material, also known as aluminum foil surface linoleum. It uses glass fiber felt for the matrix, dipping coated with oxidized petroleum asphalt, the surface with embossed aluminum foil sticky surface, its surface scattered fine particles of mineral materials or covered with polyethylene film and made of waterproof materials. Scan the code to learn about the grade and application of aluminium foil surface asphalt waterproofing membrane knowledge.

ii. Polymer modified asphalt coil

Petroleum asphalt can not meet the performance requirements of civil engineering on its own. It has defects in low-temperature flexibility, high-temperature stability, anti-aging, adhesion capacity, fatigue resistance and adaptability to component deformation. Therefore, commonly used some polymers, mineral fillers to modify petroleum asphalt, such as SBS modified asphalt, APP modified asphalt, PVC modified asphalt, recycled rubber modified asphalt, rubber modified asphalt and aluminum foil rubber modified asphalt, etc.

1. Elastomer modified bitumen waterproofing roll-roofing material

Elastomer modified asphalt waterproofing roll-roofing is impregnated with asphalt or thermoplastic elastomer (such as SBS) modified asphalt base, both sides coated with elastomeric asphalt coating layer, the upper surface scattered with fine sand, mineral grains (tablets) or covered with polyethylene film, the lower surface scattered with fine sand or covered with polyethylene film made of a class of waterproofing roll-roofing.

SBS waterproof rolling material is a kind of elastomer modified asphalt waterproof rolling material which is widely used. SBS (styrene-butadiene-styrene) polymer is a block polymer, using a special polymerization method to make butadiene two styrene, without vulcanization molding can be obtained elastic rich copolymer. Among all the modified asphalt, SBS modified asphalt has the best performance at present. The modified waterproof coil has not only high tensile strength of polystyrene, good high temperature resistance, but also high elasticity, fatigue resistance and softness of polybutadiene. It is suitable for roofing and underground waterproof engineering of industrial and civil buildings, especially for building waterproof in low temperature environment. Scan the code to learn the classification knowledge of SBS waterproof coil.

2. Plastic body modified asphalt waterproof coil

Plastic body modified asphalt waterproof coil is impregnated with asphalt or thermoplastic elastomer (such as atactic polypropylene APP or polyolefin polymer APAO, APO) modified asphalt matrix. Then, both sides of the matrix are coated with a plastic body asphalt coating layer, the upper surface is sprinkled with fine sand, mineral particles (pieces) or covered with polyethylene film, and the lower surface is sprinkled with fine sand or covered with polyethylene film.

APP waterproof coil is a kind of plastic body modified asphalt waterproof coil is widely used. APP coil has excellent heat resistance, water resistance, corrosion resistance, and low temperature flexibility (but not as good as SBS coil). Polyester felt has excellent mechanical properties, water resistance and corrosion resistance. Glass fiber felt has low price, but low strength and no extensibility. It is suitable for roofing and underground waterproof engineering of industrial and civil buildings, as well as waterproofing of roads, Bridges and other buildings, especially for building waterproofing in higher temperature environment. Scan the code to learn the classification knowledge of APP waterproof coil.

3. Modified asphalt polyethylene based waterproof coil

Scan code tolearn the knowledge of modified asphalt polyethylene tire waterproof coil.

4. Self-adhesive polymer modified asphalt waterproof coil

Scan code tolearn the knowledge of polymer modified asphalt waterproof coil.

iii. Synthetic polymer waterproof coil

Synthetic polymer waterproof rolling material is a kind of waterproof rolling material which is developed vigorously in recent years except asphalt based waterproof rolling material. Synthetic polymer waterproof coil is based on synthetic rubber, synthetic resin or two blends as the base material, adding appropriate amount of chemical additives, filling materials, and then through mixing, calendering or extrusion made of waterproof coil or sheet.

Synthetic polymer waterproof coil has good heat resistance and flexibility at low temperature, high tensile strength, tear strength, elongation at break, aging resistance, corrosion resistance, good weather resistance, suitable for cold construction.

There are many kinds of synthetic polymer waterproof rolling materials, the most representative of which are synthetic rubber EPDM waterproof rolling materials, polyvinyl chloride waterproof rolling materials and chlorinated polyethylene-rubber blend waterproof rolling materials.

1. EPDM rubber waterproof coil

EPDM rubber waterproof coil is made of EPDM rubber as the main body, adding a certain amount of butyl rubber, softener, reinforcing agent, filler, accelerator and vulcanizing agent. EPDM coil has the characteristics of aging resistance, good heat resistance ( > 160℃), long service life (more than 30 to 50 years), high tensile strength, large elongation, cold construction, strong adaptability to base cracking deformation, light weight, single-layer construction and so on. In the United States and Japan, about one third of new roofing and repair waterproof projects are used EPDM waterproof coil. Epdm waterproof coil is suitable for projects with exposed roof, large span, large vibration, long life and high waterproof quality requirements.

2. PVC waterproof coil

Polyvinyl chloride (PVC) waterproof roll is made of polyvinyl chloride resin as the main raw material, mixed with filler (such as bauxite) and an appropriate amount of modifier, plasticizer (such as dioctyl phthalate) and other additives (such as coal tar). PVC waterproof coil has good aging resistance (durable life of more than 25 years), high tensile strength, great elongation at break, abundant raw materials and cheap prices. Hot air welding is convenient and does not pollute the environment. It is applicable to the industry and civil construction of the north and south areas with high waterproof requirements and long durable life. When used for roof waterproof, it can be made into a single exposed waterproof. Scan code to learn the classification of PVC waterproof coil knowledge.

3. Chlorinated polyethylene-rubber blend waterproof coil

It is a waterproof material mainly composed of chlorinated polyethylene and synthetic rubber blends. The waterproof coil not only has the high strength, excellent ozone resistance and aging resistance of chlorinated polyethylene, but also has the high elasticity, high extensibility and good low temperature flexibility of rubber and plastic. From the point of view of physical properties, chlorinated polyethylene-rubber blend waterproof coil is close to the performance of EPDM waterproof coil, the most suitable for the roof single layer exposed waterproof. Scan code to learn the knowledge of chlorinated polyethylene-rubber blend waterproof coil.

## Ⅱ. Waterproofing coatings

Coatings that protect building components from water penetration or wetting and that form impermeable coatings are called waterproof coatings. According to the different dispersants can be divided into solvent coatings, water emulsion type coatings. With the development of science and technology, coating products not only require easy construction, film speed, good repair effect, but also to extend the service life, adapt to the needs of a variety of complex projects.

ⅰ. Asphalt waterproof coating

Asphalt waterproof coating is based on asphalt, through the dissolution or forming a water dispersion composition of waterproof coatings. Asphalt waterproof coating in addition to the basic performance of waterproofing rolls, but also has a simple construction, easy maintenance, for special building features.

Directly unmodified or modified asphalt dissolved in organic solvents and formulated coatings, known as solvent-based asphalt coatings. Dispersion of petroleum asphalt in water, the formation of a stable water dispersion and the composition of the coating, known as water emulsion asphalt waterproof coating. Scan the code to learn about the classification of asphalt-based waterproofing coatings related knowledge.

1. SL-2 solvent-type SBS rubber modified asphalt waterproof coating

This type of waterproof coating is currently the third generation at home and abroad to SBS rubber as a modified material production of a cold construction waterproofing materials. It has excellent corrosion resistance, high elasticity, ductility, adhesion, good adaptability to grass-roots cracking, high temperature does not flow, low temperature does not crack, stable product performance, can be in the negative temperature( $-20℃$ ) under construction, cold construction, labor-saving, convenient construction. It is mainly used for waterproofing and damp-proofing projects of roofs, basements, gutters and culverts of various buildings; it can be used as a cold construction binder for waterproofing rolls; it can also be used as a material for repairing and patching of waterproofing systems of buildings and anti-corrosion works of pipes.

## 2. Asbestos emulsified asphalt waterproofing coating

Asbestos emulsified asphalt waterproof coating with petroleum asphalt as the base material, asbestos as a dispersant, in the forced mixing made under the thick waterproof coating.

This waterproof coating is non-toxic, non-combustible, water-based cold construction, non-polluting, can be laid on the wet base. It has better water resistance, heat resistance, weather resistance, crack resistance and stability than general emulsified asphalt. Due to the inorganic fibre minerals used in the filler, it has a stronger emulsified film than the chemical film. The shortcoming is that the construction environment temperature generally has to be above 15℃, but the temperature is too high easy to sticky feet. Asbestos emulsified asphalt waterproof coating is widely used in China, the effect is good. With glass fiber cloth, non-woven cloth, it can be applied to reinforced concrete roof, basement, kitchen pool waterproof layer.

## 3. Latex reclaimed rubber asphalt waterproof coating

Water-emulsion reclaimed rubber asphalt waterproof coating is a water-based waterproof coating which is composed of petroleum asphalt as base material and reclaimed rubber as modified material. The waterproof coating is a kind of common waterproof coating at home and abroad. Compared with similar solvent-based products, it replaces gasoline with water, and its safety and environment are superior. This kind of paint is expensive because it is made from synthetic latex.

Water-emulsion recycled rubber bituminous waterproof coating can form waterproof film on various complex surfaces, with certain flexibility and durability.

Water as a dispersant is non-toxic, non-flammable, no smell, safe and reliable. It can be cold construction at room temperature, simple operation, convenient maintenance, and can be constructed on the wet surface without water accumulation. The disadvantage is that the film is thin, the quality of the product is easily affected by the production conditions, and the temperature is less than 5℃, which is not easy to construct. The product is suitable for industrial and civil concrete base roofing, toilet, kitchen waterproof, asphalt perlite insulation roofing waterproof, underground concrete building moisture-proof, old linoleum roofing renovation and rigid self-waterproof roofing maintenance.

## ii. Other kinds of waterproof coatings

Our country in the 1970s mainly produced chloroprene rubber and rubber modified asphalt waterproof coating, to the 1980s launched tar polyurethane waterproof coating, waterproof coating of all kinds of macromolecule materials emerge in an endless stream. Liquid, powder, solvent-based, water-emulsion, reactive, nano-type, fast type, art type and other new products continue to debut in the construction of the project.

## 1. Polyurethane waterproof coating

Polyurethane waterproof coating is a kind of chemical reactive coating, it is composed of isocyanate based polyurethane prepolymer and containing polyhydroxyl or amine group curing agent, and other additives in a certain proportion. Polyurethane waterproof coating belongs to the high-grade synthetic polymer waterproof coating, it has many outstanding advantages. It is easy to form a thick waterproof film. It can be constructed on complex base surface, and its end is easy to handle. Strong integrity, coating layer without seams. Cold construction, safe operation. The film has rubber elasticity, good extensibility, high tensile and tear strength. Waterproof life can reach more than 10 years, etc. Polyurethane waterproof coating is suitable for all kinds of underground, toilet, kitchen and other waterproof engineering, leakage prevention of sewage tank, underground pipeline waterproof, anti-corrosion engineering, etc.

2. Silicone rubber waterproof coating

Silicone rubber waterproof coating is a composite emulsion made of silicone rubber emulsion and other high polymer emulsion, and then add a certain amount of admixtures to prepare the emulsion waterproof coating. Silicone rubber waterproof coating has the dual characteristics of coating waterproof and penetration waterproof, suitable for the construction of complex components surface, non-toxic, tasteless, non-combustible, safe, cold construction, simple operation, can be prepared into various colors have a certain decorative effect. The main disadvantages of silicone rubber waterproof coating are high raw material price and high cost. The base is required to have a better flatness. The solid content is low, a coating layer is thin. The temperature is lower than 5℃ is not suitable for construction. Silicone rubber waterproof coating is suitable for waterproof and moisture-proof engineering of roof, underground and water storage structures.

3. Cement-based permeable crystalline waterproof coating

Cement-based permeable crystalline waterproof coating is composed of Portland cement, quartz sand, special active substances and some additives of inorganic powder waterproof material, referred to as CCCW. Cement-based permeable crystalline waterproof coating is a rigid waterproof material. After the action of water to produce silicate active chemical ions, through the carrier to the concrete internal penetration, diffusion, and concrete pores in the calcium ion chemical reaction, the formation of insoluble in water calcium silicate crystals filled pores, so that the concrete structure is compact and waterproof. This kind of waterproof coating is suitable for the waterproofing treatment of the facing and back water surface of underground engineering, pool, water tower and other concrete structure engineering. It provides reliable waterproof materials and technology for concrete engineering.

## Ⅲ. Building sealing material

In order to ensure the water tightness and air tightness of the building in civil engineering, the sealing material with waterproof function and prevent liquid, gas and solid intrusion is called waterproof sealing material. Its basic function is to fill the gap of complex configuration, through the deformation of sealing materials or flow wetting, so that the uneven surface of the gap, joint close contact or bonding, so as to achieve the role of waterproof sealing.

Building sealing materials can be used in building doors and Windows sealing, caulking, concrete, brick walls, Bridges, road expansion caulking, docking sealing of water supply and drainage pipes, electrical equipment manufacturing and installation of insulation, sealing, aerospace, transportation equipment, mechanical equipment connection parts of the sealing and repair of various component cracks, etc.

The base material of building sealing material is mainly oil base, rubber, resin, inorganic class and so on. Among them, rubber, resin and other polymer materials with excellent performance are the main body of building sealing materials, so they are called polymer building sealing materials. Building sealing materials are paste, liquid and powder. Scan code to understand the classification of waterproof sealing materials related knowledge.

ⅰ. Amorphous waterproof sealing materials

Amorphous sealing materials are field-formed sealing materials, mostly made of rubber, resin, synthetic materials as the base material, it is filled in the gap to play a sealing role. Engineering commonly used non-stereotyped sealing materials are silicone rubber waterproof sealing materials and acrylate waterproof sealing materials.

Silicone rubber waterproof sealant is a polysiloxane as the main component of the amorphous sealing materials, it can be cured at room temperature or heating curing liquid rubber. These materials are heat-resistant, cold-resistant, insulating, waterproof, shockproof, resistant to chemical media, ozone resistance, UV resistance, ageing resistance, resistance to some organic solvents and dilute acids, stable storage, durable sealing performance, after vulcanisation of the sealant in the range from −50℃ to 250℃ to maintain long-term elasticity, widely used in prefabricated components of the construction of seam sealing, waterproof plugging.

Acrylic ester waterproof sealing material is an amorphous sealing material with acrylic ester polymer as the main component, with rubber elasticity and flexibility, good water resistance, solvent resistance, etc.. Due to the poor cold mobility, it cannot be used for deformation joints with large expansion and contraction, and hot construction methods should be used when embedded joints. These materials are suitable for sealing the joints between steel, aluminium, wooden windows and doors, walls and glass; and sealing the joints of rigid roofs, internal and external walls, pipes and concrete elements.

ⅱ. Styling waterproof sealing material

Building sealing materials have good water tightness, air tightness and durability. With good elasticity, plasticity and strength, heat resistance, low temperature resistance, corrosion resistance, the building sealing materials meet the requirements. High dimensional precision is required to prevent cracking and falling off in the process of vibration and deformation of components. The common type of non-shaped sealing materials in the project is polyurethane water expansion rubber sealing materials and rigid water-resistant belts.

Polyurethane water-expanded rubber material, hydrophilic polyurethane prepolymer mainly made of polyether polyol as raw material. There are a lot of polar chain joints in polyurethane material which are easy to rotate. The proper cross-linking curing can produce better resilience. When it meets water, its links and water form hydrogen chains, which causes the material to expand in volume. Domestic 821AF and 821BF water-swelling sealing materials belong to this kind of products. Scan code to understand domestic 821AF and 821BF profile related knowledge.

The rigid water-resistant belt is also called the metal water stop belt. It is made of steel, copper, aluminum, alloy steel plate and so on. Steel water stop belt and copper water stop belt are mainly used in dams and large constructions. Metal waterproof materials are welded together, and the quality of metal welding is very important.

## Section IV  Thermal insulation material

Buildings often have requirements in terms of insulation and thermal insulation in use, and insulation materials can be used to meet these building function requirements. In civil engineering, it is customary to call materials used to control the outflow of heat from the room insulation materials, and materials used to prevent heat from entering the room insulation materials. Insulation and thermal insulation materials are collectively referred to as thermal insulation materials.

The reasonable use of thermal insulation materials in buildings can improve the efficiency of the building's use, better meet the energy-saving requirements and ensure normal production, work and life. A well insulated building can also greatly reduce the energy consumption of heating and air conditioning, and in the context of the "double carbon era", the application of thermal insulation materials is of increasing importance.

### I. Thermal conductivity

i. Heat transfer mode

There are three ways of heat transfer: conduction, convection and radiation.

Heat is transferred in the building and environment in the above three ways, mainly through heat conduction, but also convection and thermal radiation. The main areas of heat dissipation are walls, ceilings and roofs, floors, doors and windows. Gaps in buildings and open doors and windows greatly increase heat dissipation.

In the north of China, heat dissipation is a serious economic problem. The use of insulation can prevent buildings from absorbing too much heat in summer and losing too much heat in winter.

ii. Thermal resistance and thermal conductivity

When there is a temperature difference between the two surfaces of a material, heat is automatically transferred from the hotter side to the cooler side. In a stable state, the thermal resistance of the material can be calculated byEquation(7-1) by measuring the heat flow, the temperature of the two surfaces of the material and the effective heat transfer area:

$$R = \frac{A(T_1 - T_2)}{Q} \qquad (7\text{-}1)$$

Where　$R$——thermal resistance($m^2 \cdot K/W$);
　　　　$Q$——average heat flow(W);
　　　　$T_1$——average temperature of the hot surface of the specimen(K);
　　　　$T_2$——average cold surface temperature of specimen(K);
　　　　$A$——effective heat transfer area of the specimen($m^2$).

If the thermal resistance has a linear relationship with temperature, and the specimen can represent the whole material, and the specimen has sufficient thickness, then the thermal conductivity of the material can be calculated by equation(7-2):

$$\lambda = \frac{d}{R} = \frac{Q \cdot d}{A(T_1 - T_2)} \tag{7-2}$$

Where　$\lambda$——thermal conductivity[$W/(m \cdot K)$];
　　　　$d$——average thickness of specimens(m).

The physical meaning of material thermal conductivity $\lambda$ is that when the temperature difference is 1K, the material with a thickness of 1m will pass through the heat of $1m^2$ area within 1s. The smaller the thermal conductivity of the material, the better its adiabatic performance.

Actual materials often contain pores, and there are heat conduction, heat convection and heat radiation. What is measured is not the real thermal conductivity, but the apparent thermal conductivity, or equivalent thermal conductivity and equivalent thermal conductivity. The thermal conductivity of materials is an important parameter in the design of building maintenance structure and thermal calculation. Choosing materials with small thermal conductivity and large specific heat capacity can improve the insulation performance of the maintenance structure and keep the indoor temperature stable.

In civil engineering, the thermal conductivity of less than $0.175W/(m \cdot K)$ is often referred to as thermal insulation materials. The selection of thermal insulation material, the general requirements of its thermal conductivity is not more than $0.175W/(m \cdot K)$, apparent density is less than $600kg/m^3$, compressive strength is not less than 0.30MPa. In practical applications, due to the low compressive strength of adiabatic materials, they are often combined with load-bearing materials. In addition, because most insulation materials have a certain water absorption, moisture absorption capacity, so in the actual use should pay attention to moisture and waterproof, need to add a waterproof layer or air insulation layer in its surface.

iii. Factors affecting thermal conductivity

The heat conductivity of the material is determined by the size of the thermal conductivity of the material. The smaller the thermal conductivity, the better the thermal insulation performance. The main factors affecting the thermal conductivity of materials are material composition, microstructure, pore structure, temperature, humidity and heat flow direction.

1. Material composition: the thermal conductivity of different materials is different. Generally speaking, the thermal conductivity of metal materials is the largest, followed by inorganic non-metallic materials, and the thermal conductivity of organic materials is the smallest.

2. Microstructure: For materials with the same chemical composition, the thermal conductivity of crystalline structure is the highest, followed by microcrystalline structure, and the thermal conductivity of vitreous structure is the lowest.

3. Pore structure: Because the thermal conductivity of solid substances is much larger than that of air, so in general, the greater the porosity of the material, the smaller the thermal conductivity. In the case of similar porosity, the larger the pore size, the porosity communication will improve the thermal conductivity of the material, which is the result of air circulation and convection in the pore. For fibrous materials, it also depends on the degree of compaction. When compaction reaches a certain apparent density, its thermal conductivity is the minimum, which is called the optimal apparent density. When the apparent density is less than the optimal, the space in the material is too large, due to the air convection, the thermal conductivity will be improved.

4. Humidity: Because solid heat conduction is the best, followed by liquid, gas heat conduction is the worst, therefore, the material moisture will increase the thermal conductivity, if the water freezes, the thermal conductivity will further increase. In order to ensure the thermal insulation effect, special attention should be paid to moisture insulation materials.

5. Temperature: the thermal conductivity of the material increases with the increase of temperature. Therefore, the use of thermal insulation material at low temperature is better.

6. Direction of heat flow: For fibrous materials such as wood, when the direction of heat flow is perpendicular to the direction of fiber arrangement, the thermal conductivity of the material is smaller than that of parallel.

## II. Types of thermal insulation materials

i. Porous type

For flat porous materials, when heat is transferred from the high-temperature side to the low-temperature side, heat transfer is mainly carried out through the following heat transfer methods.

1. Heat conduction in the solid phase.

2. Radiation and convection from the high temperature solid surface to the gas in the pore space.

3. Convection and conduction of the gas itself in the pore space.

4. Radiation and convection from a hot gas to a low temperature solid surface.

5. Radiation between the surface of the hot solid and the surface of the cold solid.

These several modes of heat transfer co-exist in porous materials.

At room temperature, convection and radiation stand for a small proportion of the total heat transfer and mainly heat conduction. At room temperature, convection and radiation in the total heat transfer is a small proportion, mainly heat conduction. The thermal conductivity of closed air is only $0.025 W/(m \cdot K)$, far less than that of solid, so the resistance of heat transfer through closed pores is relatively large, and the existence of closed pores greatly reduces the heat transfer process in solid, thus greatly slowing down the heat transfer speed. This is why a material with a large number of closed pores acts as an adiabatic.

ⅱ. Fibre type

The thermal insulation mechanism of fibre type insulation material is basically similar to that of porous material, when the heat transfer direction is perpendicular to the fibre direction, its thermal insulation performance is better than when it is parallel.

ⅲ. Reflective type

When external heat radiation energy is projected onto an object, part of the energy is usually reflected off and another part is absorbed. According to the principle of conservation of energy, the sum of the absorbed energy and the reflected energy is the total radiant energy. It can be seen that any material with a strong reflective capacity has a small ability to absorb thermal radiation. Therefore, by using the reflection effect of some materials on thermal radiation and pasting this material on the surface of the parts that need insulation, most of the external thermal radiation can be reflected off, so as to play the role of insulation.

## Ⅲ. Commonly used insulation materials

According to the chemical composition, the thermal insulation materials can be divided into organic and inorganic materials. According to the structural form, the insulation materials can be divided into porous tissue materials, fibrous materials and loose granular materials. It is usually made into plate, sheet, coil or shell and other forms of products. Generally speaking, the apparent density of inorganic adiabatic materials is larger, but it is not easy to corrosion, will not burn, and some can withstand high temperature. Organic thermal insulation material is light with good thermal insulation performance and poor heat resistance.

ⅰ. Porous insulation material

Commonly used porous insulation materials are expanded vermiculite, expanded perlite, microporous calcium silicate, foaming silicate, foam glass, foam plastic and aerated concrete, etc.

ⅱ. Inorganic fibrous insulation material

Commonly used inorganic fiber insulation materials mainly refer to rock wool, mineral wool, glass wool and other artificial inorganic fiber materials. This kind of material has the same fibrous shape and structure in appearance, with low density, good thermal insulation performance, no combustion, corrosion resistance, strong chemical stability, good sound absorption performance, non-toxic, no pollution, moth-proof, cheap and other advantages, widely used in residential construction and thermal equipment, pipes and other thermalinsulation, insulation, insulation and sound absorption materials. Scan code to understand the knowledge of common inorganic fiber insulation materials.

ⅲ. Reflective insulation material

At present in China, porous materials are commonly used as the thermal insulation materials in building engineering. Vacuum insulation panel has a good effect on the performance of maintenance structure. But for the thin maintenance structure, it is difficult to set the insulation layer and air layer, and the reflective insulation material often has the ideal insulation effect. Scan code to learn about common reflective insulation materials.

ⅳ. Other insulation materials

Cork board, honeycomb board, fibre board and other materials are also commonly used as thermal insulation materials, scan the code for related knowledge.

## Section V  Sound-absorbing and sound-insulating materials

Sound-absorbing materials are civil engineering materials that can absorb the energy of sound waves transmitted by the air to a greater extent. In concert halls, theatres, halls, studios and other indoor walls, floors and ceilings, the use of appropriate sound-absorbing materials can improve the quality of sound waves transmitted in the room and maintain a good acoustic effect.

### Ⅰ. Principles of sound-absorbing materials

Sound comes from the vibration of an object. It forces neighbouring air to follow the vibration and becomes a sound wave, and spreads around in the air. Sound waves rely on the molecules of the medium to vibrate and spread energy outwards, the molecules of the medium can only vibrate but not move, so sound waves are fluctuations. Humans can hear sounds at frequencies ranging between 128Hz and 10kHz.

As sound travels, some of it diffuses with distance and some of it weakens as air molecules absorb it. This weakening phenomenon is more obvious in the outdoor space, and in the small room, the weakening of sound energy mainly depends on the absorption of sound energy by the material surface of the four walls in the room.

When the sound wave encounters the surface of the material, part of it is reflected, the other part penetrates the material, and the rest is transmitted to the material, causing friction and viscous resistance between the air molecules in the pore of the material and the pore wall, during which a considerable part of the sound energy is converted into heat energy and absorbed. The ratio of the absorbed energy ($E$) (including the sound energy partially penetrating the material) to the total sound energy transmitted to the material ($E_0$) is called the sound absorption coefficient ($\alpha$), which is the main index to evaluate the sound absorption performance of the material, expressed by the formula:

$$\alpha = \frac{E}{E_0} \tag{7-3}$$

The absorption coefficient is related to the frequency of sound and the direction of incident sound. Therefore, the absorption coefficient is the average absorption of the sound incident from all directions, and should indicate the absorption to which frequency. Six frequencies are usually used, namely 125Hz, 250Hz, 500Hz, 1000Hz, 2000Hz and 4000Hz.

Any material can absorb sound, but the degree of absorption varies greatly. Generally, the material with the average sound absorption coefficient of the above six frequencies greater than 0.20 is classified as sound absorbing material. The larger the sound absorption coefficient, the better the sound absorption effect.

## II. The characteristics of sound-absorbing materials

Sound-absorbing materials can be divided into two categories according to the different sound-absorbing mechanisms: the first is loose porous materials; the second is flexible materials, membrane-like materials, plate materials, perforated plates and so on.

Porous sound-absorbing materials have a large number of micro-pores and continuous bubbles that are connected inside and outside. When sound waves are incident on the surface of the material, the sound waves can quickly enter the interior of the material along the micro-pores, causing the air inside the pores or bubbles to vibrate. Due to friction, the viscous resistance of the air and the internal heat conduction of the material, a considerable part of the sound energy is converted into heat energy and absorbed. Flexible materials, membrane materials, plate materials, perforated plates and other sound absorption principle is the resonance of materials and sound waves, so that the sound energy into mechanical energy.

## III. Commonly used sound-absorbing materials

Please scan the code to learn more about the commonly used acoustic materials and their settings in buildings.

Many porous acoustic materials are made of the same material as porous insulation materials, but the requirements for the characteristics of the pores are different. Insulating materials require closed pores to prevent heat convection from taking place, while acoustic materials require open, interconnected pores, and the more pores there are, the better the acoustic effect. This material is the same but the porous material with different pore structure, mainly depends on the difference of a component of the raw material and the different thermal regime and pressure in the production process to achieve.

## IV. Acoustic insulation materials

The materials that play a major role in sound insulation in buildings are collectively referred to as sound-insulating materials. Sound insulation materials are mainly used for external walls, doors and windows, partition walls, partitions, etc.

Sound insulation can be divided into isolation of air sound (sound transmitted through the air) and isolation of solid sound (sound transmitted through impact or vibration) two. The principles of sound insulation are very different. Sound insulation is not only related to materials, but also has a close relationship with the structure of the building.

ⅰ. Airborne sound insulation

The ability of materials to isolate airborne sound can be measured by the transmission coefficient of sound waves or by the amount of sound insulation of the material.

$$\tau = \frac{E_t}{E_0} \tag{7-4}$$

$$R = 10\lg \frac{1}{\tau} \tag{7-5}$$

Where  $\tau$——the acoustic transmission coefficient;

$E_t$——sound energy of the transmitting material;

$E_0$——total incident sound energy;

$R$——sound insulation of the material(dB)。

The smaller the $\tau$ of the material, the greater the $R$ and the better the sound insulation performance of the material. The sound insulation performance of the material is related to the frequency of the incident sound waves and is often expressed in terms of the amount of sound insulation at six frequencies from 125Hz to 4000Hz. For ordinary classrooms between the partition wall and the floor, the requirement to achieve not less than 40dB of sound insulation, which means the transmitted sound energy is less than one ten thousandth of the incident sound energy.

Insulation of airborne sound, mainly obey the law of mass, which means the greater the bulk density of the material, the greater the mass, the better the sound insulation performance, so dense materials should be used as sound insulation materials, such as: brick, concrete, steel plates, etc. If lightweight materials or thin-walled materials are used as sound insulation materials, they need to be supplemented with porous acoustic materials or use laminated structwre, such as laminated glass which is a very good sound insulottion material.

ⅱ. Isolation of solid sound

The ability of a material to isolate solid sound is measured by the impact sound pressure level of the material. When measuring, the specimen is installed in the hole between the upper sound source chamber and the lower sound receiving chamber. There is no rigid connection between the sound source chamber and the sound receiving chamber. The surface of the specimen is hit with a standard percussive device, and the sound pressure level received by the sound receiving chamber is subtracted from the environmental constant to obtain the impact sound pressure level of the material. The standardized impact pressure level of the floor between ordinary classrooms should be less than 75dB.

The most effective measure to isolate solid sound is to apply a discontinuous structural treatment, which means adding elastic padding, such as felt, cork, rubber, between walls and load-bearing beams, between the frame and wall panels of the house, or adding elastic carpeting to the floor slab.

## Section VI  Fire resistant material

With the development of urban construction in China, the number of fire hazards in cities has increased and disaster accidents occur frequently. People put forward higher requirements for the safety of fire prevention in buildings. Fireproof material refers to the material that can keep its function for a certain period of time under fire conditions. Including fireproof plate, fireproof paint, fireproof fabric and so on.

### I. Fire Resistant Panels

They are mostly non-combustible materials and can be used as partitions, ceilings and other components of buildings. They are easy to process, lightweight, decorative, and can be constructed in a dry way, so they are widely used in modern buildings. Commonly used fire resistant panels include cement particle board, inorganic fibre reinforced cement board, steel mesh frame cement sandwich composite board, fire resistant paper faced gypsum board, calcium silicate board, flame-retardant aluminium-plastic building decorative panels, and so on. Please scan the code to learn more.

### II. Fireproof Coating

Fireproof paint is a type of coating that has both the basic function of a coating and the function of fire protection. Fireproof coating is a functional coating that is applied to the surface of combustible substrates and can reduce the flame propagation rate of combustible substrates or prevent heat transfer to the protected member, thus delaying or eliminating the ignition process of combustible substrates, or delaying the destabilisation of the member or the reduction of the mechanical strength of the member. In case of fire, the fireproof coating can effectively delay the change of the physical and mechanical properties of the fire substrate material, so that people have sufficient time to evacuate and fight the fire, and achieve the purpose of protecting people's lives and property.

### III. Fire separation facilities

Holes in buildings for cables, oil pipes, air ducts, gas ducts, and so on, become channels for fire and toxic gases in case of fire, leading to the spread of fire. Fire separation facilities refer to a series of separation facilities that can control the fire within a certain space within a certain period of time and prevent it from spreading and expanding. Fire separation facilities must meet the different levels of fire protection requirements in *China's Code of Practice for Fire Protection in Building Design* (GB 50016—2021). Commonly used fire separation facilities are fire doors, fire shutters, fire nets, fire bags, fire rings. Please scan the code to learn more.

## IV. Flame retardants

Wood, plastic and textiles are all combustible or flammable materials. In order to hinder the spread of fire, some polymers are often added to the above materials for flame retardant treatment, so that the flammable materials become non-combustible or non-combustible materials. Generally speaking, flame retardant polymer materials can be flame retarded through several types of flame retardant mechanisms such as gas phase flame retardant, condensed phase flame retardant and interrupted heat exchange flame retardant. By preventing the combustible gas products from the decomposition of polymers or flame reaction to prevent the role of gas-phase flame retardant; by preventing the thermal decomposition of polymers and the release of combustible gases belong to the condensed phase flame retardant; by taking away part of the heat generated by the combustion of polymers to achieve flame retardant is to interrupt the heat exchange flame retardant. The halogen-antimony positive combustion system commonly used in industry is a gas-phase flame retardant; aluminium hydroxide, a condensed-phase flame retardant; and gasified paraffin, an interrupted heat exchange flame retardant.

# Questions for review

1. The adiabatic performance of an adiabatic material decreases obviously after it is exposed to moisture. Try to analyze the reason.

2. There are two high-grade high-rise building under construction in Guangzhou and Harbin, respectively. Which one need to build glass curtain walls? There are two kinds of materials available: endothermic glass and heat-reflecting galss. Please choose and explain your reasons.

3. Please analyze the difference in the main functions of the architectural decoration materials used outdoor and indoor.

# Reference

[1] ZHAO Q X. Civil Engineerig Materials (in Chinese) [M]. Beijing: China Electric Power Press, 2010.

[2] ZHANG Y M. Civil Engineering Materials (in Chinese) [M]. Sixth Eidtion. Nanjing: Southeast Univerisity Press, 2021.

[3] LIAO G S, ZENG S H. Civil Engineering Materials (in Chinese) [M]. Second Edition. Beijing: Metallugical Industry Press, 2022.

[4] MEHTA P K, MONTEIRO P J M. Concrete: Microstructure, Properties, and Materials [M]. 4th edition. New York: McGraw-Hill Education, 2014.

[5] CLAISSE P A. Civil Engineering Materials [M]. Oxford: Butterworth-Heinemann, 2016.